Regionalization of Watersheds

Water Science and Technology Library

VOLUME 58

The titles published in this series are listed at the end of this volume.

Regionalization of Watersheds

An Approach Based on Cluster Analysis

A. Ramachandra Rao
School of Civil Engineering, Purdue University, West Lafayette, IN, USA

and

V.V. Srinivas
Department of Civil Engineering, Indian Institute of Science (IISc), Bangalore, India

Springer

A. Ramachandra Rao
Purdue University
West Lafayette, IN, USA

V.V. Srinivas
Indian Institute of Science
Bangalore, India

ISBN: 978-90-481-7737-0 e-ISBN: 978-1-4020-6852-2

Printed on acid-free paper.

9 8 7 6 5 4 3 2 1

springer.com

Dedication

This book is respectfully dedicated to
the unique yogini of the twentieth century
Maatha Jayalakshmi
and to her son the great Siddha Purusha
Sri Sri Sri Ganapathi Sachchidananda Swamiji
of
Avadhootha Datta Peetham
Shri Ganapathi Sachchidananda Ashrama,
Mysore 570 025, India
with namaskarams

Preface

Design of water control structures, reservoir management, economic evaluation of flood protection projects, land use planning and management, flood insurance assessment, all rely on knowledge of the magnitude and frequency of floods. Often, estimation of this information is not easy because of paucity of flood records at the target sites. Regional flood frequency analysis (RFFA) entails estimating the flood frequency distribution at a target site by utilizing flood records pooled from several other watersheds, which are similar to the watershed of the site in flood producing mechanisms. The process of identifying similar watersheds for pooling peak flow information is known as regionalization. Research in this area is active over past four decades with new and intriguing findings constantly being reported.

Clustering techniques are useful to identify groups of watersheds which have similar flood producing mechanisms. This book deals with regionalization of watersheds. It provides a detailed account of several recently developed clustering techniques, including those based on fuzzy set theory and artificial neural networks. It also documents research findings on application of clustering techniques to RFFA. An attempt is made to make the technical level of explanation simple and comprehensive for the benefit of practitioners.

In regional frequency analysis, the optimal number of regions is based on cluster validation measures and visual interpretation. The potential of various cluster validity measures in identifying optimal set of regions is investigated. The L-moment based homogeneity tests form the basis to check the regions for homogeneity. The regions formed by any regionalization method are, in general, heterogeneous and they need adjustments to make them homogeneous. It is demonstrated that the subjectivity involved and the effort needed to identify homogeneous groups of watersheds with conventional approaches are greatly reduced by using efficient clustering techniques. To achieve better results, some modifications are suggested to conventional fuzzy clustering approach to regionalization in Chapter 3. Further, a novel two-level self-organizing feature map based clustering approach is developed in Chapter 4. The theoretical background of the proposed approaches is provided and their performance in practical situation is assessed.

In RFFA, a distribution such as log-Pearson type III or Generalized extreme value is recommended as default choice to fit peak flows in different parts of the world. The stipulation that a particular flood frequency distribution can be preferred to fit peak flow data in all the regions is investigated in Chapter 5. Further the validity of simple scaling methods that have been developed in RFFA is tested in practical situation. It is demonstrated that better flood estimates with smaller confidence intervals are obtained by analysis of data from homogeneous watersheds. The importance of regionalization in flood frequency analysis is demonstrated.

It is suggested that Chapter 1 be read before proceeding to other chapters. There are some repetitions for the sake of completeness. The words: site, watershed, and catchment are used interchangeably. Feature vectors that are formed using site characteristics are referred to by the words object, data point, and site at several locations. Similarly, the words: model, method, procedure, algorithm and technique are used interchangeably to refer to clustering algorithm.

Professor Rao would like to thank the members of his family, Smt. Mamatha Rao his wife, Dr. Malini Rao Prasad his daughter, Dr. Sathya Prasad his son-in-law, Karthik A. Rao and Siddhartha S. Rao his sons and especially Kum. Shambhavi N. Prasad, his delightful grand daughter for their support. Dr. Srinivas would like to acknowledge Sri. Sharma and Smt. Katyayani his parents, Smt. Srivatsala his wife, and families of his sisters Smt. Radhika and Smt. Sudha for their continuous encouragement and affection. He would like to express his gratitude to Prof. Rao S Govindaraju and his student Mr. Shivam Tripathi, School of Civil Engineering Purdue University, for useful discussions and input to Chapter 4.

Some of the material in the first four chapters of this book is reprinted from three papers published by the authors in Journal of Hydrology (Rao and Srinivas (2006a), Rao and Srinivas (2006b), and Srinivas et al. (2008)), with permission from Elsevier. Authors express their special thanks to Elsevier Sciences for granting the permission.

We would like to thank a number of people for both direct and indirect support during the period that we worked on this book project. We would like to thank Dr. V.P. Singh for his support. We thank the Publishing Editorial and Production staff at Springer Publishers (Dordrecht, The Netherlands) who helped in bringing this book project to a successful conclusion. Our special thanks are due to Petra D. van Steenbergen (Publishing Editor). We also thank authors whose work is cited in this volume, for their contributions to this research.

West Lafayette, IN — *A. Ramachandra Rao*
Bangalore — *V. V. Srinivas*
September, 2007

Contents

Chapter 1
Introduction

1.1 Regionalization for Flood Frequency Analysis

Floods cause widespread damage to property and life in different parts of the world. Determination of the plausible magnitude and frequency of these hydrologic extreme events is necessary in the design of various flow control structures such as levees, culverts, bridges, barrages and dams. Flood frequency analysis (FFA) procedures are in use for a long time in hydrology to relate the magnitude of floods to their frequency of occurrence. In traditional procedures of FFA, site-specific (at-site) hydrologic information in the form of annual maximum flow series or peak-over-threshold series is needed to estimate the flooding potential at the site of interest (which is also referred to as 'target site' or 'subject site'). However, because of the paucity of flood data at target sites, it is not always possible to use at-site frequency analysis to estimate flood of required frequency for hydrologic design. To contend with this situation, hydrologists use regional flood frequency analysis (RFFA) methods that are based on pooling flood information from several watersheds which are similar to the watershed of target site in flood producing mechanisms. A group of watersheds with sufficient homogeneity in flood generating mechanisms constitutes a *homogeneous region* or *pooling-group* for RFFA, and the procedure to identify the homogeneous regions is traditionally referred to as regionalization.

A region for FFA is sized to provide at least $5T$ peak flow values, where T is referred to as target 'return period' or 'recurrence interval' in years (Reed et al., 1999). The choice between at-site and regional frequency analysis methods depends on both the length of gauged record at the 'target site' and the 'target return period'. Design of water control structures such as highway culverts, bridges, urban storm sewers, airfields and small dams may require estimates of flood quantiles corresponding to 50 to 100-year recurrence interval. Design of levees around cities and intermediate to large dams may require quantile estimates corresponding to 100 to 200-year recurrence interval (Chow et al., 1988, p. 419). Floodway channels are designed for flood events corresponding to 500-year or even higher return period. Various regionalization approaches have been developed in the past to form regions for pooling the required information for reliable estimation of flood quantiles corresponding to the T-year return period.

A.R. Rao, V.V. Srinivas, *Regionalization of Watersheds*,

1.2 Approaches to Regionalization

Regions for RFFA are often chosen to be groups of geographically contiguous watersheds based on political, administrative, or physiographic boundaries. However, this practice is criticized because delineation of regions using these factors does not guarantee hydrological homogeneity. Consequently, several approaches to regionalization have been developed which seek similarity between sites by examining catchment attributes such as physiographic characteristics, geographical location, and at-site flood statistics.

The approaches to regionalization of watersheds include: (i) the method of residuals (Thomas and Benson, 1970; Wandle, 1977; Glatfelter, 1984; Choquette, 1988); (ii) the canonical correlation analysis (Ribeiro-Corréa et al., 1995; Ouarda et al., 2000, 2001; Cavadias, 1989, 1990; Cavadias et al., 2001); (iii) the region of influence (ROI) approach and its extensions (Burn, 1990a,b; Zrinji and Burn, 1994; Cunderlik and Burn, 2006a); (iv) the hierarchical approach and its extension to ROI framework (Gabriele and Arnell, 1991; Zrinji and Burn, 1996); and (v) the cluster analysis (Mosley, 1981; Tasker, 1982; Acreman and Sinclair, 1986; Wiltshire, 1986; Bhaskar and O'Connor, 1989; Burn, 1989; Nathan and McMahon, 1990; Hosking and Wallis, 1997; Hall and Minns, 1999; Burn and Goel, 2000; Hall et al., 2002; Jingyi and Hall, 2004; Rao and Srinivas, 2006a,b). Javelle et al. (2002) developed regional flood-duration-frequency (QdF) curves based on the index-flood method (Dalrymple, 1960) for describing the flood regime for a basin. Shu and Burn (2004) used a fuzzy expert system with genetic enhancement for RFFA. A detailed comparison of some of the approaches generally used for regionalization of watersheds is found in Cunnane (1988) and GREHYS (1996), whereas Bobée and Rasmussen (1995) provide a review of the relevant literature.

In the method of residuals (MOR) approach to RFFA, regions are formed using the positive and the negative signs of residuals extracted from a regional regression model relating flood quantile at each gauged site to the characteristics of watersheds. This approach is widely used by the United States Geological Survey (USGS) for regionalization. This method delineates flood regions in a rather arbitrary manner and the regions are often arranged to be coincident with recognized geographic and/or hydrologic boundaries, political or administrative areas. Therefore, the regions delineated by this approach are likely to contain watersheds whose flood-frequency characteristics may not be similar (Wiltshire, 1986; Bhaskar and O'Connor, 1989).

In the canonical correlation analysis (CCA) based approach to RFFA (Cavadias, 1989, 1990), drainage basins are represented as points in the spaces of pairs of uncorrelated flood-related canonical variables and pairs of uncorrelated basin-related canonical variables to examine similarity in the corresponding point patterns in these spaces. If the point patterns are sufficiently similar, regions are formed in the space of the flood-related canonical variables. The approach was originally based on subjective visual judgement of clustering patterns that may be available. Ribeiro-Corréa et al. (1995) and Cavadias (1995) extended the approach to determining homogeneous hydrological neighborhoods and applied it to regionalization of flood flows. The problem with this approach to regionalization is that similarity in point patterns may not be found (Bobée and Rasmussen, 1995). Ouarda et al. (2001) present theoretical framework for the use of canonical correlations in RFFA. Chokmani and Ouarda (2004) proposed a physiographical space-based kriging method for regional flood frequency estimation at ungauged sites. The physiographical space was defined based on physiographical and meteorological characteristics of gauged catchments, using CCA or principal component analysis. Ordinary kriging was then used to interpolate flow quantiles in the physiographical space.

The RFFA is sought for reliable estimation of flood quantiles at target sites having inadequate flood records. Whereas the results of CCA method depend on at-site estimates of extreme quantiles that cannot be reliably estimated due to paucity of flood records. Hence it seems unlikely that CCA method can give dependable results (Hosking and Wallis, 1997, p. 147).

The ROI approach of Burn (1990a,b) allows each site to have its own region. The ROI of a target site consists of those sites in the study region whose distance to the target site in a weighted multi-dimensional attribute space does not exceed a chosen threshold value. In the estimation of a regional growth curve, each site could be weighted according to its proximity to the target site. The selection and weighting of attributes and sites is one of the problems where no strict mathematical solution is available (Bobée and Rasmussen, 1995). It becomes a point of concern as the number of attributes available for the analysis increase. Recently Cunderlik and Burn (2006a) recommend using Mahalanobis distance measure for assessing similarity between sites, instead of the conventionally used Euclidean measure, to account for the correlation between watershed attributes used for regionalization. The Mahalanobis distance takes into account the variance and covariance of the variables, which was not possible with the Euclidean distance. The proposed approach allows considering estimation uncertainty due to sampling variability in measures describing flood seasonality of watersheds.

For fixed regions, Gabriele and Arnell (1991) proposed hierarchical approach to RFFA which explicitly accounts for spatial variability in different flood characteristics. The skewness of annual maximum flood data is assumed to be constant over a larger area than the coefficient of variation (CV), which in turn is assumed to vary more slowly over space than the mean annual flood. Therefore, more sites are used to estimate the distribution parameters controlling the skewness than are used to derive the parameters determining the CV. Zrinji and Burn (1996) incorporated the concept of hierarchical approach into the ROI framework by defining a set of ROIs for each site as opposed to a single ROI. Similarities between catchments were computed by using directional statistics (i.e., measures of average time of occurrence and seasonality of flood events in the catchments). Further, three measures proposed by Hosking and Wallis (1993) for testing regional homogeneity are used to obtain three different ROIs for each site.

Each regionalization approach has its strengths and limitations. However, because of the constraints imposed by scarcity of data and the subjectivity involved in the selection of attributes, weights, threshold values and distance measures, there are no established criteria by which the superiority of any particular regionalization method can be clearly established.

Recently, increase in awareness of the use of hydro-climatic data seems to have prompted several agencies to work towards creating databanks of a variety of variables that influence hydrological processes. To effectively use the data archives in regionalization studies, there is a need to develop potential approaches that are useful to identify and interpret patterns inherent in hydrologic data. For this task, clustering algorithms that are effective in recognizing patterns in both large and small data sets appear promising. These techniques are referred to by different names in different disciplines, including unsupervised learning in pattern recognition, numerical taxonomy in biology and ecology, typology in social sciences and partition in graph theory (Theodoridis and Koutroubas, 1999). Introductory material on cluster analysis is found in Hartigan (1975), Aldenderfer and Blashfield (1984), Jain and Dubes (1988), Kaufman and Rousseeuw (1990), Everitt (1993), Gordon (1999), Jain et al. (1999) and others.

1.3 Cluster Analysis in Regionalization

Cluster analysis is the generic name of a variety of multivariate statistical procedures that are used to investigate, interpret and classify given data into similar groups or clusters, which may or may not be overlapping. The data points within a cluster should be as similar as possible and the data points of different clusters should be as dissimilar as possible.

In this section, first a brief description of various attributes used in regionalization by cluster analysis is provided. Following this, broad classification of existing clustering algorithms is presented. Subsequently, steps in regionalization are described. Finally the section is concluded with a discussion of issues in cluster analysis.

1.3.1 Attributes Used in Regionalization

A cluster consists of one or more *feature vectors*. A feature vector (also referred to as ‘data point’, ‘data vector’ or ‘object’) comprises of several attributes or variables. The attributes that have been used for regionalization of watersheds include: (i) physiographic characteristics such as drainage area, length of longest stream, main stream slope, average basin slope, storage index, fraction of the basin covered by lakes, reservoirs, and swamps; (ii) soil cover characteristics such as infiltration potential, effective mean soil moisture deficit, and runoff coefficient; (iii) characteristics associated with the land use pattern such as fraction of the basin covered by forests, agricultural, suburban or urban land; (iv) drainage characteristics of the basin such as drainage density; (v) geographical location attributes such as latitude, longitude and altitude of the gauging station, and the centroid of the catchment containing the site; (vi) meteorological characteristics such as storm direction in the catchments, mean annual number of days below certain temperature (Ouarda et al., 2006); (vii) Geologic features of the basin such as fraction of catchment underlain by various types of rock formations (Nathan and McMahon, 1990); (viii) a measure of basin response time such as basin lag or time to peak (Potter and Faulkner, 1987); and (ix) flood seasonality descriptors such as directional statistics (Mardia, 1972; Fisher, 1993) and relative frequency of flood occurrence (Black and Werritty, 1997; Lecce, 2000; Cunderlik and Burn, 2002b; Cunderlik et al., 2004a,b). Directional statistics include measures that describe the average time of occurrence of floods and its variability in the catchment, and mean delay between precipitation and floods (Burn, 1997; Castellarin et al., 2001; Cunderlik and Burn, 2002a; Cunderlik et al., 2004a).

Shape indicators of catchments such as form factor, compactness coefficient, elongation ratio, or circularity ratio may also be used as attributes to form regions for flood frequency analysis. The form factor of a basin is defined as the ratio of area of the basin to square of its axial length, where axial length is the distance from the outlet of the basin to the most remote point in the basin. The compactness coefficient of a basin is the ratio of the perimeter of the basin to the circumference of a circle of area equal to the basin area. The elongation ratio of a basin is the ratio of the diameter of a circle having area the same as the basin area, to the maximum length of the basin. The circularity ratio of a basin is the ratio of the basin area to the area of a circle of same perimeter as that of the basin.

At-site flood statistics have also been used as attributes for regionalization in the past. Examples include mean, coefficient of variation and skewness of annual flood series, plotting position estimate of T-year flood event interpolated from the annual flood series (Burn, 1990b), flood magnitude corresponding to a T-year recurrence interval (Tasker, 1980).

In practice, the homogeneity of regions formed using a regionalization approach is tested by using flood statistics. Hence, these statistics are not supposed to be used as attributes to form regions. A drawback in using flood statistics as attributes to form regions is that the resulting regions may appear homogeneous but are not necessarily effective for RFFA (Burn et al., 1997). Moreover, using flood statistics in forming regions for FFA precludes the use of information from the derived regions to estimate flood quantiles at ungaged sites in the study region. Similarly, the formation of regions should not be based entirely on physiographic characteristics of catchments. This is because similarity in only physiographic characteristics does not necessarily imply similarity in catchment hydrologic response. Therefore in forming regions it is reasonable to include some attributes that are estimated from data measured at the sites, provided that these measurements are not highly correlated with the flood values themselves (Hosking and Wallis, 1997, pp. 54–55). Examples include flood seasonality descriptors (Hosking and Wallis, 1997; Burn et al., 1997; Castellarin et al., 2001). The flood seasonality descriptors are less prone to errors and are more robust than measures based on flood magnitude data. However, they are subject to estimation uncertainty resulting from sampling variability. Other site characteristics to form a region may be based on estimates that are sufficiently accurate to be treated as though they are deterministic quantities. For example, mean annual precipitation can be reliably estimated from isohyetal maps.

1.3.2 Classification of Clustering Algorithms

Most existing clustering algorithms can be classified into two categories (Jain and Dubes, 1988): hierarchical clustering and partitional clustering. Hierarchical clustering procedures provide a nested sequence of partitions, whereas partitional clustering procedures generate a single partition of the data in an attempt to recover the natural grouping present in the data. In this subsection, a brief description of these clustering procedures is presented. Further details on these procedures are provided in subsequent chapters.

Hierarchical clustering algorithms can be subdivided into two categories: Agglomerative and Divisive. For a given set of N feature vectors, the agglomerative hierarchical clustering procedures begin with N *singleton clusters*. The singleton clusters are those that consist of only one feature vector. A distance measure such as the Euclidean is chosen to evaluate the dissimilarity between any two clusters. The clusters that are least dissimilar are found and merged. This provides N-2 singleton clusters and a cluster with two feature vectors. The process of identifying and merging two closest clusters is repeated till the desired number of clusters is obtained. On the other hand, the divisive hierarchical clustering procedures begin with a single cluster consisting of all the N feature vectors. The feature vector that has the greatest dissimilarity to other vectors of the cluster is then identified and separated to form a splinter group. The dissimilarity values of the remaining feature vectors in the original cluster are then examined to determine if any additional vectors are to be added to the splinter group. This step divides the original cluster into two parts. The larger cluster is subjected to the aforementioned procedure in the next step. The algorithm terminates when the desired number of clusters is obtained.

The hierarchical clustering process (both agglomerative and divisive) can be represented as a nested sequence or tree, called *dendrogram*, which shows how the clusters that are formed at the various steps of the process are related. The drawback of hierarchical clustering algorithms is that the resulting clusters are usually not optimal because the feature vectors committed to a cluster in the early stages cannot move to another cluster. Divisive hierarchical clustering algorithms always split clusters. In contrast, agglomerative algorithms always merge clusters.

Partitional clustering procedures attempt to recover the natural grouping present in the data through a single partition. Prototype-based clustering algorithms are the most popular class of partitional clustering methods which consider the prototype, such as cluster centroid, as representive of the cluster.

Clustering algorithms can also be classified as hard clustering and fuzzy clustering. In hard clustering, each feature vector is assigned to one of the clusters with a degree of membership equal to one. This is based on the assumption that feature vectors can be divided into non-overlapping clusters with well defined boundaries between them. This is natural for compact and well-separated groups of data. Nevertheless, in many realistic situations feature vectors bear partial resemblance to several clusters and therefore one cannot justify fully assigning a feature vector to one cluster or the other. The fuzzy set theory (Zadeh, 1965) is a natural way to represent such a situation. Fuzzy clustering allows a feature vector to belong to all the clusters simultaneously with a certain degree of membership in the interval [0, 1].

While hierarchical clustering procedures are not influenced by initialization and local minima, partitional clustering procedures are influenced by initial guesses (number of clusters, cluster centers, etc.). The partitional clustering procedures are dynamic in the sense that feature vectors can move from one cluster to another to minimize the objective function. In contrast, the feature vectors committed to a cluster in the early stages cannot move to another in hierarchical clustering procedures.

In the past decade, a special class of artificial neural network called Self-Organizing Feature Map (SOFM) has been applied as a clustering technique. However, SOFM is not a clustering method because it is seldom possible to interpret clusters from the output of an SOFM. In Chapter 4 we demonstrate that SOFMs may, however, serve as a useful precursor to clustering algorithms.

1.3.3 Steps in Regionalization by Cluster Analysis

1. *Selection of attributes*: The goal of this step is to analyze data of various variables to identify attributes influencing the flood response of watersheds in the study region.
2. *Preparing feature vectors*: The data available for each attribute are rescaled to nullify differences in their variance and relative magnitude. The rescaling may involve transforming the values of attributes by appropriate transformation function (such as logarithmic) and dividing the transformed values by standard deviation. Each feature vector consists of rescaled (dimensionless) attributes of a watershed.
3. *Forming clusters*: This step involves selection of a clustering algorithm to partition feature vectors prepared in step 2 into disjoint or overlapping clusters. The watersheds represented by feature vectors in a cluster constitute a region for flood frequency analysis. In general, distance (or dissimilarity) measure and a clustering criterion characterize a clustering algorithm.
4. *Selecting optimum number of regions*: The clusters formed in step 3 are interpreted visually and by using cluster validity indices to determine optimum number of regions.

Visual interpretation: Clusters obtained in step 3 are visually interpreted by plotting them in geographical space of the study region to identify stable regions. The stable regions do not change their configuration drastically with change in the number of clusters formed by clustering algorithm.

Cluster validity indices: These indices are used to identify compact and well separated clusters. A variety of validity indices are in use with hard and fuzzy clustering algorithms. These will be discussed in the following chapters.

5. *Testing the regions for homogeneity*: The regions determined in step 4 are tested for homogeneity by using statistical homogeneity tests that are described in Section 1.4.
6. *Adjustment of heterogeneous regions*: The regions that are closer to being homogeneous are adjusted to improve their homogeneity. The various plausible options available for adjusting the regions are presented in Section 1.4.1.
7. *Estimation of flood quantiles*: The aim of this step is to perform regional goodness-of-fit tests to identify and fit a suitable flood frequency distribution to flood data of sites in a region. The fitted distribution is then used to obtain flood quantile estimates for hydrologic design (Fig. 1.3.1).

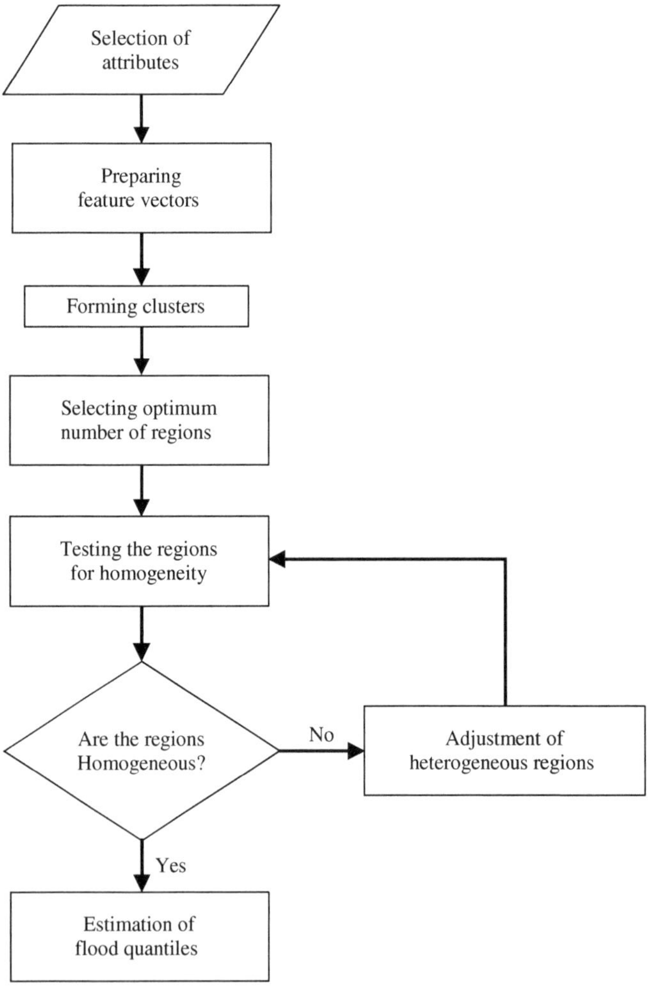

Fig. 1.3.1 Steps in regionalization by cluster analysis

1.3.4 Issues in Cluster Analysis

Clustering algorithms attempt to partition a given data set based on certain assumptions and criteria that will be discussed in the following chapters. Important issues that arise in cluster analysis include: (i) choice of clustering algorithm; (ii) choice of appropriate attributes for clustering; (iii) selection of an objective function; (iv) choice of dissimilarity (or distance) measure; (v) appropriate initialization of the clustering algorithm; and (vi) selection of appropriate number of clusters in the data.

There are several clustering algorithms which may be chosen for regionalization. It is generally expected that these algorithms help in identifying homogeneous groups of watersheds by exploring structure hidden in given data. The performance of a clustering algorithm depends on the definition of similarity considered for identifying neighboring watersheds in attribute space. Domain knowledge will be useful in choosing an appropriate algorithm.

Independent attributes which affect flood response of catchments in the study region must be selected for cluster analysis from a set of causal variables. However, in reality, it is impossible to identify and prepare an exhaustive set of causal variables. Hence the resulting clusters may need adjustment to improve their statistical homogeneity. The ability of a clustering algorithm to produce a partition that represents a meaningful interpretation of structure in the data depends on the chosen objective function. Therefore the objective function should be chosen judiciously.

Furthermore, in cluster analysis the shape of clusters is determined by the distance measure. For instance, the use of Euclidian distance measure is suitable for identification of clusters with a spherical shape (Dunn, 1973). If information is available regarding the shape of expected clusters, a suitable distance metric can be chosen to form the clusters. However, in RFFA, information on shape of expected clusters is not known *a priori*.

In a partitional clustering procedure, the optimal value attained by an objective function depends on cluster centers (also called cluster centroids or cluster seeds) used to initialize the algorithm. As no single procedure of initializing the cluster centers is theoretically proven to yield global optimum value for an objective function, several methods of initializing cluster seeds are in use. A detailed description of various options in vogue in hydrologic literature to initialize partitional clustering algorithms is provided in Chapter 2.

Optimal number of clusters can be chosen by visual interpretation of clusters in geographical space. However, since this method is subjective and cumbersome, a number of cluster validity indices have been developed to aid in their selection. Yet no single index has proven to be efficient in identifying appropriate clusters for a wide variety of datasets. In general, for a given set of feature vectors and clustering algorithm, different cluster validity measures are likely to suggest different values for optimal number of clusters. Therefore it is suggested that selection of optimal number of clusters should be based on a set of validity measures and visual interpretation for better confidence in the result.

1.4 Testing Regional Homogeneity

The homogeneity of regions obtained from cluster analysis is assessed statistically using homogeneity tests Examples of regional homogeneity tests include those proposed by Acreman and Sinclair (1986), Wiltshire (1986), Buishand (1989), Chowdhury et al. (1991), Lu and Stedinger (1992), Hosking and Wallis (1993, 1997), Fill and Stedinger (1995), Cunderlik and Burn (2006b), and Viglione et al. (2007). The L-moment based homogeneity test of Hosking and Wallis (1993) that is widely used by practicing hydrologists is described in this section.

Hosking and Wallis (1993) proposed heterogeneity measures that use the advantages offered by sampling properties of L-moment ratios. A discussion of L-moments is found in Hosking and Wallis (1997). One of the prime advantages of using L-moment based methods for testing homogeneity is that they avoid assumptions about the form of the underlying probability distribution of the observed data.

In a homogeneous region all sites are supposed to have the same population L-moment ratios. However, their sample L-moment ratios (LMRs: L-coefficient of variation (L-CV), L-skewness and L-kurtosis) may be different due to sampling variability. The regional homogeneity tests are developed to examine whether the between-site dispersion of the sample LMRs for the group of sites under consideration is larger than the dispersion expected in a homogeneous region.

Suppose that the region to be tested for homogeneity has N_R sites, with site i having record length of peak flows n_i. Further, let $t^{(i)}$, $t_3^{(i)}$ and $t_4^{(i)}$ denote L-CV, L-skewness and L-kurtosis respectively at site i. The regional average L-CV, L-skewness and L-kurtosis, represented by t^R, t_3^R and t_4^R respectively, are computed as:

$$\left.\begin{aligned} t^R &= \sum_{i=1}^{N_R} n_i t^{(i)} \Big/ \sum_{i=1}^{N_R} n_i \\ t_3^R &= \sum_{i=1}^{N_R} n_i t_3^{(i)} \Big/ \sum_{i=1}^{N_R} n_i \\ t_4^R &= \sum_{i=1}^{N_R} n_i t_4^{(i)} \Big/ \sum_{i=1}^{N_R} n_i \end{aligned}\right\} \tag{1.4.1}$$

where, $n_i \Big/ \sum_{i=1}^{N_R} n_i$ denotes the weight applied to sample LMRs at site i, which is proportional to the sites' record length. The regional average mean l_1^R is set to 1.

Heterogeneity measures (*HMs*) are based on three measures of dispersion: (i) weighted standard deviation of the at-site sample L-CVs (V); (ii) weighted average distance from the site to the group weighted mean in the two dimensional space of L-CV and L-skewness (V_2); and (iii) weighted average distance from the site to the group weighted mean in the two-dimensional space of L-skewness and L-kurtosis (V_3).

$$
\left.
\begin{aligned}
V &= \left\{ \sum_{i=1}^{N_R} n_i (t^{(i)} - t^R)^2 \Big/ \sum_{i=1}^{N_R} n_i \right\}^{1/2} \\
V_2 &= \sum_{i=1}^{N_R} n_i \left\{ (t^{(i)} - t^R)^2 + (t_3^{(i)} - t_3^R)^2 \right\}^{1/2} \Big/ \sum_{i=1}^{N_R} n_i \\
V_3 &= \sum_{i=1}^{N_R} n_i \left\{ (t_3^{(i)} - t_3^R)^2 + (t_4^{(i)} - t_4^R)^2 \right\}^{1/2} \Big/ \sum_{i=1}^{N_R} n_i
\end{aligned}
\right\} \qquad (1.4.2)
$$

In these dispersion measures, distance of sample LMRs for site i from the regional average LMRs is weighted proportionally to the record length of the site, thus allowing greater variability of LMRs for sites having small sample size in a region.

A large number of realizations ($N_{sim} = 500$) of the region are simulated from a kappa distribution fitted to regional average LMRs: l_1^R, t^R, t_3^R and t_4^R. Each realization constitutes a homogeneous region, with N_R sites having same record length as their real-world counterparts. Further, in each realization, the data simulated at any site in the region is serially independent and the data simulated at different sites in the region are not cross-correlated. For each simulated realization, V, V_2 and V_3 are computed.

Let μ_V, μ_{V_2} and μ_{V_3} denote the mean and σ_V, σ_{V_2} and σ_{V_3} the standard deviation of the N_{sim} values of V, V_2 and V_3 respectively. These statistics are used to estimate the following three heterogeneity measures (*HMs*):

$$H_1 = \frac{(V - \mu_V)}{\sigma_V} \qquad (1.4.3)$$

$$H_2 = \frac{(V_2 - \mu_{V_2})}{\sigma_{V_2}} \qquad (1.4.4)$$

$$H_3 = \frac{(V_3 - \mu_{V_3})}{\sigma_{V_3}} \qquad (1.4.5)$$

A region can be regarded as 'acceptably homogeneous' if $HM < 1$, 'possibly homogeneous' if $1 \leq HM < 2$, and 'definitely heterogeneous' if $HM \geq 2$. Further details of the homogeneity test are found in Hosking and Wallis (1997). The values of H_2 and H_3 rarely exceed 2 even for grossly heterogeneous regions and hence lack power to discriminate between homogeneous and heterogeneous regions. Consequently, H_1 is considered to be superior to H_2 and H_3.

1.4.1 Adjusting the Regions

If the regions obtained from the cluster analysis are not statistically homogeneous, they are adjusted to improve their homogeneity. This step of regionalization is justified because the regions are not generally expected to be homogeneous when they are formed based on a set of attributes which is not exhaustive. Nevertheless, the modifications should not be substantial if the attributes used for cluster analysis include a reasonable number of causal variables affecting flood response of catchments and if an efficient clustering algorithm is used for regionalization.

The options suggested by Hosking and Wallis (1997) for adjusting the regions resulting from clustering algorithm include: (i) eliminating (or deleting) one or more sites from the data set; (ii) transferring (or moving) one or more sites from a region to other regions; (iii) dividing a region to form two or more new regions; (iv) allowing a site to be shared by two or more regions; (v) dissolving regions by transferring their sites to other regions; (vi) merging a region with another or others; (vii) merging two or more regions and redefining groups; and (viii) obtaining more data and redefining regions. Among these, the first three options are useful in reducing the values of heterogeneity measures of a region, whereas the options (iv) to (vii) help in ensuring that each region is sufficiently large. In the following, the size of a region, which is estimated as sum of record lengths of peak flow data at all the sites in a region, is expressed in station-years. The effort required for the task of merging or splitting a region is minimal when cluster analysis is used to form regions. This is brought out in the following chapters.

The primary option considered for revising a region is to eliminate one or more sites that are grossly discordant with respect to other sites within the region. In hard clustering, the site eliminated from a region is transferred to another region (recipient) that is nearest to the eliminated site in multi-dimensional attribute space, provided the transfer does not affect the homogeneity of the recipient region adversely. In contract, in fuzzy clustering, a site simultaneously belongs to all the regions and hence there is no need to transfer the eliminated site to another region.

1.4.2 Discordancy Measure

A discordancy measure is useful to identify sites with gross errors in their data or those that are grossly discordant with the region as a whole. In practice, discordancy measure suggested by Hosking and Wallis (1997) is widely used by hydrologists. To estimate discordancy values for sites in a region, the sites are considered as points in three-dimensional space of sample L-moment ratios (L-CV, L-Skewness, and L-Kurtosis). Centroid of the region is regarded as a point depicting average of sample L-moment ratios of the sites in the region. Any point that is far from the centroid of the region is flagged as discordant.

Let N_R represent the number of sites in a region. Further, let $\boldsymbol{u}_{\mathrm{i}} = \left[t^{(i)} t_3^{(i)} t_4^{(i)}\right]^{\mathrm{T}}$ be a vector containing the t, t_3, and t_4 values of site i in the region, where the superscript T denotes transpose of a vector. The discordancy statistic for site i is defined as:

$$\mathbf{D}_i = \frac{1}{3} N_R (\mathbf{u}_i - \bar{\mathbf{u}})^{\mathrm{T}} \mathbf{S}^{-1} (\mathbf{u}_i - \bar{\mathbf{u}}) \qquad (1.4.6)$$

Table 1.4.1 Critical values for the discordancy statistic $\mathbf{D}_i$ (Hosking and Wallis, 1997)

N_k	Critical value of $\mathbf{D}_i$
5	1.333
6	1.648
7	1.917
8	2.140
9	2.329
10	2.491
11	2.632
12	2.757
13	2.869
14	2.971
≥ 15	3.000

where $\bar{\mathbf{u}}$ is the unweighted group average of the L-moment ratios computed using Eq. (1.4.7) and $\mathbf{S}$ is a covariance matrix computed using Eq. (1.4.8).

$$\bar{\boldsymbol{u}} = \frac{\sum_{i=1}^{N_R} \boldsymbol{u}_i}{N_R} \tag{1.4.7}$$

$$\boldsymbol{S} = \sum_{i=1}^{N_R} (\boldsymbol{u}_i - \bar{\boldsymbol{u}})(\boldsymbol{u}_i - \bar{\boldsymbol{u}})^{\mathrm{T}} \tag{1.4.8}$$

Hosking and Wallis (1993) suggested 3 as the critical value for the discordancy statistic for regions containing any number of sites. Later it was found that critical value of $\mathbf{D}_i$ for a region depends on its size. Hosking and Wallis (1997) provide critical values of $\mathbf{D}_i$ for regions of various sizes, which are presented in Table 1.4.1. In many instances the site discordancy may arise out of sampling variability. Therefore, the data at all sites with large values of $\mathbf{D}_i$ should be carefully scrutinized before deciding whether the sites are discordant.

1.5 Data Used in Examples

In the examples discussed in the following chapters, flow records from 245 gauging stations in and around the state of Indiana, USA, are used. The stations are the same as those considered by Glatfelter (1984) in an earlier work of regionalization of Indiana watersheds. The location of these stations in the study region is shown in Fig. 1.5.1.

Information related to the magnitude of peak flows and the date and time of occurrence of the flood events at the gauging stations in Indiana is extracted from the electronic file of Indiana Department of Natural Resources (IDNR) Division of Water (2001). The latitude and longitude values of all the 245 stations and the peak flow records of the stations located outside Indiana are extracted from

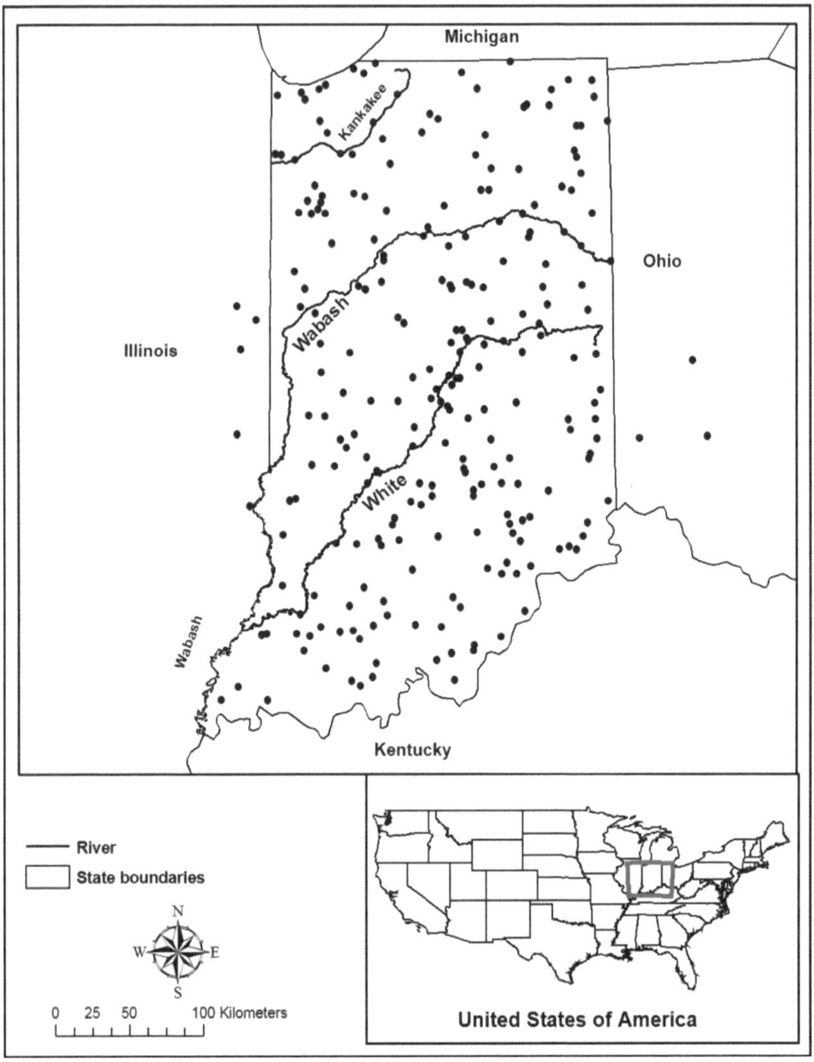

Fig. 1.5.1 Location of gauging stations in the study region considered for regionalization of watersheds

USGS (United States Geological Survey) national water information system web site *http://water.usgs.gov/nwis/peak.* Details of nine attributes used to assess the degree of similarity between drainage basins in Indiana are available for the 245 stations from Glatfelter (1984). The range of each of these attributes is presented in Table 1.5.1. The attributes are subjected to a screening process with a view to extract independent attributes for use in cluster analysis. It is also important to note that these attributes are not used in testing the statistically homogeneity of the regions.

Table 1.5.1 Attributes considered for regionalization of Indiana watersheds

Attribute	Range
Drainage Area	0.28–28,813 km^2
Mean Annual Precipitation	86.36–116.84 cm
Main channel Slope	0.17–50.57 m/km
Main channel Length	0.48–506.94 km
Basin Elevation	125.6–362.7 m
Latitude†	38.00–41.75
Longitude†	84.08–87.90
Storage*	0 %–11 %
Soil Runoff coefficient	0.30–1.00
Forest cover in Drainage Area	0.0–88.4 %
I(24,2)**	6.60–8.51 cm

* Storage – percentage of the contributing drainage area covered by lakes, ponds or wetlands.
** I(24,2) – 24-hour rainfall having a recurrence interval of 2 years.
†Latitude and longitude are in decimal degrees.

1.6 Organization of the Text

The following part of the book is organized into five chapters. It expands on the concepts, issues and approaches described in the foregoing sections, using the database prepared for watersheds in Indiana, USA.

Regionalization of watersheds by hard cluster analysis is discussed in Chapter 2. Various methods of forming regions using hard clustering procedures are reviewed. Following this, the idea of hybrid cluster analysis is presented. In hard cluster analysis, initial set of regions has to be identified by using hard cluster validity measures. These measures are presented and discussed. This is followed by a description of the procedure used for feature extraction, identification of optimal set of clusters, validating the regions and testing the regions for robustness.

Regionalization of watersheds by fuzzy cluster analysis is discussed in Chapter 3. First the classification of fuzzy clustering algorithms is presented. Subsequently fuzzy c-means algorithm is presented and fuzzy cluster validity measures which are useful to identify optimal number of fuzzy clusters are described. This is followed by a description of the procedure used for identification and validation of regions and testing them for robustness.

Chapter 4 is concerned with regionalization of watersheds using a special class of artificial neural networks called Self-Organizing Feature Maps (SOFMs). Various issues concerning use of the SOFMs for RFFA are discussed. This is followed by a discussion of hard and fuzzy clustering of SOFMs. Subsequently a novel two-level clustering algorithm based on SOFMs and Fuzzy clustering is suggested to derive effective clusters for flood frequency analysis. An example of using SOFM and two-level clustering algorithm is given along with the results. The results from validation and testing the regions for robustness conclude this chapter.

Various aspects of the effects of regionalization on flood frequency analysis are discussed in the fifth chapter. Improvement in flood quantile estimation brought about by regionalization of watersheds is investigated. For this purpose, different flood quantile estimation methods are considered because conclusions based on only one method could be misleading. The methods discussed include Index flood method, generalized least square (GLS) regional regression method, and a method based on combination of these two methods. Further, since past three decades, practicing hydrologists in United States resorted to log-Pearson type III (LP3) distribution to estimate frequency of floods in the country following recommendations of the U.S. Water Resources Council (1976, 1977, 1981) in Bulletin 17. The premise of whether a single frequency distribution can be used to fit peak flows in all the regions formed in Indiana State is examined.

Furthermore, in the past decade, simple scaling methods have been developed to show that within hydrologically homogeneous regions moments of peak flows scale with drainage area of watersheds according to log-log linear relations (Gupta and Waymire, 1990; Smith, 1992; Kumar et al., 1994; Gupta and Dawdy, 1995; Ribeiro and Rousselle, 1996). To verify this assertion, the behavior of scaling methods in regionalized watersheds of Indiana is investigated and the results are presented. This is followed by selection of probability distributions for use in regionalized watersheds.

A set of concluding remarks is presented in Chapter 6. Some of the recent approaches to non-stationary at-site and regional frequency analysis of floods are discussed. These approaches desire attention in the scenario of climate change.

Chapter 2
Regionalization by Hybrid Cluster Analysis

2.1 Introduction to Hybrid Cluster Analysis

Hybrid cluster analysis is discussed in this chapter. The hybrid clustering algorithms form clusters by combining two hard clustering algorithms, namely hierarchical and partitional clustering. A brief description of these hard clustering algorithms is provided. Subsequently, it is shown how these algorithms are combined to perform hybrid clustering. Following this, performance of the hybrid clustering algorithms is demonstrated through application to a real world data set.

The hard clustering algorithms partition watersheds in the area of interest into non-overlapping clusters such that each watershed is in one of the clusters. The results are good if clusters are well separated. The hard clustering algorithms have been widely used in hydrology for regional analysis.

2.2 Classification of Hard Clustering Algorithms

Hard clustering algorithms have been classified in Chapter one as hierarchical and partitional clustering algorithms. Hierarchical clustering algorithms provide a nested sequence of partitions, whereas partitional clustering algorithms generate a single partition of the data to recover the natural grouping present in it.

The hierarchical clustering algorithms can be broadly classified into two categories: Agglomerative and Divisive. The agglomerative hierarchical clustering begins with singleton clusters and proceeds successively by merging smaller clusters into larger ones. On the other hand, the divisive hierarchical clustering begins with one large cluster comprising all feature vectors and proceeds by splitting them into smaller clusters.

The partitional clustering algorithms require an initial guess of the number of clusters and cluster centers. They can be classified based on the technique used to initiate clusters, clustering criteria, and the type of data for which they are applicable. The classification of hard clustering procedures is shown in Fig. 2.2.1. The K-means algorithm and agglomerative hierarchical clustering algorithms have been used for regionalization in hydrology.

A.R. Rao, V.V. Srinivas, *Regionalization of Watersheds*,

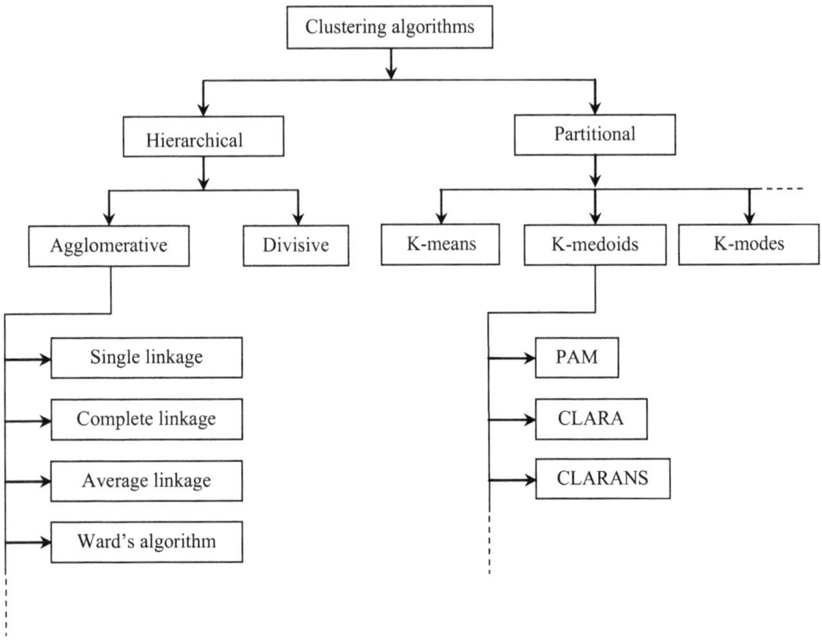

Fig. 2.2.1 Classification of hard clustering algorithms

2.2.1 Hierarchical Clustering Methods

2.2.1.1 Agglomerative Hierarchical Clustering

For a given set of N feature vectors, the agglomerative hierarchical clustering procedures begin with N singleton clusters. A distance measure such as those shown in Table 2.2.1 is chosen to evaluate the dissimilarity between any two cluster centroids, or feature vectors. The clusters that are least dissimilar are found and merged. This results in N–2 singleton clusters and a cluster with two feature vectors. The process of identifying and merging two closest clusters is repeated till a single cluster is left. In general, the number of clusters left at the end of n merges is equal to N-n. The entire process may be represented as a nested sequence, called the dendrogram, which shows how the clusters that are formed at various steps of the process are related.

Algorithms that are representative of the agglomerative hierarchical method of clustering include: (i) single linkage or nearest neighbor; (ii) complete linkage or furthest neighbor; (iii) average linkage; and (iv) Ward's algorithm. These algorithms differ from each other by the strategy used for defining nearest neighbor to a chosen cluster. Clusters with the smallest distance between them are merged.

In the *Single linkage* algorithm, distance between two clusters is the distance between the closest pair of feature vectors, each of which is in one of the two clusters. This algorithm tends to form a small number of large clusters, with remaining

Table 2.2.1 Dissimilarity measures for computing distance between cluster centroids, or feature vectors

Distance measure	Equation
Euclidean	$\sqrt{\sum_{k=1}^{n}(x_{ik}-x_{jk})^2}$
Squared Euclidean	$\sum_{k=1}^{n}(x_{ik}-x_{jk})^2$
Mahalanobis distance	$\sqrt{(\boldsymbol{x}_i-\boldsymbol{x}_j)^{\mathrm{T}}\sum^{-1}(\boldsymbol{x}_i-\boldsymbol{x}_j)}$
Manhattan or City block	$\sum_{k=1}^{n}\lvert x_{ik}-x_{jk}\rvert$
Canberra	$\sum_{k=1}^{n}\frac{\lvert x_{ik}-x_{jk}\rvert}{\lvert x_{ik}\rvert+\lvert x_{jk}\rvert}$
Chebychev	$\max_{1\le k\le n}\lvert x_{ik}-x_{jk}\rvert$
Cosine	$1-\frac{\sum_{k=1}^{n}x_{ik}x_{jk}}{\sqrt{\sum_{k=1}^{n}x_{ik}^2\sum_{k=1}^{n}x_{jk}^2}}$
Minkowski	$\left(\sum_{k=1}^{n}\lvert x_{ik}-x_{jk}\rvert^t\right)^{1/t}$

n: number of attributes; x_{ik}: attribute k of feature vector $\boldsymbol{x}_i$ in cluster-1; x_{jk}: attribute k of feature vector $\boldsymbol{x}_j$ in cluster-2; In Mahalanobis distance measure, T is transpose of matrix, and Σ is covariance matrix. If the covariance matrix is the identity matrix, the Mahalanobis distance reduces to the Euclidean distance. t denotes the order of Minkowski distance.

small outlying clusters on the fringes of the space of site characteristics and is not likely to yield good regions for regional flood frequency analysis (Hosking and Wallis, 1997, pp. 58–59; Rao and Srinivas, 2006a).

In the *Complete linkage* algorithm, distance between two clusters is defined as the greatest distance between a pair of feature vectors, each of which is in one of the two clusters. This algorithm tends to form small, tightly bound clusters. It is usually not suitable for application to large data sets.

In the *Average linkage* algorithm, the distance between two clusters is defined as average distance between them. There are several methods available for computing the average distance. These include unweighted pair-group average, weighted pair group average, unweighted pair group centroid and weighted pair group centroid.

- *Unweighted pair-group average* (*UPGA*): The distance between two clusters is defined as average distance between all pairs of feature vectors, each of which is in one of the two clusters.
- *Weighted pair-group average* (*WPGA*): This method is identical to the *UPGA*, except that in the computations, the size of the respective clusters (i.e., the

number of feature vectors contained in them) is used as a weight. This method is preferred when the cluster sizes are suspected to be greatly uneven.

- *Unweighted pair-group centroid* (*UPGC*): The distance between two clusters is defined as the distance between their centroids. The centroid of a cluster is the mean vector of all the feature vectors contained in the cluster. In this method, if two clusters to be merged are very different in their size, the centroid of the cluster resulting from the merger tends to be closer to the centroid of the larger cluster.
- *Weighted pair-group centroid* (*WPGC*): This method is identical to the *UPGC*, except that feature vectors are weighted in proportion to the size of clusters.

Ward's algorithm (Ward, 1963) is a frequently used technique for regionalization studies in hydrology and climatology (Willmott and Vernon, 1980; Winkler, 1985; Kalkstein and Corrigan, 1986; Acreman and Sinclair, 1986; Nathan and McMahon, 1990; Hosking and Wallis, 1997). It is based on the assumption that if two clusters are merged, the resulting loss of information, or change in the value of objective function, will depend only on the relationship between the two merged clusters and not on the relationships with any other clusters. The governing equation and a detailed explanation of Ward's algorithm are provided in Section 2.3.3.

In regional flood frequency analysis, Mosley (1981) used agglomerative hierarchical clustering available with BioMeDical computer Program 2M (BMDP2M, Dixon, 1975) for regionalization of catchments in New Zealand. Tasker (1982) applied complete linkage algorithm of Sokal and Sneath (1963) for regionalization of watersheds in Arizona, USA.

Nathan and McMahon (1990) compared the performance of single linkage, complete linkage, average linkage, centroid, median and Ward's algorithms of agglomerative hierarchical clustering available with Statistical Package for the Social Sciences (SPSS, 1988). Euclidean, squared Euclidean, Manhattan, Chebychev and Cosine distance measures were considered in their study. Burn et al. (1997) have used agglomerative hierarchical clustering algorithm for regionalization of watersheds in Canada. Their study used the dissimilarity measure shown in Eq. (2.2.1), which was extracted from Webster and Burrough (1972):

$$D_{ij}^{d} = \frac{D_{ij} + \dfrac{d_{ij}}{d_{\max}} w}{1 + w} \tag{2.2.1}$$

where D_{ij} is the Canberra dissimilarity measure of Lance and Williams (1966), whose expression is found in Table 2.2.1; d_{ij} represents the geographic distance between catchments i and j; $d_{\max}$ denotes the maximum geographic distance between catchment pairs, each of which is in one of the two clusters; w is the weighting factor that reflects the relative importance of scaled geographic separation ($d_{ij}/d_{\max}$) and the dissimilarity term in the combined dissimilarity metric.

Catchment seasonality measures, called mean date of occurrence of flood events and the regularity of the phenomenon at each gauging station have been considered by Burn et al. (1997) as attributes in the Canberra dissimilarity measure. The seasonality measures based on flood seasonality may not be useful attributes when the catchments in the area of interest do not show strong flood seasonality, or if they all have similar flood seasonality.

2.2.1.2 Divisive Hierarchical Clustering

The divisive hierarchical clustering procedures begin with a single cluster consisting of all the N feature vectors. The feature vector that has the greatest dissimilarity to other vectors of the cluster is then identified and separated to form a splinter group. The dissimilarity values of the remaining feature vectors in the original cluster are then examined to determine if any additional vectors are to be added to the splinter group. This step divides the original cluster into two parts. The larger of the two clusters is subjected to the same procedure in the next step. The process continues until a stopping criterion (such as, the requested number of clusters) is achieved. If no stopping criterion is specified, the algorithm terminates when clusters resulting from the analysis are all singleton clusters. Description of divisive clustering algorithms can be found in Murtagh (1983) and Guenoche et al. (1991). Savaresi et al. (2002) discuss strategies for selection of a cluster to be split in divisive clustering algorithms. The divisive clustering methods are yet to be applied in regionalization studies.

2.2.2 Partitional Clustering Methods

In partitional clustering procedures, an attempt is made to recover the natural grouping present in the data through a single partition. These procedures are subdivided into K-means and K-medoids methods.

In K-means method (Ball and Hall 1967; MacQueen, 1967), each cluster is represented by its centroid, which is mean (weighted or unweighted average) of feature vectors within the cluster. This method is known for its efficiency in clustering large data sets with numerical attributes. However, it has limitations in clustering categorical data (Ralambondrainy, 1995; Huang and Ng, 2003). Further, the method is sensitive to the presence of outliers.

In K-medoids method, median of each cluster is considered as its representative. This has two advantages. First, the method can be used with both numerical and categorical attributes, and, second, the choice of medoids is dictated by the location of a predominant fraction of data points inside a cluster and, therefore, it is less sensitive to the presence of outliers (Berkhin, 2002). Examples of algorithms that can be grouped under K-medoids method include PAM (Partitioning around medoids, Kaufman and Rousseeuw, 1990), CLARA (Clustering Large Application, Ng and Han, 1994), CLARANS (Clustering Large Applications based on Randomized Search). Among these the PAM is effective with small data sets.

Huang (1997, 1998) proposed K-modes algorithm for clustering large categorical data sets by modifying the K-means algorithm. Each cluster is represented by its mode and a frequency-based method is used to update modes in the clustering process to minimize the clustering cost function.

Algorithms such as K-means and PAM are suitable for regionalization studies in hydrology. The K-means algorithm and its modifications have been used for RFFA by Wiltshire (1986), Burn (1989), Bhaskar and O'Connor (1989) and Burn and Goel (2000). Burn (1989) used the K-means clustering algorithm to determine appropriate grouping of a network of streamflow gauging stations in southern Manitoba, Canada. Flood statistics (coefficient of variation of peak flows, mean annual flow divided by the drainage area) and geographic position of catchments (latitude and longitude) were used as attributes in the feature vector. Traditionally, flood statistics such as the coefficient of variation are used to test the homogeneity of the derived regions. The use of the same flood related variables to form regions and subsequently to evaluate the homogeneity of the derived regions leads to formation of regions that are homogeneous but may not be effective for regional flood frequency analysis (Burn et al., 1997). If at-site flood statistics are used as attributes in the feature vector, one has to ensure that they do not exhibit a high degree of correlation with the flood quantiles of interest. Moreover, the use of flood statistics in a similarity (or dissimilarity) measure constrains the use of the derived regions for estimating extreme flow quantiles at ungaged sites in the study region.

When cluster analysis is based on site characteristics, the at-site statistics are available for use as the basis of an independent test of the homogeneity of the final regions (Hosking and Wallis, 1997). Burn and Goel (2000) applied the K-means algorithm to site characteristics (catchment area, length and slope of the main stream of river) of a collection of catchments in India to derive regions for flood frequency analysis. The drawback in using only physiographic characteristics for forming regions is that similarity in physiographic characteristics does not necessarily imply similarity in catchment hydrologic response (Burn et al., 1997, p. 76).

Wiltshire (1986) adopted the iterative relocation algorithm of Gordon (1981), whereas Bhaskar and O'Connor (1989) used the FASTCLUS clustering procedure of SAS package. While the former work made use of random partition of data to initiate their clustering algorithm, the latter work specified a limiting value for the minimum distance between initial cluster centers.

2.2.3 *Hybrid Clustering*

Hierarchical clustering algorithms are not influenced by initialization and local minima, whereas the partitional clustering algorithms are greatly influenced by initial guesses about number of clusters, cluster centers, etc. The partitional clustering algorithms are dynamic in the sense that feature vectors can move from one cluster to another to minimize the objective function. In contrast, in hierarchical clustering algorithms, the feature vectors committed to a cluster in the early stages cannot move from one cluster to another. The relative merits of using both the hierarchical and partitional clustering algorithms spurred the development of hybrid clustering algorithm. In the Hybrid clustering algorithm, the cluster centers resulting from a hierarchical clustering algorithm are used to initialize a partitional clustering algorithm.

2.3 Clustering Algorithms and Performance Assessment

In this section, the hybrid-clustering algorithm for regionalization of watersheds is presented. It uses K-means algorithm (a partitional clustering algorithm) to identify groups of homogeneous watersheds by adjusting the clusters derived from agglomerative hierarchical clustering algorithm. The K-means algorithm and three agglomerative hierarchical clustering algorithms namely single linkage, complete linkage and Ward's algorithm are presented and discussed. Subsequently, hard cluster validity indices which are useful to identify optimal partition of watersheds provided by the hybrid clustering algorithm are described.

2.3.1 Hybrid Algorithm

Let $\boldsymbol{Y} = \{\boldsymbol{y}_i / i = 1, \ldots, N\}$ denote a set of N feature vectors in n-dimensional attribute space (i.e., $\boldsymbol{y}_i = [y_{i1}, \ldots, y_{in}] \in \Re^n$), each of which characterizes one of the N sites. Further, let $\boldsymbol{x}_i$ denote the i-th rescaled feature vector in the n-dimensional attribute space ($\boldsymbol{x}_i = [x_{i1}, \ldots, x_{in}] \in \Re^n$) obtained by rescaling $\boldsymbol{y}_i$ using Eq. (2.3.1).

$$x_{ij} = \frac{w_j}{\sigma_j}\left[f(y_{ij})\right] \quad \text{for} \quad j = 1, \ldots, n \tag{2.3.1}$$

where $f(\cdot)$ represents the transformation function; y_{ij} denotes the value of attribute j in the n-dimensional feature vector $\boldsymbol{y}_i$; x_{ij} denotes the rescaled value of y_{ij}; w_j is the weight assigned to attribute j; σ_j is the standard deviation of attribute j. Rescaling the attributes may be necessary because of the differences in their variance, relative magnitude and importance.

The K clusters formed in the step 'N-K' of an agglomerative hierarchical clustering algorithm are used to initialize the K-means algorithm (Hartigan and Wong, 1979). The K-means algorithm (KMA) is an iterative procedure in which the feature vectors move from one cluster to another to minimize the value of objective function, F, defined in Eq. (2.3.2).

$$F = \sum_{k=1}^{K} \sum_{j=1}^{n} \sum_{i=1}^{N_k} d^2(x_{ij}^k - x_{\bullet j}^k) \tag{2.3.2}$$

where K denotes the number of clusters, N_k represents the number of feature vectors in cluster k; $\boldsymbol{x}_{ij}^k$ denotes the rescaled value of attribute j in the feature vector i assigned to cluster k; $x_{\bullet j}^k$ is the mean value of attribute j for cluster k, computed as:

$$x_{\bullet j}^{k} = \frac{\sum_{i=1}^{N_k} x_{ij}^{k}}{N_k} \tag{2.3.3}$$

By minimizing F in Eq. (2.3.2), the distance of each feature vector from the center of the cluster to which it belongs is minimized. We have the option to incorporate the knowledge about the global shape or size of clusters by using an appropriate distance measure $d(\cdot)$, such as Euclidean or Mahalanobis. Euclidean distance measure that is suitable for clusters with spherical shape is used in the case study presented in this chapter.

The optimal value attained by the objective function, F, depends on cluster centers used to initialize the KMA. As no single procedure of initializing the cluster centers has been proven to yield a global minimum value for the objective function F, several methods of initialization are in use. Wiltshire (1986) randomly partitioned data to initiate the clustering algorithm. Bhaskar and O'Connor (1989) considered initial cluster centers as feature vectors that are separated by at least a specified minimum distance (Bhaskar and O'Connor, 1989, p. 795). Burn (1989) suggested choosing K of the N feature vectors as the starting centroids to ensure that each cluster has at least one member (Burn, 1989, p. 569). In the case study to follow, results from the hierarchical clustering algorithms namely, single linkage, complete linkage and Ward's algorithm are used to provide initial cluster centers for the KMA.

Every feature vector is assigned to a cluster center that is nearest to it among the K-clusters. After assigning the feature vectors to the K-cluster centers, the center of each of the K-clusters is updated and the value of the objective function, F, is computed. This completes an iteration of K-means algorithm. The procedure of assigning feature vectors to nearest cluster centers and updating the cluster centers is repeated in each of the subsequent iterations. The algorithm is stopped at a point when change in the value of objective function between two successive iterations becomes sufficiently small. The L-moments package of Hosking (2005) contains Fortran routines for regional frequency analysis using the hybrid algorithm. The source code and documentation are available from the StatLib software repository at Carnegie Mellon University.

2.3.2 Single Linkage and Complete Linkage Algorithms

The algorithms begin with N singleton clusters each comprising a rescaled feature vector. Among the N singleton clusters, two closest clusters $\boldsymbol{x}_i$ and $\boldsymbol{x}_j$ are identified and merged to form a new cluster $[\boldsymbol{x}_i, \boldsymbol{x}_j]$.

In the single linkage algorithm the distance between the new cluster $[\boldsymbol{x}_i, \boldsymbol{x}_j]$ and any other singleton cluster $\boldsymbol{x}_k$ is the smaller of the distances between $\boldsymbol{x}_i$ and $\boldsymbol{x}_k$, or $\boldsymbol{x}_j$ and $\boldsymbol{x}_k$. In general, the distance between two non-singleton clusters is the smallest of the distances between all possible pairs of feature vectors in the two clusters (Fig. 2.3.1a). On the other hand, in complete linkage algorithm the distance

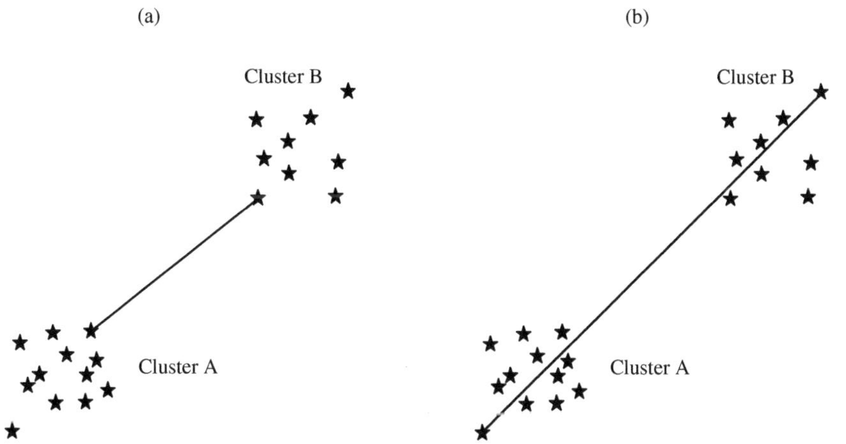

Fig. 2.3.1 Illustration of the definition of distance (or dissimilarity) between two clusters in case of (**a**) Single linkage and (**b**) Complete linkage agglomerative hierarchical clustering algorithms. The stars shown against each cluster denote feature vectors and the line joining the stars refers to the distance between them

between the new cluster $[\boldsymbol{x}_i, \boldsymbol{x}_j]$ and any other singleton cluster $\boldsymbol{x}_k$ is the greater of the distances between $\boldsymbol{x}_i$ and $\boldsymbol{x}_k$, or $\boldsymbol{x}_j$ and $\boldsymbol{x}_k$. In general, the distance between two non-singleton clusters is the largest of the distances between all possible pairs of feature vectors in the two clusters (Fig. 2.3.1b). The pair consists of one feature vector from each cluster.

At each step, two closest clusters are identified and merged. As a consequence, the number of available clusters decreases by one with each additional step. The algorithms are terminated at the step when the number of clusters is equal to the specified value K.

2.3.3 Ward's Algorithm

The objective function, W, of Ward's algorithm (Ward, 1963) minimizes the sum of squares of deviations of the feature vectors from the centroid of their respective clusters.

$$W = \sum_{k=1}^{K} \sum_{j=1}^{n} \sum_{i=1}^{N_k} (x_{ij}^k - x_{\bullet j}^k)^2 \tag{2.3.4}$$

The Ward's algorithm starts with singleton clusters. At this point the cluster centers are the same as feature vectors. Therefore, the value of the objective function is zero. At each step in the analysis, union of every possible pair of clusters is considered and two clusters whose fusion results in the smallest increase in W are merged. The change in the value of objective function, W, due to merger depends only on the relationship between the two merged clusters and not on the relationships with other clusters. To understand this point Eq. (2.3.4) can be rewritten as:

$$W = \sum_{k=1}^{K} \sum_{j=1}^{n} \sum_{i=1}^{N_k} [(x_{ij}^k - x_{\bullet j}) + (x_{\bullet j} - x_{\bullet j}^k)]^2 \tag{2.3.5}$$

where, $x_{\bullet j}$ denotes the mean value of j-th attribute over all feature vectors.

$$x_{\bullet j} = \frac{\sum_{k=1}^{K} \sum_{i=1}^{N_k} x_{ij}^k}{\sum_{k=1}^{K} N_k} \tag{2.3.6}$$

$$\begin{aligned} W = {} & \sum_{k=1}^{K} \sum_{j=1}^{n} \sum_{i=1}^{N_k} (x_{ij}^k - x_{\bullet j})^2 + \sum_{k=1}^{K} \sum_{j=1}^{n} N_k (x_{\bullet j} - x_{\bullet j}^k)^2 \\ & + 2 \sum_{k=1}^{K} \sum_{j=1}^{n} \sum_{i=1}^{N_k} (x_{ij}^k - x_{\bullet j})(x_{\bullet j} - x_{\bullet j}^k) \end{aligned} \tag{2.3.7}$$

From Eq. (2.3.3):

$$\sum_{i=1}^{N_k} x_{ij}^k = N_k (x_{\bullet j}^k) \tag{2.3.8}$$

The value $x_{\bullet j}$ is unique for a given set of feature vectors. Therefore,

$$\sum_{i=1}^{N_k} x_{\bullet j} = N_k (x_{\bullet j}) \tag{2.3.9}$$

Substituting the values of Eqs. (2.3.8) and (2.3.9) in (2.3.7), we have

$$\begin{aligned} W = {} & \sum_{k=1}^{K} \sum_{j=1}^{n} \sum_{i=1}^{N_k} (x_{ij}^k - x_{\bullet j})^2 + \sum_{k=1}^{K} \sum_{j=1}^{n} N_k (x_{\bullet j} - x_{\bullet j}^k)^2 \\ & + 2 \sum_{k=1}^{K} \sum_{j=1}^{n} N_k (x_{\bullet j}^k - x_{\bullet j})(x_{\bullet j} - x_{\bullet j}^k) \\ \Rightarrow W = {} & \sum_{k=1}^{K} \sum_{j=1}^{n} \sum_{i=1}^{N_k} (x_{ij}^k - x_{\bullet j})^2 + \sum_{k=1}^{K} \sum_{j=1}^{n} N_k (x_{\bullet j} - x_{\bullet j}^k)^2 \\ & - 2 \sum_{k=1}^{K} N_k \sum_{j=1}^{n} (x_{\bullet j} - x_{\bullet j}^k)^2 \\ \Rightarrow W = {} & \sum_{k=1}^{K} \sum_{j=1}^{n} \sum_{i=1}^{N_k} (x_{ij}^k - x_{\bullet j})^2 - \sum_{k=1}^{K} \mathrm{N}_k \sum_{j=1}^{n} (x_{\bullet j} - x_{\bullet j}^k)^2 \end{aligned} \tag{2.3.10}$$

Let us consider two clusters labeled 1 and 2, before and after merger. Let N_1 and N_2 denote the number of feature vectors in clusters 1 and 2 respectively. The value of objective function before merger is obtained by substituting K=2 in Eq. (2.3.10).

$$\begin{aligned} W &= \sum_{k=1}^{2}\sum_{j=1}^{n}\sum_{i=1}^{\mathrm{N_k}}(x_{ij}^{k} - x_{\bullet j})^2 - \sum_{k=1}^{2}\mathrm{N}_k\sum_{j=1}^{n}(x_{\bullet j} - x_{\bullet j}^{k})^2 \\ &= \sum_{j=1}^{n}\sum_{i=1}^{\mathrm{N_1}}(x_{ij}^{1} - x_{\bullet j})^2 + \sum_{j=1}^{n}\sum_{i=1}^{\mathrm{N_2}}(x_{ij}^{2} - x_{\bullet j})^2 - \sum_{k=1}^{2}\mathrm{N}_k\sum_{j=1}^{n}(x_{\bullet j} - x_{\bullet j}^{k})^2 \end{aligned} \tag{2.3.11}$$

Since there are only two clusters, the centroid for the cluster resulting from their merger is the same as the centroid of data set $x_{\bullet j}$. The value of objective function after merger is given as

$$\begin{aligned} W &= \sum_{k=1}^{K=2}\sum_{j=1}^{n}\sum_{i=1}^{N_k}(x_{ij}^{k} - x_{\bullet j})^2 \\ &= \sum_{j=1}^{n}\sum_{i=1}^{N_1}(x_{ij}^{1} - x_{\bullet j})^2 + \sum_{j=1}^{n}\sum_{i=1}^{N_2}(x_{ij}^{2} - x_{\bullet j})^2 \end{aligned} \tag{2.3.12}$$

Comparing Eqs. (2.3.11) and (2.3.12), increase in the value of W due to merger is:

$$\begin{aligned} \Delta W &= \sum_{k=1}^{2} N_k \sum_{j=1}^{n} (x_{\bullet j} - x_{\bullet j}^{k})^2 \\ &= N_1 \sum_{j=1}^{n} (x_{\bullet j} - x_{\bullet j}^{1})^2 + N_2 \sum_{j=1}^{n} (x_{\bullet j} - x_{\bullet j}^{2})^2 \end{aligned} \tag{2.3.13}$$

In Eq. (2.3.13), $x_{\bullet j}$ is the centroid co-ordinate of the cluster resulting from the merger, whereas $x_{\bullet j}^{1}$ and $x_{\bullet j}^{2}$ are the centroid coordinates of clusters 1 and 2 before merger. The equation indicates that in Ward's algorithm, if two clusters are merged, the resulting loss of information or change in the value of objective function ΔW depends only on the relationship between the two merged clusters and not on the relationships with other clusters. At each step, the contribution to the objective function by the clusters that are not merged remains the same as their contribution to the objective function before the merger.

Ward's algorithm is good at recovering the cluster structure and it tends to form spherical clusters of nearly equal size. This characteristic of the Ward's algorithm makes it useful for identification of homogeneous regions for regionalization. However, like other hierarchical clustering techniques, there is no provision in Ward's algorithm for reallocation of feature vectors that may have been poorly classified at an early stage in the analysis.

2.3.4 Hard Cluster Validity Measures

A number of hard cluster validity measures have been in use to determine optimal number of clusters in a data set (Romesburg, 1984; Everitt, 1993; Theodoridis and Koutroubas, 1999; Halkidi et al., 2001). Cluster validity constitutes the procedure of evaluating the results of a clustering algorithm. In general, the approaches in vogue to investigate cluster validity can be broadly classified into three categories (Theodoridis and Koutroubas, 1999). The first approach, which is based on external criteria, evaluates the results of a clustering algorithm based on a pre-specified structure, which is imposed on the data set and reflects our intuition about the clustering structure of the data set. In the second approach, which is based on internal criteria, results of clustering are evaluated in terms of quantities that involve the vectors of the data set themselves (e.g., proximity matrix). The third approach to validate clusters is based on relative criteria, which involves evaluation of cluster structure by comparing it to other clustering schemes, resulting with different input parameter values.

In this section, six cluster validity indices, namely cophenetic correlation coefficient (Sokal and Rohlf, 1962), average silhouette width (Rousseeuw, 1987), Dunn's index (Dunn, 1973), Davies-Bouldin index (Davies and Bouldin, 1979), Calinski Harabasz index (Calinski and Harabasz, 1974), and Minimum Description Length (Qin and Suganthan, 2004) are presented and discussed. These measures, which evaluate the clustering result by using internal criteria, find use in identification of optimal partition of watersheds provided by the hybrid clustering algorithm.

2.3.4.1 Cophenetic Correlation Coefficient

The cophenetic correlation coefficient, abbreviated as CPCC by Farris (1969), is a validity measure for hierarchical clustering algorithms. A hierarchical clustering process can be represented as a nested sequence or tree, called dendrogram, which shows how the clusters that are formed at the various steps of the process are related. The CPCC is used to measure how well the hierarchical structure from the dendrogram represents in two Dimensions the multi-dimensional relationships within input data. The CPCC is defined as the correlation between the $M = N(N-1)/2$ original pairwise dissimilarities (proximity) between the feature vectors and their cophenetic dissimilarities from the dendrogram. The cophenetic dissimilarity, c_{ij}, between two feature vectors i and j is the intercluster distance at which the two feature vectors are first merged in the same cluster.

$$CPCC = \frac{\left(1/M \sum_{i=1}^{N-1} \sum_{j=i+1}^{N} d_{ij}^{p} c_{ij} - \mu_p \mu_c\right)}{\sqrt{\left[(1/M) \sum_{i=1}^{N-1} \sum_{j=i+1}^{N} \left(d_{ij}^{p}\right)^2 - \mu_p^2\right]\left[(1/M) \sum_{i=1}^{N-1} \sum_{j=i+1}^{N} c_{ij}^2 - \mu_c^2\right]}} \tag{2.3.14}$$

where μ_p and μ_c are the means of elements in proximity and cophenetic matrices respectively, whereas d_{ij}^p and c_{ij} are respectively the (i, j)th elements of proximity and cophenetic matrices. The concordance between the input data and the dendrogram is close if value of the index is close to 1.0. A high value for CPCC is regarded as a measure of successful classification. A value of 0.8 or above indicates that the dendrogram does not greatly distort the original structure in the input data (Romesburg, 1984). Nonetheless, the CPCC is not always a reliable measure of the distortion due to a hierarchical model (Holgersson, 1978; Romesburg, 1984; Everitt, 1993).

2.3.4.2 Silhouette Width

The silhouette width (Rousseeuw, 1987) for a feature vector is a measure of how similar that feature vector is to feature vectors in its own cluster compared to feature vectors in other clusters. The silhouette width $s(i)$ for i-th feature vector in cluster k is defined as

$$s(i) = \frac{b(i) - a(i)}{\max\{a(i), b(i)\}} \tag{2.3.15}$$

where $a(i)$ is the average distance from the i-th feature vector to all other feature vectors in the cluster k; $b(i)$ is the minimum average distance from the i-th feature vector to all the feature vectors in another cluster j $(j = 1, \ldots, K;\ j \neq k)$. From this formula it follows that $-1 \leq s(i) \leq 1$.

If $s(i)$ is close to 1, we may infer that the i-th feature vector has been assigned to an appropriate cluster. On the other hand, when $s(i)$ is close to –1, we may conclude that the i-th feature vector has been misclassified. When $s(i)$ is approximately zero, it indicates that the i-th feature vector lies equally far away from the two clusters. For the given K clusters, the overall average silhouette width is the average of the silhouette widths for all the feature vectors in the dataset. The partition with the maximum overall average silhouette width is taken as the optimal partition.

2.3.4.3 Dunn's and Davies-Bouldin Indices

Dunn's index (Dunn, 1973) and Davies-Bouldin index (Davies and Bouldin, 1979) are widely recognized for their ability to identify sets of clusters that are compact and well separated. The Davies-Bouldin index is a function of the ratio of the sum of within-cluster scatter to between-cluster separation.

Suppose that the given set of N-feature vectors (in n-dimensional space) has been partitioned into K clusters $\{C_1, C_2, \ldots, C_K\}$ such that cluster C_k has N_k feature vectors and each feature vector is in exactly one cluster, so that $\sum_{k=1}^{K} N_k = N$. The scatter within the k-th cluster, $S_{k,q}$, is computed using Eq. (2.3.16) and the Minkowski distance of order 't' between the centroids that characterize clusters C_j and C_k is defined by Eq. (2.3.17).

$$S_{k,q} = \left(\frac{1}{N_k} \sum_{\boldsymbol{x}_i \in C_k} \| \boldsymbol{x}_i - z_k \| \,|_2^q \right)^{1/q} \tag{2.3.16}$$

$$d_{jk,t} = \left\| z_j - z_k \right\|_t \tag{2.3.17}$$

where z_k represents the centroid of cluster k and $S_{k,q}$ is the q-th root of the q-th moment of the Euclidean distance of points in cluster k with respect to their mean. First moment (i.e., q=1) and Minkowski distance of order 2 (i.e., t = 2) which are commonly adopted by practitioners (Pakhira et al., 2004), have been used in examples presented in this book. The Davies-Bouldin index is computed using Eq. (2.3.18). A small value for DB indicates good partition, which corresponds to compact clusters with their centers far apart.

$$DB = \frac{1}{K} \sum_{k=1}^{K} \max_{j, j \neq k} \left\{ \frac{S_{k,q} + S_{j,q}}{d_{jk,t}} \right\} \tag{2.3.18}$$

Dunn's index is computed by using Eq. (2.3.19),

$$D = \min_{1 \leq i \leq K} \left\{ \min_{1 \leq j \leq K, j \neq i} \left\{ \frac{\delta \left(C_i, C_j \right)}{\max\limits_{1 \leq k \leq K} \Delta \left(C_k \right)} \right\} \right\} \tag{2.3.19}$$

where $\delta(C_i, C_j)$ denotes the distance between clusters C_i and C_j (intercluster distance) computed using Eq. (2.3.20); $\Delta(C_k)$ represents the intracluster distance of cluster C_k defined by Eq. (2.3.21). The value of K for which D is maximized is taken as the optimal number of clusters.

$$\delta(C_i, C_j) = \max_{\boldsymbol{x}_i \in C_i, \boldsymbol{x}_j \in C_j} \left\{ d(\boldsymbol{x}_i, \boldsymbol{x}_j) \right\} \tag{2.3.20}$$

$$\Delta(C_k) = \max_{\boldsymbol{x}_i, \boldsymbol{x}_j \in C_k} \left\{ d(\boldsymbol{x}_i, \boldsymbol{x}_j) \right\} \tag{2.3.21}$$

where $d(\boldsymbol{x}_i, \boldsymbol{x}_j)$ is the distance between rescaled feature vectors $\boldsymbol{x}_i$ and $\boldsymbol{x}_j$.

2.3.4.4 Calinski-Harabasz Index

Calinski-Harabasz Index (V_{CH}) of a partition $G = \{C_1, \ldots, C_K\}$ comprising K clusters is computed as

$$V_{\text{CH}} = \frac{\left[trace\, \mathbf{B} / (K - 1) \right]}{\left[trace\, \mathbf{W} / (N - K) \right]} \tag{2.3.22}$$

where $\boldsymbol{B}$ and $\boldsymbol{W}$ are matrices. The matrix $\boldsymbol{B}$ describes dispersion of cluster centroids and $\boldsymbol{W}$ represents within cluster dispersion. The traces of the matrices $\boldsymbol{B}$ and $\boldsymbol{W}$ can be written as

$$trace\,\boldsymbol{B} = \sum_{k=1}^{K} N_k \left\| z_k - \bar{\boldsymbol{x}} \right\|^2 \tag{2.3.23}$$

$$trace\,\mathbf{W} = \sum_{k=1}^{K} \sum_{\boldsymbol{x}_i \in C_k} \left\| \boldsymbol{x}_i - z_k \right\|^2 \tag{2.3.24}$$

where N_k denotes the number of feature vectors in k-th cluster C_k, z_k is centroid of C_k, and $\bar{\boldsymbol{x}}$ is centroid of the entire set of rescaled feature vectors $\{\boldsymbol{x}_i / i = 1, \ldots, N\}$. Maximum value of V_{CH} denotes optimal partition.

2.3.4.5 Minimum Description Length

Minimum Description Length (*MDL*) principle, which was originally proposed by Rissanen (1989), can be used to determine the optimum number of clusters by finding prototype vectors that can minimize the length of description of the set **X** containing N feature vectors. The feature vectors of **X** are divided into two subsets **I** and **O**, which are composed of inliers and outliers, respectively. Inliers are feature vectors assigned to clusters, whereas outliers are those which are not allocated to any cluster. The expression of *MDL* criterion is formulated as follows:

$$MDL(\mathbf{X}, \mathbf{Z}) = \text{mod } L(\mathbf{I}, \mathbf{Z}) + \text{error } L(\mathbf{I}, \mathbf{Z}) + \text{mod } L(\mathbf{O}) \tag{2.3.25}$$

where **Z** represents the set of cluster centroids $\mathbf{Z} = \{z_1, z_2, \ldots, z_K\}$. The complexity of the entire model is evaluated by the term mod $L(\mathbf{I}, \mathbf{Z})$.

The length of encoding mod $L(\mathbf{I}, \mathbf{Z})$ is given by the sum of: (i) the length of encoding **Z**, denoted by $L(\mathbf{Z})$; and (ii) the length of encoding all the indices of **I**, given by $L(\mathbf{I}(\mathbf{Z}))$. Herein, an index is the identity of cluster to which feature vector of **I** is assigned.

The encoding length of 'error $L(\mathbf{I}, \mathbf{Z})$' represents residual errors generated in describing all inlier data points **I** with prototype set **Z**. The description length of the outlier set **O**, denoted by mod $L(\mathbf{O})$, is usually encoded in the same way as the prototype vectors. The capability of the model to describe the whole data set $\mathbf{X} = \mathbf{I} + \mathbf{O}$ is reflected by the last two terms in Eq. (2.3.25).

Let b denote the number of bits needed for encoding a single data vector. Then, $L(\mathbf{Z}) = Kb$ and mod $L(\mathbf{O}) = |\mathbf{O}|b$, where K is the number of prototypes and $|\mathbf{O}|$ represents the cardinality (i.e., number of feature vectors) of the outlier set. The b is computed using the average value range of rescaled feature vectors and the resolution (or accuracy) of data η as $b = [\log_2(\text{range}/\eta)]$. Each inlier feature vector $\boldsymbol{x} \in \mathbf{I}$ is encoded with $\log_2 K$ bits following the fixed length encoding scheme of Bischof et al. (1999) and thus $L(\mathbf{I}(\mathbf{Z})) = |\mathbf{I}| \log_2 K$. The *MDL* value is instantiated as:

$$MDL(\mathbf{X},\mathbf{Z}) = Kb + |\mathbf{I}|\log_2 K + \kappa \sum_{i=1}^{K}\sum_{x\notin S_i}\sum_{j=1}^{n} max\left(\log_2\left(\frac{\|x_j - z_{ij}\|}{\eta}\right), 1\right) + |\mathbf{O}|b \quad (2.3.26)$$

where K, n, and S_i represent the current number of prototypes, the dimension of input vectors and the inlier receptive field of prototype z_i, respectively. Parameter κ is used to balance the contribution of model complexity mod $L(\mathbf{I}, \mathbf{Z})$ and model efficiency error $L(\mathbf{I}, \mathbf{Z})$ in the computation of MDL value. Herein, it is assumed that the length of encoding the error term $L(\mathbf{I}, \mathbf{Z})$ is proportional to its magnitude, and that data accuracy is η in all n dimensions. Minimum value of *MDL* denotes optimal partition. Further details of this validity measure can be found in Qin and Suganthan (2004).

2.4 Application of Hybrid Clustering Algorithms to Regionalization

2.4.1 Feature Extraction

Sensitivity of flood response of drainage basins to variations in the values of attributes is examined by plotting each of the attributes discussed earlier (Section 1.5) against a flood-related variable. The flood related variables considered herein include: (i) mean value of annual peak flows (or mean annual flood, MAF); (ii) median value of annual peak flows (or median annual flood, MEF); (iii) mean annual flood per unit area of drainage basin (MAF/A); (iv) median annual flood per unit area of drainage basin (MEF/A); (v) mean annual flood divided by the mean annual precipitation (MAF/P); and (vi) median annual flood divided by the mean annual precipitation (MEF/P).

The magnitude of flood flow increases with increase in drainage basin area (Fig. 2.4.1a). The contribution to flood magnitude from unit area of a drainage basin increases, in general, with increase in the slope and length of the main channel (Figs. 2.4.1b,c) and soil runoff coefficient of the drainage basin. Also, the magnitude of MAF from a drainage basin for unit depth of precipitation increases with runoff coefficient values of the contributing drainage areas.

Figure 2.4.1d and Table 2.4.1 show that the main channel length is highly correlated with the area of drainage basin (correlation coefficient = 0.850). Since the objective of the feature extraction is to identify independent attributes, either the main channel length or the drainage area could be considered as a physiographic attribute for cluster analysis. The correlation of flood-related statistics (MAF, MEF, and MAF/P) is significant with drainage area than with channel length (Table 2.4.1). Therefore, drainage basin area is selected as an attribute for further analysis.

The MAF and MEF are expected to increase with increase in the mean annual precipitation volume for a given watershed. This is evident from Table 2.4.1 which

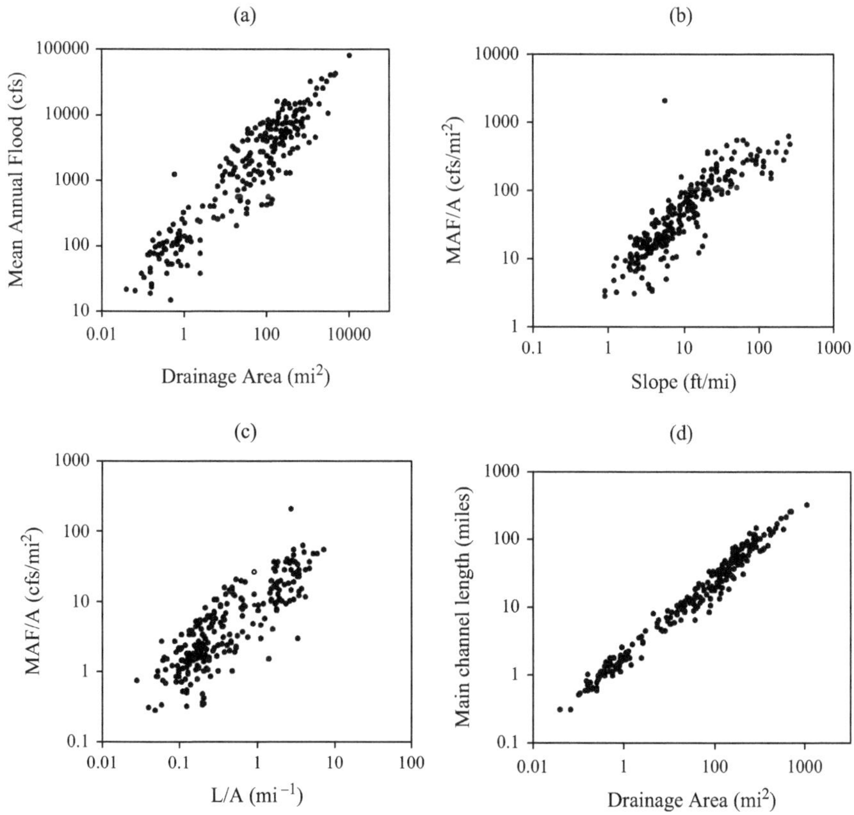

Fig. 2.4.1 Typical plots prepared for extraction of independent attributes for cluster analysis. MAF denotes mean annual flood, A is drainage area, and L refers to length of main channel

shows the correlation coefficient of precipitation volume (P×A) with MAF and MEF as 0.906 and 0.912, respectively. However, the dependence of MAF/A and MEF/A on depth of mean annual precipitation is insignificant. This is because mean annual precipitation does not reflect the exact volume of effective rainfall contributing to flood events experienced in a water year. Investigation of temporal variation of precipitation and its relationship with the date and time of occurrence of the resulting flood events would provide further insight in this regard. The meteorological attributes: mean annual precipitation and I(24,2) are correlated with correlation coefficient of 0.804 (Table 2.4.1). Mean annual precipitation is selected for inclusion in the feature vector for cluster analysis.

It is apparent from Table 2.4.1 that storage, which is percentage of the contributing drainage area covered by lakes, ponds, or wetlands is weakly correlated with MAF, MEF and MAF/P. Further, the dependence of flood related statistics on the attributes, elevation and forest cover in a drainage area is found to be insignificant. These two attributes are therefore not used for cluster analysis.

Table 2.4.1 Correlations between catchment attributes and flood related variables used to extract attributes to form feature vector. A: Area; L: Length of Main channel; ELEV: Average basin Elevation; STOR: Storage; FOR: Forest cover; P: Precipitation; RC: Runoff coefficient; MAF: Mean Annual Flood; MEF: Median Annual Flood; I(24,2): 24-hour rainfall having a recurrence interval of 2 years

Attribute	Catchment attributes											Flood related variables				
	A	Slope	L	ELEV	STOR	FOR	P	P × A	I(24,2)	I×A	RC	MAF	MAF/A	MEF	MEF/A	MAF/P
A	1.000	−0.169	0.850	−0.062	−0.001	−0.026	0.031	0.999	−0.016	0.999	0.029	0.903	−0.174	0.910	−0.168	0.901
Slope		1.000	−0.324	−0.246	−0.108	0.135	0.362	−0.163	0.347	−0.166	0.202	−0.239	0.511	−0.236	0.467	−0.247
L			1.000	0.034	0.009	−0.015	−0.007	0.839	−0.067	0.847	0.070	0.874	−0.322	0.875	−0.312	0.884
ELEV				1.000	0.145	−0.194	−0.444	−0.070	−0.593	−0.069	−0.004	0.005	−0.154	0.000	−0.412	0.020
STOR					1.000	0.024	−0.119	−0.006	−0.254	−0.005	−0.198	−0.051	0.272	−0.047	0.299	−0.049
FOR						1.000	0.453	−0.019	0.355	−0.020	0.215	0.056	0.026	0.050	0.024	0.043
P							1.000	0.051	0.804	0.044	0.528	0.145	0.141	0.136	0.129	0.119
P×A								1.000	0.001	0.999	0.040	0.906	−0.167	0.912	−0.162	0.901
I(24,2)									1.000	−0.001	0.539	0.086	0.226	0.080	0.209	0.066
I×A										1.000	0.039	0.907	−0.170	0.914	−0.164	0.904
RC											1.000	0.192	0.119	0.183	0.117	0.184

Note: Not all correlations are meaningful. They have been provided for completeness.

Directional statistics (measures of average time of occurrence and seasonality of flood events in catchments) could also be considered as attributes for cluster analysis (Burn, 1997). However, the catchments in Indiana did not show strong seasonal response (Rao et al., 2002).

Finally, the attributes selected for cluster analysis are: (i) four physiographic attributes: *drainage area*, *slope of the main channel in the drainage basin*, *soil runoff coefficient* and *storage*; and (ii) one meteorological attribute: *mean annual precipitation*. The geographic location attributes *latitude* and *longitude* are included in the feature vector to identify regions that are geographically contiguous. Geographically nearby sites could exhibit similar extreme flow responses due to similarities in the causative precipitation events that act as inputs to the flow generation process (Burn, 1990a).

Among the seven attributes, only drainage area was transformed by using logarithmic transformation. Then, each of the seven attributes was rescaled by using Eq. (2.3.1). Equal weight was assigned to all the attributes, implying equal importance to all the features.

2.4.2 Results from Clustering Algorithms

The K clusters obtained from the agglomerative hierarchical clustering component of the hybrid model after (245-K) merges are used to initiate the K-means algorithm. The value of the objective function (Eq. (2.3.2)), in general, decreases with increase in the number of clusters. It has maximum value when all the feature vectors are lumped in a single cluster and has a minimum value of zero when K equals the number of feature vectors considered for cluster analysis. This is because in the latter case the centroid of a cluster coincides with its feature vector. As a consequence, the objective function, which is the sum of squares of deviations of feature vectors about their respective cluster centers, would be zero.

The sites in a region should collectively supply five times as many station-years of record as the target return period (Reed et al., 1999). Several of the clusters obtained with the Indiana data for the choice of K greater than 10 are found to be quite small in size. Hence the results obtained for K greater than 10 are not presented and discussed further.

Variation in the optimal value of objective function for K ranging from 1 to 10 is presented in Table 2.4.2 for each of the three hybrid clustering models and their respective hierarchical clustering constituents, namely single linkage, complete linkage and Ward's hierarchical clustering algorithms. Further, to understand the effectiveness of Hybrid clustering over the K-means algorithm (KMA), the KMA was initialized with three different options. Option-1 initializes KMA with the first K feature vectors in the data set; Option-2 initializes KMA with centroids of the K clusters formed by uniform partitioning of the data; Option-3 initializes KMA with the K farthest feature vectors in the data set.

Among the three hierarchical clustering constituents of the hybrid model, Ward's algorithm gave the minimum value for the objective function (Table 2.4.2).

Table 2.4.2 Minimization of objective function given by Eq. (2.3.2) – A comparison between Hierarchical, K-means and Hybrid clustering models (K represents number of clusters; SL denotes single linkage; CL refers to complete linkage; W denotes Ward's). Option-1 initializes K-means algorithm (KMA) with the first K feature vectors in the data set; Option-2 initializes KMA with centroids of K clusters formed by uniform partitioning of the data; Option-3 initializes KMA with the K farthest feature vectors in the data set. The optimal value of objective function for a chosen value of K is shown in bold font

K	Hierarchical			K-Means (KM)			Hybrid clustering		
	SL	CL	W	Option-1	Option-2	Option-3	SL+KMA	CL+K	W+K
1	1704.3	1704.3	1704.3	1704.3	1704.3	1704.3	1704.3	1704.3	1704.3
2	1545.2	1509.1	1314.1	1280.6	1280.6	1306	1302.3	1302.3	**1277.0**
3	1535.5	1214.9	1120.3	1067.6	1069.6	1100.8	1097.9	1097.9	**1052.2**
4	1529.6	1085	971.9	895.4	899.7	894.1	920.5	**890.6**	896.1
5	1511.5	913.5	829	788.2	800.8	760	789.8	789.8	**746.0**
6	1473.7	772.5	721.1	660.5	768.6	670.9	673.9	**656.3**	668.9
7	1465.1	696.9	648.8	**585.9**	605.8	599.4	592.7	609.8	592.7
8	1449.7	632.2	583.7	**527.6**	555.4	571.7	584.1	554.8	531.9
9	1396.5	604.0	530.3	524.3	515.1	502.3	574.4	491.3	**490.6**
10	1394.8	586.7	491.6	455.5	460.9	457.5	567.6	486.6	**452.4**

As expected, the performance of each of the three hybrid-clustering algorithms in minimizing the objective function is better than that of the hierarchical clustering algorithm used to initialize them. In particular, the blend of Ward's algorithm and KMA gave the minimum value of objective function for most values of K in the range from 1 to 10. The blend of complete linkage and KMA yielded minimum value of objective function for the choice of K equal to 4 and 6. In essence, the overall performance of hybrid models in minimizing the objective function is better than that of the hierarchical and the K-means clustering models considered separately.

As mentioned in the Section 2.3.1, the output provided by K-means clustering algorithm depends on cluster centers used to initialize the algorithm. In a hybrid clustering algorithm, the hierarchical clustering model is expected to provide more meaningful initial values to the KMA, so that the KMA provides better and meaningful output. However, one cannot guarantee a better output from KMA by hybrid clustering. This point is evident from the results presented in Table 2.4.2 for choice of number of clusters, K, equal to 7 and 8, for which the KMA initialized with the first K feature vectors in the data set provided the smallest value of the objective function. In other words, one can always consider hybrid clustering as a potential option to initialize the KMA. However it is not always the best of all options for initialising it.

Before hybridization, the plausible hydrologic regions (or clusters) obtained from single-linkage, complete-linkage and Ward's clustering algorithms were examined. The clusters obtained from single linkage algorithm consisted of one large cluster and several small clusters, indicating that the algorithm is not suitable for regionalization of watersheds. Visual interpretation of the results showed that the clusters obtained from complete linkage algorithm are not clearly distinguishable (Fig. 2.4.2),

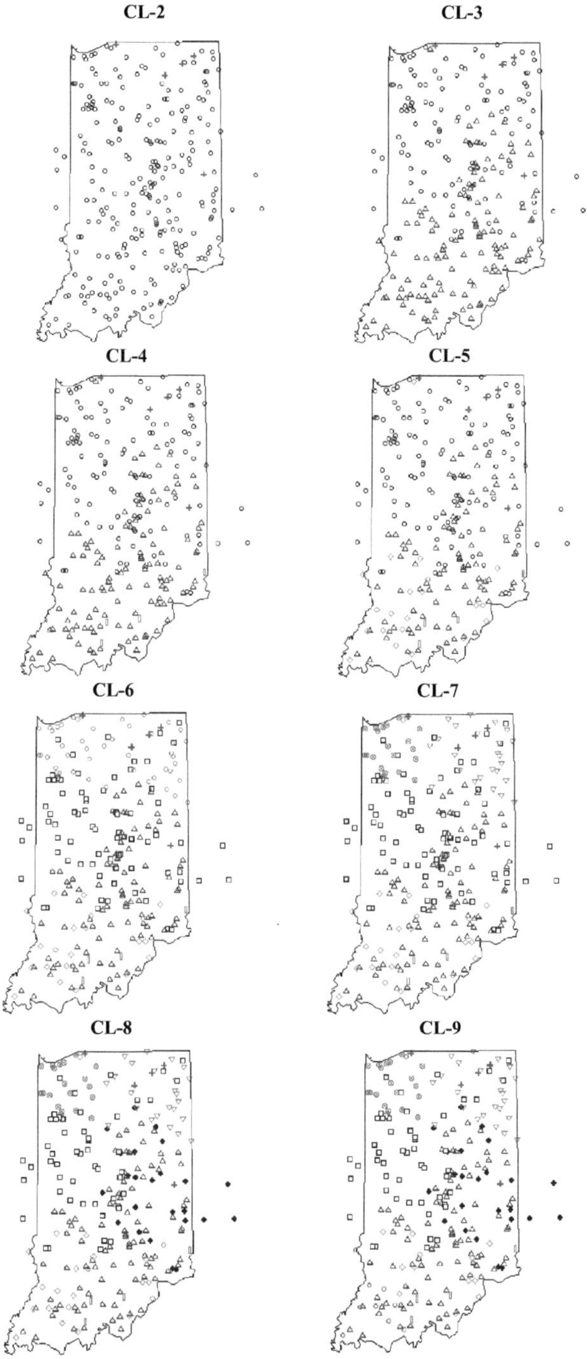

Fig. 2.4.2 Location of clusters (plausible homogeneous hydrologic regions) obtained from complete-linkage (CL) clustering algorithm. The number that follows CL denotes the number of clusters. Each of the symbols in the diagram characterizes a different cluster

while those resulting from Ward's algorithm are well separated for the values of K up to 6 (Fig. 2.4.3).

After hybridization, the clusters obtained from the three hybrid clustering algorithms were examined pictorially. In spite of considerable differences in the results from the three hierarchical clustering algorithms, the difference in results from the three hybrid clustering algorithms is found to be small (Fig. 2.4.4). The clusters obtained from the hybrid of Ward's algorithm and KMA are found to be very similar to those resulting from Ward's algorithm, indicating that the result provided by the Ward's algorithm is only slightly altered by the KMA to arrive at the final clusters. In contrast, the clusters resulting from single linkage algorithm are considerably modified by KMA.

2.4.3 Validation of the Results

The cluster validity indices, namely cophenetic correlation coefficient (CPCC), average silhouette width, Dunn's index, Davies-Bouldin index, Calinski-Harabasz Index, and minimum description length are computed for the clusters obtained from the clustering methods by Eqs. (2.3.14)–(2.3.26) to determine optimal number of clusters in the dataset. While CPCC is an index to validate results from hierarchical clustering, the other indices are useful to validate clusters obtained from hierarchical, partitional and hybrid clustering algorithms.

The CPCC is found to be considerably high for clusters obtained from single linkage algorithm than for clusters obtained from complete linkage and Ward's algorithms (Table 2.4.3). Following the definition of CPCC, one may argue that the multi-dimensional relationship within the input data is represented better in the dendrogram provided by single linkage algorithm than in the dendrograms provided by complete linkage and Ward's algorithms. This is in contradiction with our earlier findings that among the three agglomerative hierarchical clustering algorithms single linkage exhibits poor performance in optimizing the objective function and Ward's algorithm performs the best. Moreover, for the dataset considered herein, single linkage algorithm provides several singleton clusters and one very large cluster comprising more than 97% of the sites in the study region. This defeats the purpose of regionalization because such regions are highly heterogeneous. In essence, CPCC appears ineffective in suggesting a optimal partitioning scheme for the Indiana dataset. Detailed discussion on the performance of CPCC can be found in Holgersson (1978).

The average silhouette width (ASW), which has a feasible range from -1 to $+1$, varied generally within a narrow range of 0.31–0.46 for the Indiana data set over the variety of clustering options considered. The ASW is reasonably high for complete linkage clustering with K equal to 2 and for single linkage clustering with K in the range 2–4 (Table 2.4.4). However, these cases provide one large heterogeneous region (cluster) and remaining very small regions, which are not suitable for RFFA (Fig. 2.4.2 and Table 2.4.5). In general, the ASW of hybrid clusters is marginally

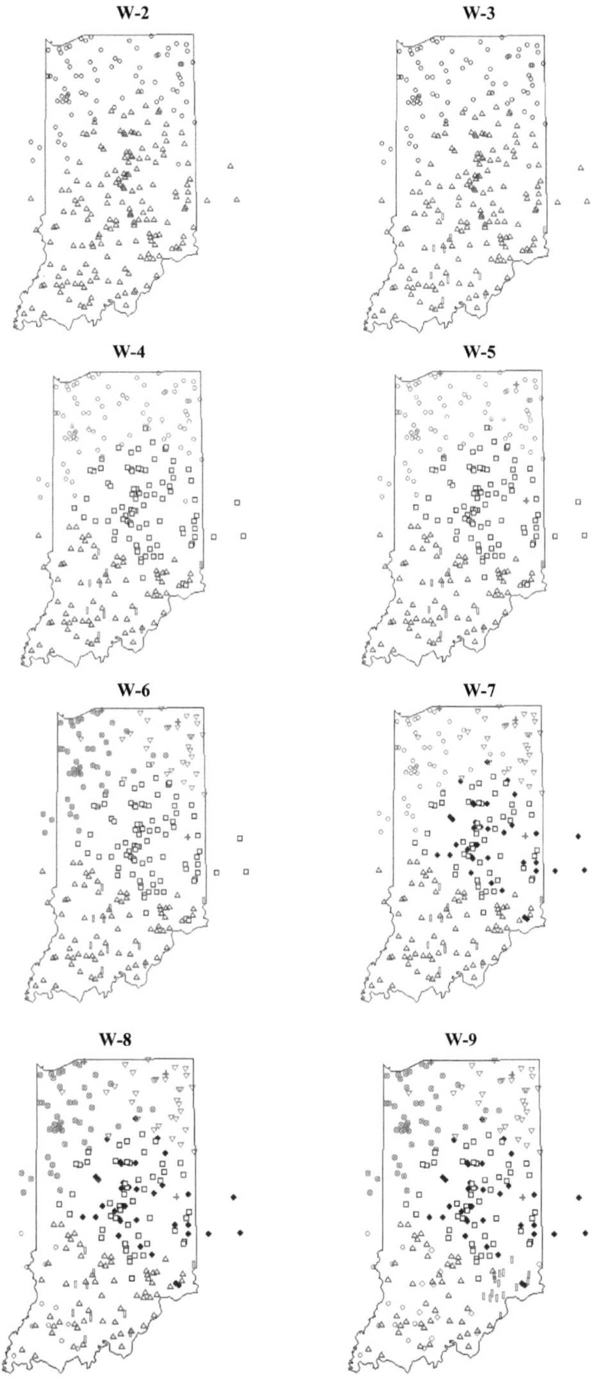

Fig. 2.4.3 Location of plausible hydrologic regions in Indiana obtained from Ward's (W) clustering algorithm. The number that follows W denotes the number of clusters. Each of the symbols in the diagram characterizes a different cluster

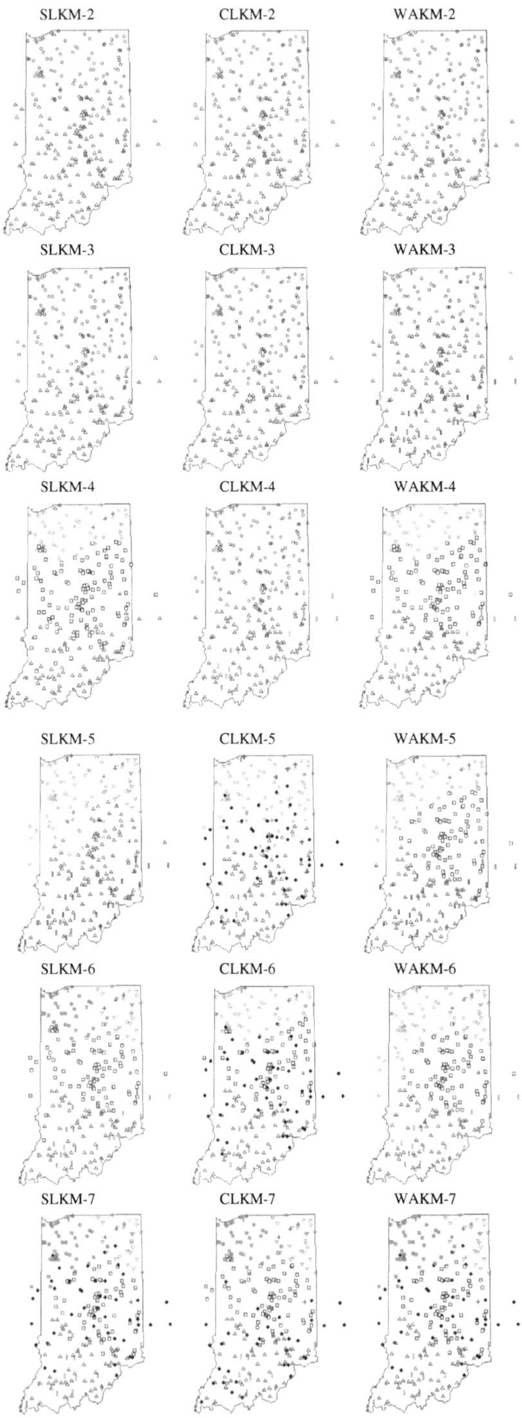

Fig. 2.4.3 (continued)

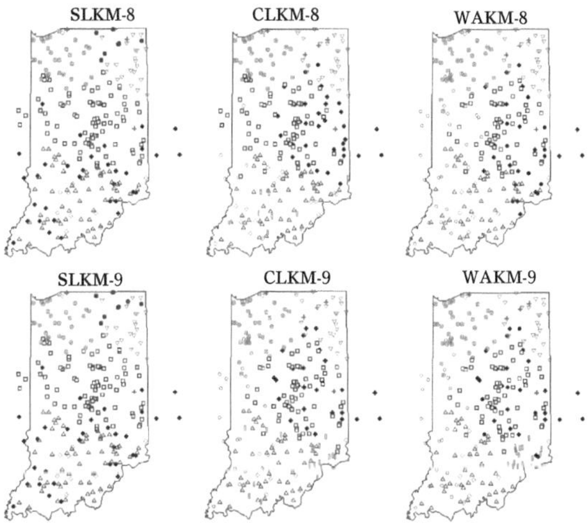

Fig. 2.4.4 Location of plausible hydrologic regions in Indiana obtained from the three hybrid clustering algorithms. SLKM – Single linkage and K-means, CLKM – Complete linkage and K-means, WAKM – Ward's and K-means. Each of the symbols in the diagram characterizes a different cluster

higher than that of the hierarchical clusters used to initialize the K-means algorithm (Table 2.4.4), suggesting improvement in performance due to hybridization.

Among the hybrid clustering models the ASW is found to be maximum (optimal) for the clusters obtained from the hybrid of Ward's and K-means algorithms with K equal to 9 (Table 2.4.4). The Dunn's index, Davies-Bouldin index, Calinski-Harabasz Index, and MDL also indicated the hybrid of Ward's and K-means algorithms to be the best (Fig. 2.4.5).

Optimal number of clusters could not be discerned using Calinski-Harabasz Index and MDL. Calinski-Harabasz Index suggested 2 clusters as optimal partition, whereas MDL suggested 4. These cases provide large heterogeneous clusters. The Dunn's index identified clusters obtained with K equal to 9 as optimal partition, whereas the Davies-Bouldin index suggested clusters obtained with K equal to 10 as optimal partition. Nevertheless, the difference in the value of Dunn's index between $K = 9$ and $K = 10$ for the hybrid of Ward's and K-means algorithms is found to be very small. Based on the foregoing analysis, the clusters provided by the hybrid of Ward's and K-means algorithms with $K = 10$ are selected at this stage for further analysis.

Table 2.4.3 Cluster validity using Cophenetic correlation coefficient (CPCC) – A measure to compare the performance of Hierarchical clustering algorithms

Algorithm	CPCC
Single linkage	0.72
Complete linkage	0.54
Ward's	0.50

Table 2.4.4 Cluster validity using Silhouette width – A comparison between Hierarchical, K-means and Hybrid clustering models (K represents number of clusters; SL denotes single linkage; CL refers to complete linkage; W denotes Ward's). Option-1 initializes K-means algorithm (KMA) with the first K feature vectors in the data set; Option-2 initializes KMA with centroids of K clusters formed by uniform partitioning of the data; Option-3 initializes KMA with the K farthest feature vectors in the data set. The optimal value of validity index for each model is shown in bold font

	Hierarchical			K-Means algorithms (KMA)			Hybrid clustering		
K	SL	CL	W	Option-1	Option-2	Option-3	SL+KMA	CL+KMA	W+KMA
2	**0.797**	**0.758**	0.397	0.382	0.382	0.392	0.392	0.392	0.382
3	0.767	0.277	**0.406**	0.329	0.365	0.411	**0.414**	0.414	0.424
4	0.748	0.279	0.250	0.380	0.318	**0.433**	0.366	0.378	0.318
5	0.455	0.302	0.295	0.404	0.326	0.377	0.395	0.375	0.366
6	0.459	0.319	0.345	0.382	0.254	0.369	0.382	0.382	0.372
7	0.067	0.327	0.274	0.414	0.330	0.387	0.382	0.380	0.381
8	0.022	0.320	0.314	0.418	0.339	0.371	0.381	0.376	0.407
9	0.020	0.319	0.356	0.392	0.338	0.426	0.377	**0.419**	**0.430**
10	0.011	0.313	0.353	**0.432**	**0.404**	0.431	0.383	0.380	0.425

The heterogeneity measures of Hosking and Wallis (1993, 1997), which are described in Section 1.4, are used as indicators to determine if the plausible regions resulting from the hybrid-clustering algorithms are homogeneous. In Figs. 2.4.6 and 2.4.7 the clusters obtained from the hybrid of Ward's algorithm and K-means algorithm for different choices of K are compared. Interestingly, increase in the number of clusters resulted in segregation of a collection of sites that are highly heterogeneous. Further, the best partition of Indiana data identified with Dunn's index and Davies-Bouldin index is found to contain clusters that are closer to being homogeneous.

It is seen from Fig. 2.4.7 that when the entire set of 245 sites was considered as a single cluster, the region is highly heterogeneous ($H_1 = 14.96$, $H_2 = 5.81$ and $H_3 = 1.10$). As the number of clusters, K, is increased beyond one, the algorithm exhibited the tendency to provide groups of sites that are relatively less heterogeneous. However, the sizes of clusters decrease, in general, with increase in K (Fig. 2.4.6). The collective record length of sites in a region should be reasonably large to make it effective for RFFA. Therefore, upper limit on K has been fixed following the recommendation of Reed et al. (1999) on data requirement for a region, as mentioned in Section 2.4.2.

Table 2.4.5 Sizes of clusters obtained by single linkage and complete linkage clustering

Number of clusters	Single linkage	Complete linkage
2	242, 3	240, 5
3	242, 2, 1	121, 5, 119
4	242, 1, 1, 1	121, 5, 115, 4
5	241, 1, 1, 1, 1	121, 5, 93, 4, 22

The groups of stations resulting from cluster analysis are, in general, heterogeneous and are therefore adjusted following the procedure described in Section 1.4.1 to improve their homogeneity. The options suggested by Hosking and Wallis (1997) for adjusting the regions resulting from clustering algorithm include: (i) eliminating (or deleting) one or more sites from the data set; (ii) transferring (or moving) one or more sites from a region to other regions; (iii) dividing a region to form two or more new regions; (iv) allowing a site to be shared by two or more regions; (v) dissolving regions by transferring their sites to other regions; (vi) merging a region with another or others; (vii) merging two or more regions and redefining groups; and (viii) obtaining more data and redefining regions. Of these, the first three options are useful in reducing the values of heterogeneity measures, whereas the options (iv) to (vii) help in ensuring that each region is sufficiently large. In this example presented, first the options (v) and (vi) are implemented to ensure that each region

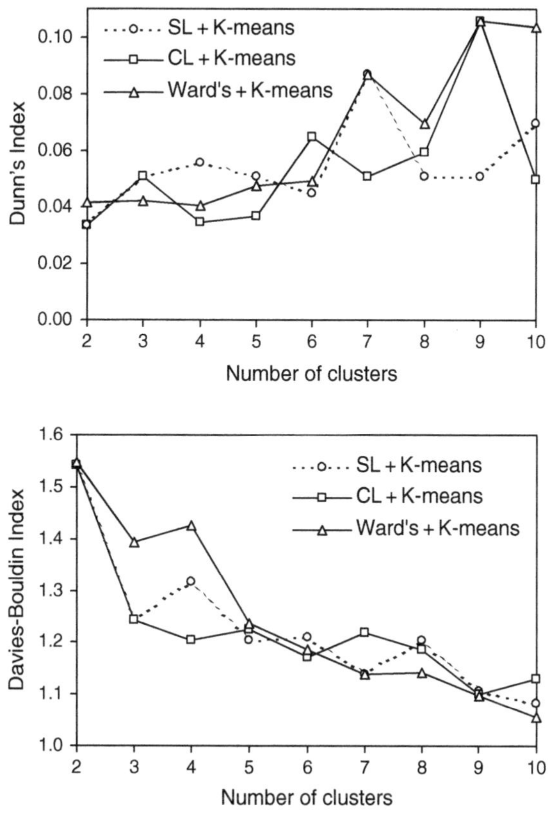

Fig. 2.4.5 Identification of optimal partition provided by the hybrid clustering algorithms using Dunn's index, Davies-Bouldin index, Calinski-Harabasz index, and minimum description length for the Indiana data set. The partition with the maximum value for Dunn's index (or Calinski-Harabasz index) and the minimum value for Davies-Bouldin index (or minimum description length) is taken as the optimal partition

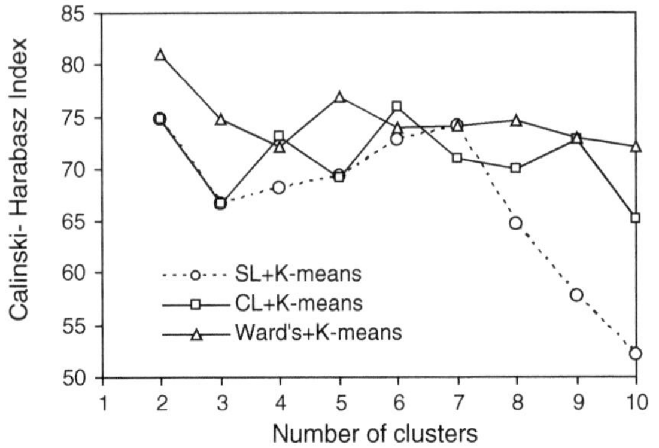

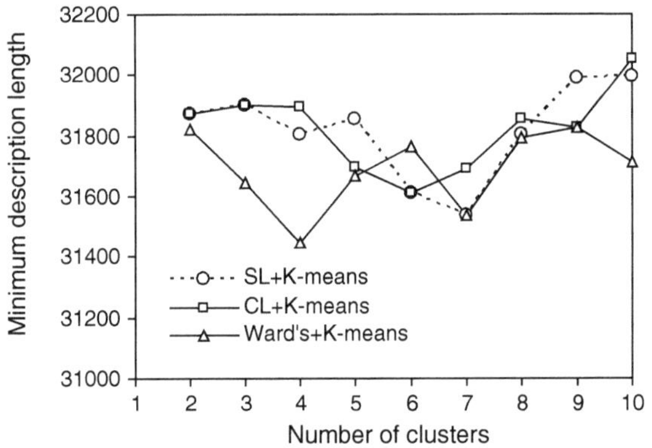

Fig. 2.4.5 (continued)

is sufficiently large. Next, the options (i)–(iii) are implemented to reduce the values of heterogeneity measures for the regions.

To adjust a region by options (i) and (ii), firstly the sites that are flagged discordant by the discordancy measure presented in Section 1.4.2 are identified. Though Hosking and Wallis (1997) provide critical values for the discordancy measure to declare a site unusual, it is worth identifying all the sites with high discordancy values. Secondly, the heterogeneity measures (H_1, H_2 and H_3) of the region to be adjusted are examined as they changed with exclusion of each site from the region. In this context, one site is eliminated at a time with replacement. Thirdly, the discordant site, whose exclusion reduces the heterogeneity measures of a region by a significant amount, is identified and removed from the region after ensuring that the site discordancy is not due to sampling variability (Fig. 2.4.8).

The Fig. 2.4.8a shows that the heterogeneity measures of region-4 improve significantly by eliminating the site having serial number 64 in the region that has high discordancy value in Fig. 2.4.8b. The Figs. 2.4.8c,d denote the scenario after elimination of the site with serial number 64 from the region. These figures further

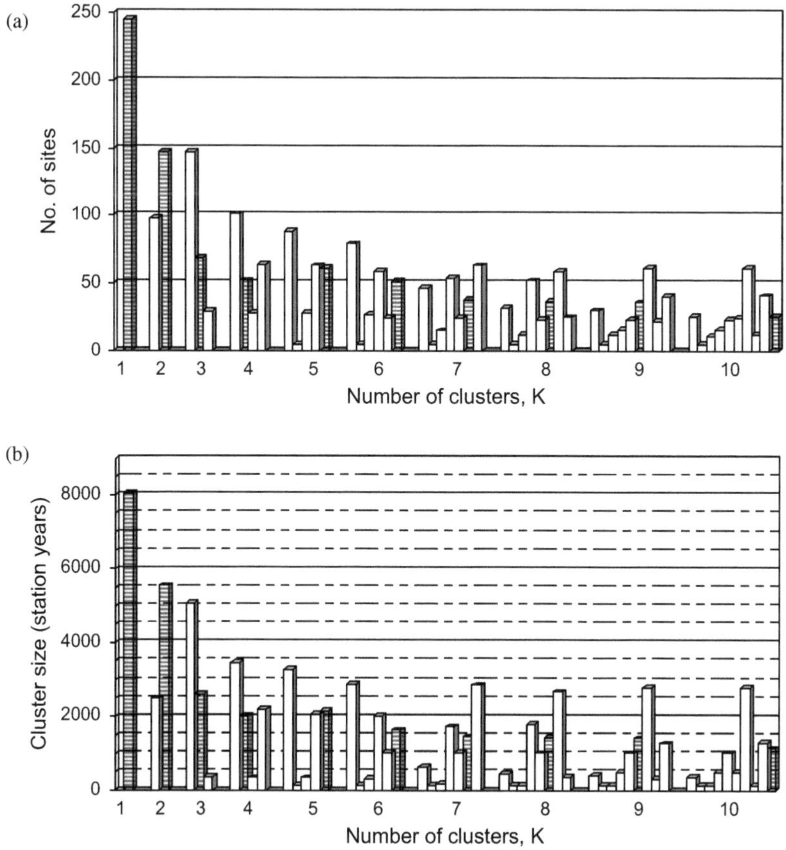

Fig. 2.4.6 Composition of clusters provided by the hybrid of Ward's and K-means algorithms with increase in the number of clusters from 1 to 10. In (**a**) the size of each cluster is expressed in terms of the number of sites or feature vectors contained in it. In (**b**) the size of each cluster is expressed as the sum of lengths of peak flow records (in years) at all the sites contained in it. In (**a**) and (**b**) the shaded bar shown for each K value denotes the most heterogeneous cluster

suggest eliminating site with serial number 7 that has high discordancy value in the updated region for improving its homogeneity.

The sites excluded from a region are examined to see whether they fit in any other region. In some instances, a site excluded from one region would fit in more than one region. Such a site is considered to be common to all the concerned regions.

Among the ten clusters identified as optimal partition for the Indiana data, the second cluster had just five sites. Following option (v) for adjusting the regions, this cluster is broken-up by transferring the sites contained in it to other regions. *Region-1* is obtained by merging clusters 6 and 8. The cluster 8 has just 12 sites and several of them are grouped with sites of cluster 6 for a lower choice of K. This

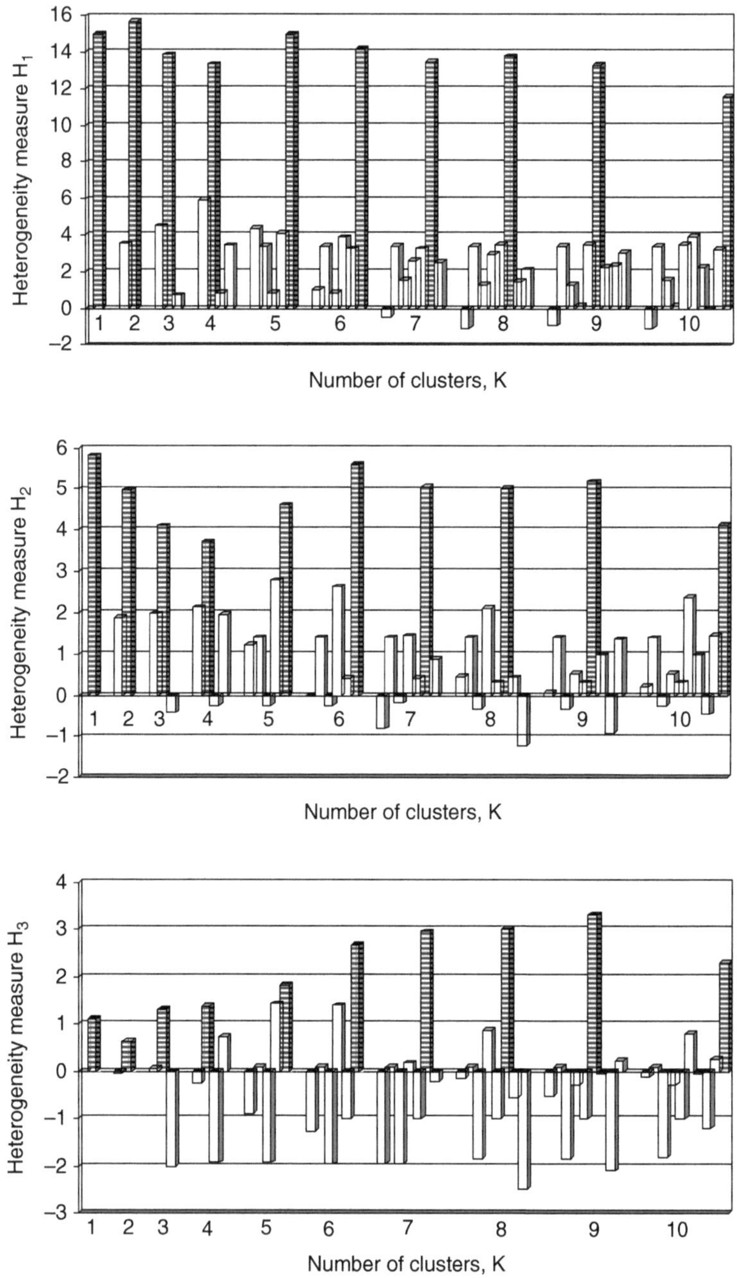

Fig. 2.4.7 Heterogeneity of clusters provided by the hybrid of Ward's and K-means algorithms with increase in the number of clusters from 1 to 10. The shaded bar shown for each K value denotes the most heterogeneous cluster

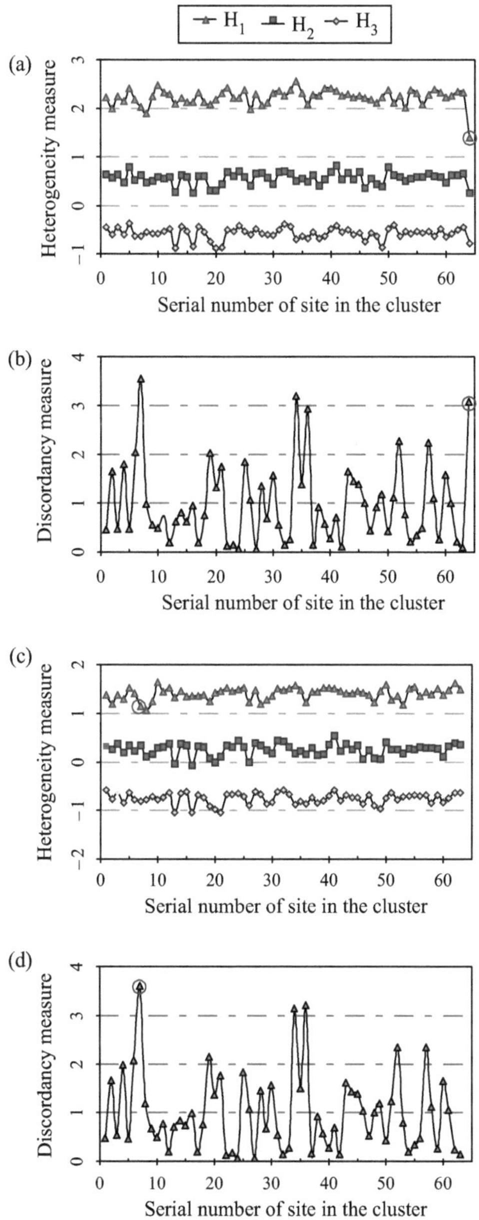

Fig. 2.4.8 Illustration of the procedure adopted for adjusting a region to improve its homogeneity. The discordancy values of sites in region-4, and heterogeneity indices that are computed after eliminating one site at a time from the region with replacement, are shown. The circle indicates the site identified for elimination from the region. (**a**) and (**b**) denote the scenario before elimination of the site identified for exclusion from the region, whereas (**c**) and (**d**) denote the scenario after elimination of the site

formed the basis for merging these two clusters and constitutes implementation of option (vi).

The cluster-3 has just 133 station-years of data making it the smallest of all clusters in terms of information. Following option (vi), the cluster-3 is merged with the cluster-9 that contains it geographically to obtain *Region-2*. The basins comprising clusters 3 and 9 appear as a single group for a smaller choice of K. The *Region 3* is formed from the fourth cluster, which characterises small drainage basins with high soil runoff coefficient. The *Region-4* is formed when cluster-1 and cluster-7 are merged following option (vi). Clusters 1 and 7 consist of geographically neighbouring drainage basins that have similar soil runoff characteristics, mean annual precipitation and surface storage features. They are, however, quite distinct in their drainage areas and slope of main streams draining the basins. Basins in clusters 1 and 7 appear as a single group for choice of K less than 7. This justifies merging these clusters.

The *Region-5* is formed from cluster 5. The tenth cluster was split into two using option (iii) for revising the regions. The first part comprising of a collection of heterogeneous sites form *Region-6*, whereas the second part with the homogeneous sites constituted *Region-7*. Following option (iv), majority of sites in the Region-7 could be considered common to both Region-7 and Region-5. All the regions obtained from the foregoing analysis are then adjusted using options (i) and (ii) to improve their homogeneity.

2.4.4 Testing the Regions for Robustness

The heterogeneity measures of Hosking and Wallis (1993, 1997) weigh information from each station in proportion to its record length. As a consequence, influence of stations with longer record length will be greater than that of stations with shorter record length. This may have adverse effects especially when some stations in a region have much longer record lengths than others. Therefore, the hydrologic regions are further examined for their robustness. By specifying various threshold values, the stations with record lengths significantly different from that of the rest of the group are removed and the region with the remaining stations was examined for homogeneity. In this step, the stations that have adverse affect on the homogeneity of the identified regions are excluded in an attempt to make the regions robust. The results of this exercise presented in Table 2.4.6 indicate that all the homogeneous regions identified are indeed robust.

2.4.5 Final Results

Fourteen sites, out of the 245 sites considered in this study, could not be allocated to any region. Four of these 14 sites were eliminated from the regions in the previous step to make them robust. Further, the remaining unallocated sites include a

Table 2.4.6 Results from the experiment performed to test the regions for robustness. R is region number, RL denotes record length, and NS represents the number of stations

R	Condition	NS	Heterogeneity measure			Region type
			H_1	H_2	H_3	
1	Entire region	48	0.60	0.03	−0.47	Homogeneous
	Sites with RL≤10 are eliminated	26	1.54	0.64	0.10	Possibly Homogeneous
2	Entire region	59	0.96	0.14	−0.86	Homogeneous
	Sites with RL<20 are eliminated	37	0.60	0.83	0.24	Homogeneous
	Sites with RL>50 are eliminated	49	0.78	0.19	−0.96	Homogeneous
	Sites with RL≤10 and RL>50 are eliminated	40	0.72	0.53	−0.29	Homogeneous
3	Entire region	32	−0.30	1.00	0.62	Homogeneous
	Sites with RL≤10 are eliminated	22	−0.31	0.72	0.53	Homogeneous
	Sites with RL<20 are eliminated	15	−0.41	0.01	−0.10	Homogeneous
4	Entire region	69	0.50	−0.28	−1.62	Homogeneous
	Sites with RL<20 are eliminated	59	0.90	0.28	−0.81	Homogeneous

catchment in Illinois and five catchments in Indiana which have less than or equal to 10 years of peak flow record. In hydrology, such catchments with short record length are often discarded in RFFA. However, they have been considered here to enable comparison with the regions derived by Glatfelter (1984).

The results presented in Table 2.4.7 indicate that regions 1 to 5 and region 7 are all acceptably homogeneous, while region-6 adjoining the Lake Michigan is highly heterogeneous and consists of 13 catchments in the Kankakee basin of Indiana.

For sites of the region-6, regional frequency analysis is not suitable as the region is highly heterogeneous. At the same time, it is not possible to reallocate (or transfer) sites from the region-6 to any other region because the heterogeneity measure of a region accepting site(s) from the region-6 increases dramatically. The average record length per station in the region-6 is 38-years, which is reasonably high.

Table 2.4.7 Characteristics of the regions formed by hybrid cluster analysis. NS represents the number of stations and RS denotes region size in station-years

Region number	NS	RS	Heterogeneity measure		
			H_1	H_2	H_3
1	48	820	0.60	0.03	−0.47
2	59	1790	0.96	0.14	−0.86
3	32	829	−0.30	1.00	0.62
4	69	2903	0.50	−0.28	−1.62
5	37	1705	0.48	−1.45	−1.56
6	13	493	12.40	5.99	2.78
7	14	543	0.32	0.08	0.00

Table 2.4.8 Characteristics of the regions formed by Glatfelter (1984). NS represents the number of stations and RS denotes region size in station-years

Region number			Heterogeneity measure		
	NS	RS	H_1	H_2	H_3
1	16	598	4.85	1.39	−0.62
2	31	1191	4.99	0.96	−0.62
3	60	3449	2.09	0.14	−0.40
4	46	1294	1.08	1.65	0.20
5	35	852	2.96	−0.24	−1.47
6	32	913	4.57	2.96	2.13
7	22	901	10.81	2.31	0.72

The regions formed by Glatfelter (1984) and those formed in this study are presented in Figs. 2.4.9 and 2.4.10, respectively. It is evident from the figures and Tables 2.4.7 and 2.4.8 that the regions identified by hybrid clustering differ significantly from those identified by Glatfelter (1984), and are better. All the homogeneous regions identified have enough pooled data (Table 2.4.7).

The region-1 identified using hybrid clustering is spread mainly along the course of Wabash river and consists predominantly of alluvial deposits of the flood plains. Region-2 contains karst formations associated with limestones of the Mississippian age, laid down 320–360 million years ago. Region-3 has a karst area consisting of older Devonian and Silurian limestones. Sinkholes, sinking streams, large springs and caves dominate the topography of these areas. For the ungauged catchments lying at the border between the regions 2 and 3, the possibility of including information from both the regions can be considered. Region-4 is in central Indiana. The soil in region 4 is predominantly loamy glacial till. Region 5 is spread over northern part of Indiana. It is composed of a wide range of soil classes (clayey glacial till, sandy and loamy deposits, loamy glacial till) overlying the Mississippian rocks of Michigan basin and Devonian and Mississippian shale. The delineated regions are found to resemble natural regions of Indiana (Figs. 2.4.11–2.4.13), thus lending credibility to this method of regionalization.

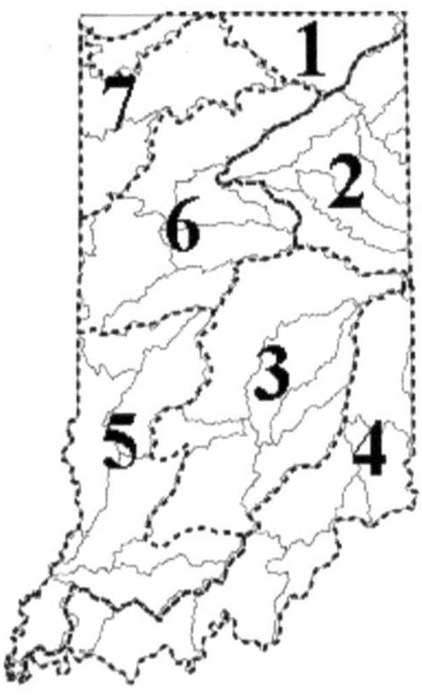

Fig. 2.4.9 The seven hydrological regions identified by Glatfelter (1984) for estimating the magnitude and frequency of floods on streams in Indiana

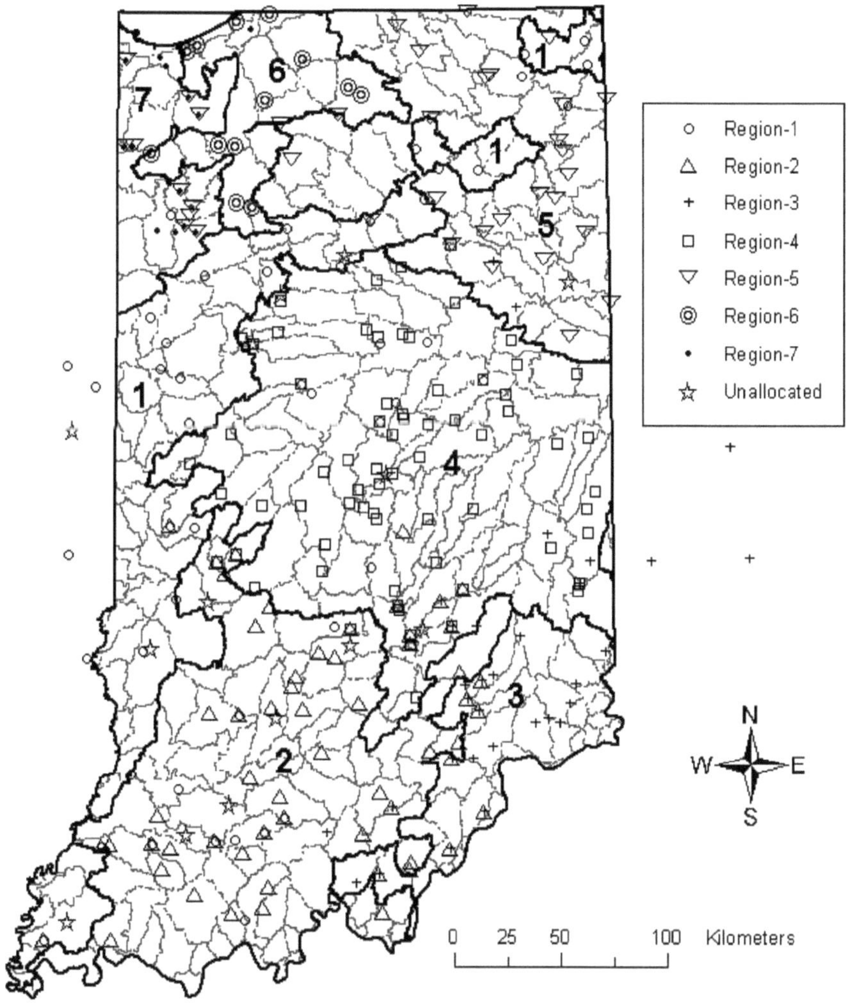

Fig. 2.4.10 Location of the regions defined by using the hybrid cluster analysis. The gray coloured lines within each region denote boundaries of 11 digit watersheds in Indiana, USA

2.5 Concluding Comments

The performance of hard clustering algorithms that are a blend of agglomerative hierarchical and partitional clustering procedures is investigated in regionalization of watersheds for flood-frequency analysis. The hierarchical clustering algorithms considered for hybridization are single linkage, complete linkage and Ward's algorithms, whereas the partitional clustering algorithm used is the K-means algorithm (KMA).

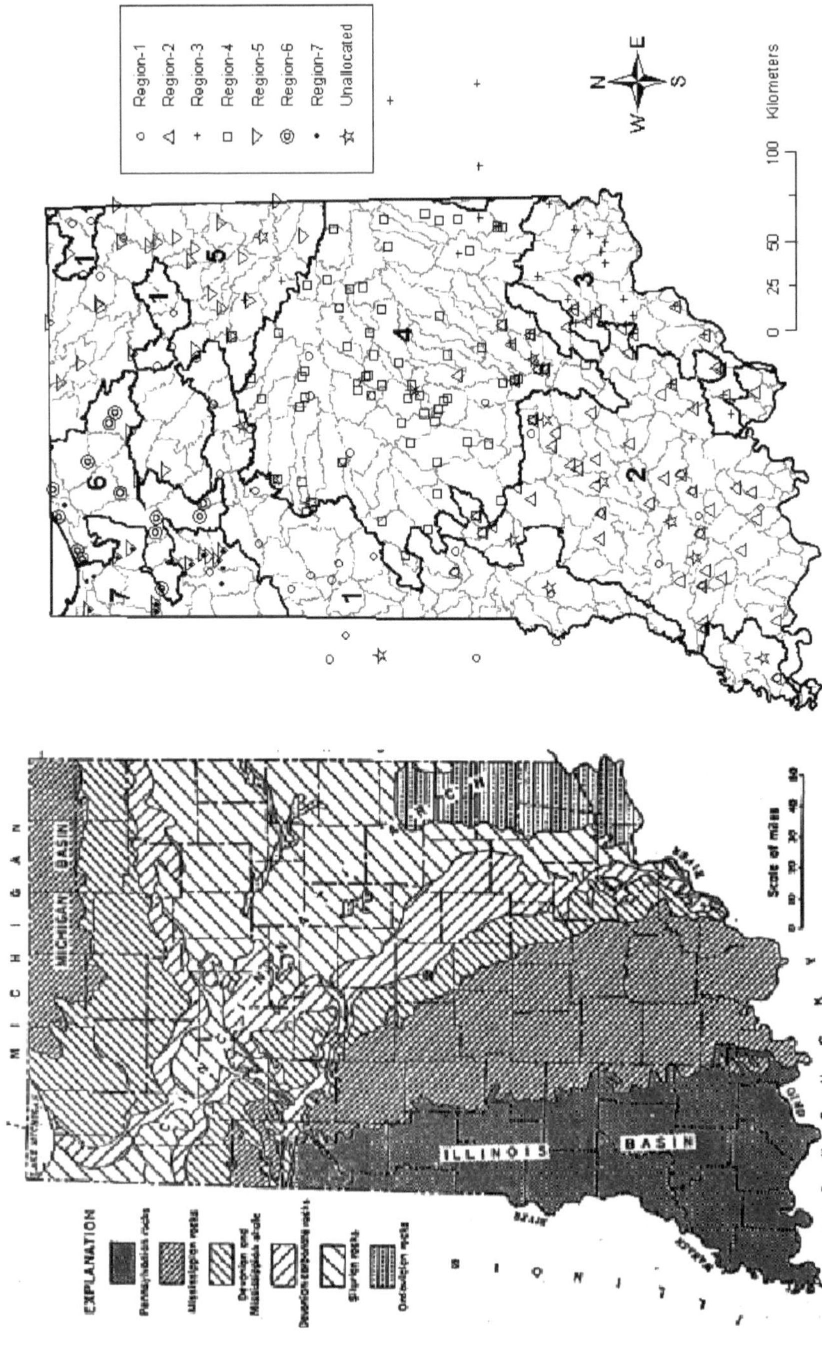

Fig. 2.4.11 Comparison of the hydrological regions identified in Indiana with geologic features of the state

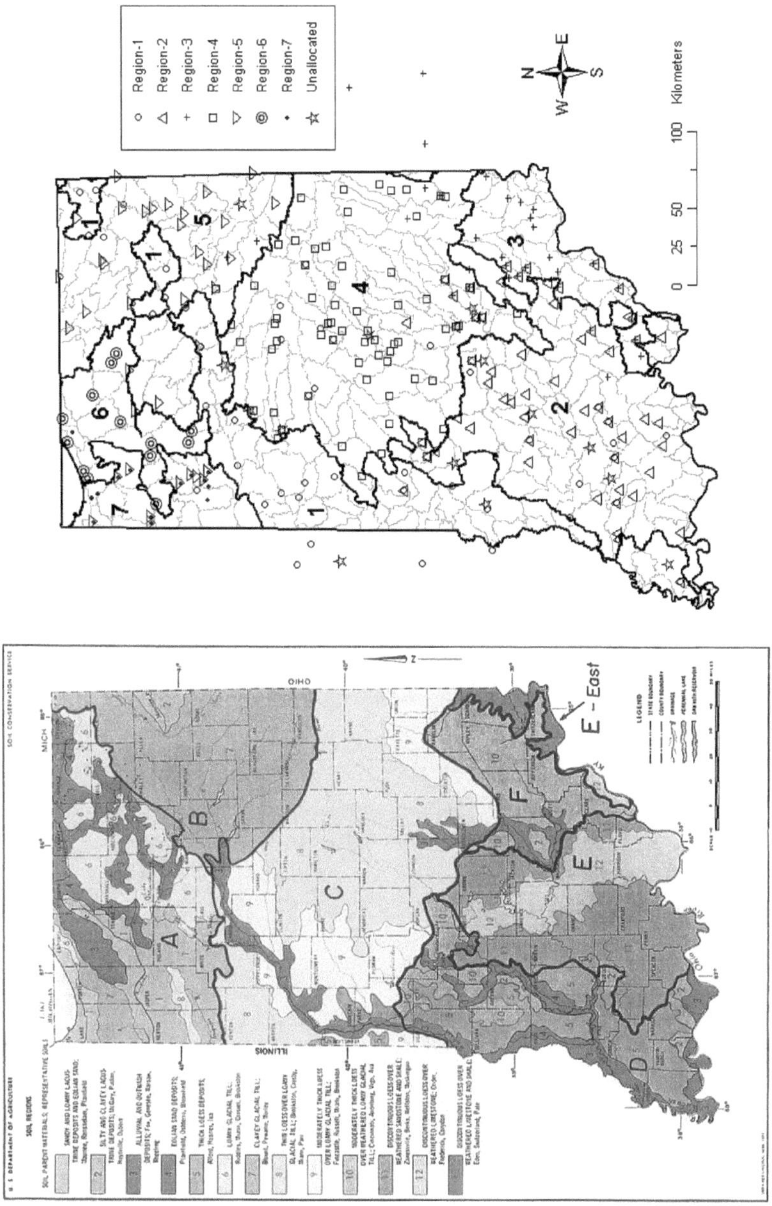

Fig. 2.4.12 Comparison of the hydrological regions identified in Indiana with soil regions in the state identified by soil conservation service, US Department of Agriculture

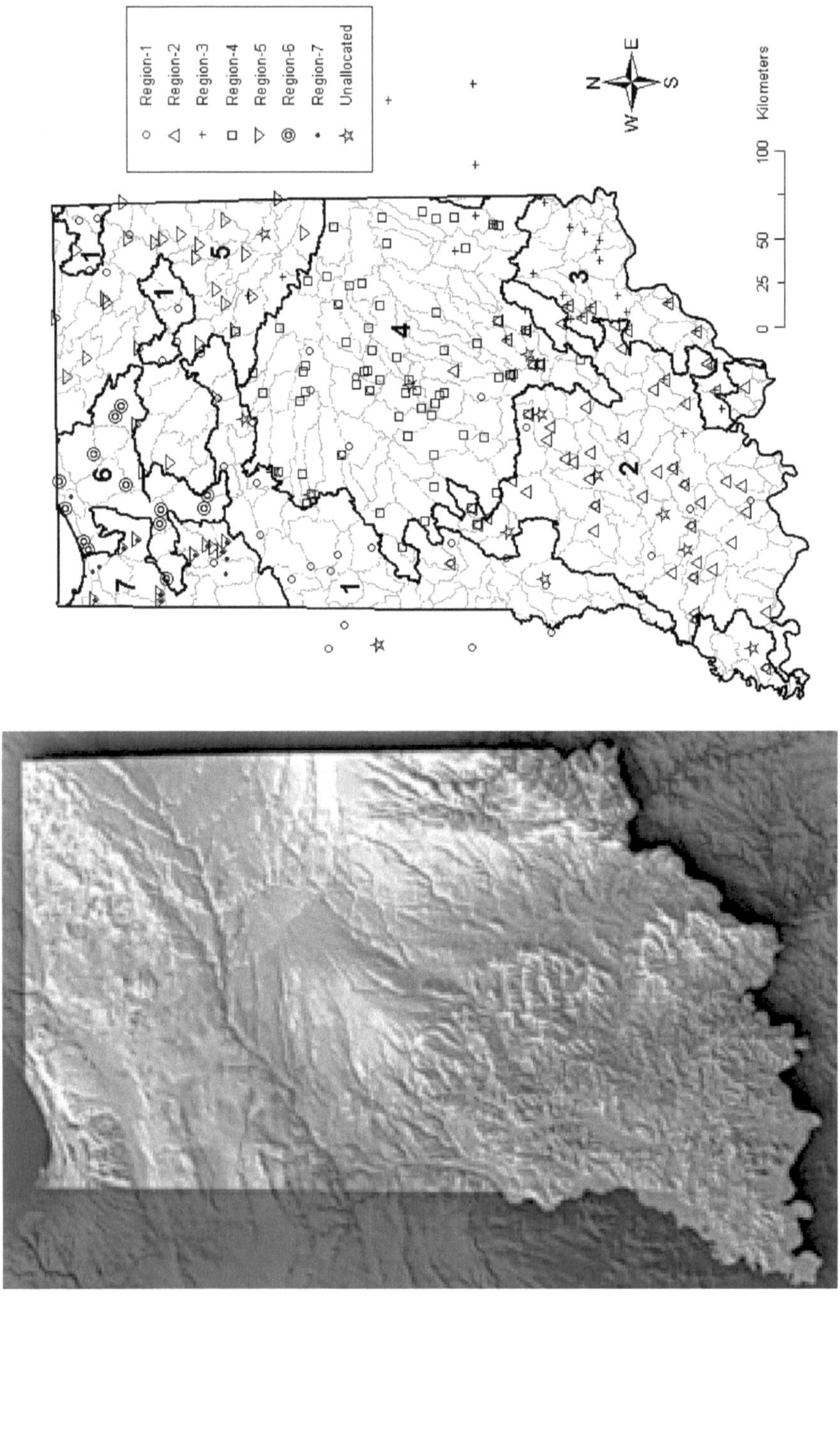

Fig. 2.4.13 Comparison of the hydrological regions identified in Indiana with tapestry produced by union of geology and topography of Indiana

In a hybrid-clustering algorithm, hierarchical clustering algorithm is expected to provide effective cluster centroids for initializing the KMA, so that the KMA provides better and meaningful output. Results obtained for Indiana watersheds showed that one cannot guarantee a better output from KMA by hybrid-clustering. The overall performance of hybrid models is found to be better than that of the hierarchical and the K-means clustering algorithms. Among the three hybrid models presented, the combination of Ward's and K-means algorithms consistently provided good initial estimates of groups of watersheds.

Six hard cluster validity indices, namely cophenetic correlation coefficient (CPCC), average silhouette width (ASW), Dunn's index, Davies-Bouldin index, Calinski-Harabasz Index, and minimum description length are tested to determine their effectiveness in identifying optimal partition provided by the hard clustering algorithms. The CPCC, Calinski-Harabasz Index, and minimum description length are found to be ineffective, whereas the ASW performed reasonably well. The Dunn's index and Davies-Bouldin index are found to be effective in identifying optimal partition containing clusters that are closer to being homogeneous. The clusters resulting from hard cluster analysis needed adjustment to improve their homogeneity.

Chapter 3
Regionalization by Fuzzy Cluster Analysis

3.1 Introduction

In the previous chapter, regionalization of watersheds using hybrid cluster analysis was discussed. The hybrid cluster analysis is a hard clustering method. In regionalization by hard clustering, a catchment is classified as belonging to or not belonging to a cluster. In reality, most catchments bear partial resemblance to several clusters. Therefore one cannot justify fully assigning a catchment to one cluster or another. In contrast, fuzzy clustering allows a catchment to have partial or distributed memberships in all the clusters. In other words, in fuzzy clustering, a catchment can belong to more than one cluster simultaneously. Thus, it results in identification of clusters with vague boundaries between them, as against crisp clusters with well-defined boundaries in the case of hard clustering. The fuzzy clustering method for regionalization, which is discussed in this chapter, is therefore expected to convey more information than hard clustering as it describes the reality better.

The fuzzy set theory (Zadeh, 1965) is a natural way to represent situations where data vectors bear partial resemblance to several clusters. Fuzzy clustering allows a feature vector to belong to all the clusters simultaneously with a certain degree of *membership* or *belonging* in the interval [0, 1]. Ruspini (1969, 1970) first introduced this idea, which was used by Dunn (1974) to construct a fuzzy clustering method.

In fuzzy clustering algorithms, there is not a total commitment of a data point to a given cluster. Therefore, they require more memory storage than hard clustering algorithms. However, advent of powerful computational facilities over the past three decades spurred the development and utility of fuzzy clustering methods for a variety of applications, including regionalization.

3.2 Classification of Fuzzy Clustering Algorithms

Clustering algorithms may be classified as supervised and unsupervised based on the uncertainty in the number of natural classes (or clusters) and hierarchies present in the data. Supervised clustering algorithms are used when the number of clusters in the input data set is known *a priori*, whereas unsupervised clustering algorithms are used when the number of clusters in the input data set is not known. In the context of regional flood frequency analysis, since the internal structure of the data is not known *a priori*, unsupervised clustering algorithms are the options for regionalization, by default.

The majority of unsupervised clustering algorithms start with two clusters. The number of clusters is increased every time after clustering is performed and cluster validity measures are computed. Cluster validity measures are independently used to evaluate and compare clustering partitions and even to determine optimal number of clusters existing in a data set. Unsupervised clustering algorithms differ from one another in their strategy of computing the new cluster center.

Fuzzy clustering methods can be divided into two types based on the strategy adopted for partitioning the data (Yang, 1993): One that uses a fuzzy relation to perform fuzzy clustering; the other that uses the objective function to determine fuzzy clustering. The groupings achieved from fuzzy relations are separate segments, whereas those resulting from the use of objective functions constitute soft segmentation. The fuzzy clustering based on fuzzy relations is proposed by Tamura et al. (1971). They presented a multistep procedure by using the composition of fuzzy relations beginning with a reflexive and symmetric relation. The description of the original fuzzy clustering algorithm based on objective function dates back to 1973 (Bezdek, 1973; Dunn, 1974). This algorithm was conceived in 1973 by Dunn (1974) and further generalized by Bezdek (1973, 1981) and Bezdek et al. (1984). Subsequently, Roubens (1982), Trauwaert (1985, 1988), Gath and Geva (1989), Gu and Dubuisson (1990), Xie and Beni (1991), Krishnapuram and Keller (1996), Frigui and Krishnapuram (1997, 1999) among others developed the approach to form fuzzy clusters. The description of these developments can be found in Bezdek and Pal (1992), Sato-Ilic and Jain (2006), and Oliveira and Pedrycz (2007).

Among the existing fuzzy clustering methods, the Fuzzy c-means (FCM) algorithm proposed by Bezdek (1981) is the simplest and is the most popular technique of clustering. It is an extension of the hard K-means algorithm to fuzzy framework. The hard K-means algorithm has been discussed in the previous chapter. The FCM algorithm has found applications in a variety of areas including agricultural engineering, astronomy, chemistry, geology, image analysis, medical diagnosis, shape analysis and target recognition (Bezdek, 1987).

In hydrology, several investigators have used hard cluster analysis for classifying watersheds into groups that are homogeneous in hydrologic response (Tasker, 1982; Burn, 1989; Bhaskar and O'Connor, 1989; Nathan and McMahon, 1990; Hosking and Wallis, 1997; Burn and Goel, 2000; Srinivas et al., 2002; Srinivas and Rao, 2002; Rao and Srinivas, 2006a). However, very few attempts have been made to explore the potential of fuzzy clustering for regionalization. Bargaoui et al. (1998) considered two fuzzy clustering methods, Iphigenie and ISODATA, for regionalization. Hall and Minns (1999) examined the utility of fuzzy c-means algorithm for regionalization by applying it to a sample of 101 gauged sites from two regions identified in the United Kingdom Flood Studies Report (NERC, 1975). The study considered catchment area, main stream length, main stream slope, mean annual rainfall and soil index as features for the analysis.

3.3 The Fuzzy C-Means Algorithm

A description of the fuzzy c-means algorithm is given in this section. Following this, the criteria used for evaluating the validity of the clusters resulting from the algorithm are discussed.

3.3.1 Description of the Algorithm

In the literature of fuzzy clustering, Fuzzy c-Means (FCM) algorithm proposed by Dunn (1974) and extended by Bezdek (1981) is popular. This algorithm, which is based on iterative optimization of a fuzzy objective function, is useful to partition N watersheds in a region into c fuzzy clusters.

Let $\boldsymbol{y}_k$ denote k-th feature vector depicting k-th watershed in n-dimensional feature space with coordinate axis labels $(y_1, \ldots, y_n)$, i.e., $\boldsymbol{y}_k = [y_{1k}, \ldots, y_{nk}] \in \Re^n$, where y_{ik} denotes the value of attribute i in $\boldsymbol{y}_k$. Various attributes, which have been considered for regionalization in flood frequency analysis, are described in Section 1.3. The attributes of the feature vector $\boldsymbol{y}_k$ are rescaled as

$$x_{ik} = \frac{w_i}{\sigma_i}\left[f(y_{ik})\right] \quad \text{for} \quad 1 \le i \le n, 1 \le k \le N \qquad (3.3.1)$$

where, x_{ik} denotes the rescaled value of y_{ik}; w_i is the weight assigned to attribute i; σ_i refers to the standard deviation of attribute i; $f(\cdot)$ represents the transformation function and N represents the number of n-dimensional feature vectors. Rescaling the attributes is necessary because of the differences in their variance, relative magnitude and importance.

The set of N rescaled feature vectors can be represented as a $n \times N$ data matrix $\boldsymbol{X}$.

$$\boldsymbol{X} = \begin{bmatrix} x_{11} & \cdots & x_{1N} \\ \cdots & \ddots & \cdots \\ x_{n1} & \cdots & x_{nN} \end{bmatrix} \qquad (3.3.2)$$

Further, let $\boldsymbol{V} = (\boldsymbol{v}_1, \ldots, \boldsymbol{v}_c)$ represent a c-tuple of prototypes $\boldsymbol{v}_i$, each of which characterizes the centroid of one of the c clusters. The FCM algorithm partitions the matrix $\boldsymbol{X}$ into c overlapping subsets (or clusters) by minimizing the following objective function.

$$\text{Minimize} \quad J(\boldsymbol{U},\boldsymbol{V} : \boldsymbol{X}) = \sum_{i=1}^{c} \sum_{k=1}^{N} (u_{ik})^{\mu} d^2(\boldsymbol{x}_k, \boldsymbol{v}_i) \qquad (3.3.3)$$

subject to the following constraints,

$$\sum_{i=1}^{c} u_{ik} = 1 \qquad \forall k \in \{1, \ldots, N\} \tag{3.3.4}$$

$$0 < \sum_{k=1}^{N} u_{ik} < N \qquad \forall i \in \{1, \ldots, c\} \tag{3.3.5}$$

where $u_{ik} \in [0, 1]$ denotes the membership (or degree of belongingness) of the k-th rescaled feature vector $\boldsymbol{x}_k$ in the i-th fuzzy cluster; $\boldsymbol{U}$ is the fuzzy partition matrix which contains the membership of each rescaled feature vector in each fuzzy cluster Eq. (3.3.6); the parameter $\mu \in [1, \infty]$ refers to the weight exponent for each fuzzy membership; $d^2(\boldsymbol{x}_k, \boldsymbol{v}_i)$ is the distance from k-th rescaled feature vector $\boldsymbol{x}_k$ to the centroid of i-th cluster $\boldsymbol{v}_i$. When point prototypes are used, the general form of the distance measure is given by Eq. (3.3.7)

$$\boldsymbol{U} = \begin{bmatrix} u_{11} & \ldots & u_{1k} & \ldots & u_{1N} \\ \vdots & & \vdots & & \vdots \\ u_{i1} & \ldots & u_{ik} & \ldots & u_{iN} \\ \vdots & & \vdots & & \vdots \\ u_{c1} & \ldots & u_{ck} & \ldots & u_{cN} \end{bmatrix}_{c \times N} \tag{3.3.6}$$

$$d^2(\mathbf{x}_k, \mathbf{v}_i) = (\mathbf{x}_k - \mathbf{v}_i)^{\mathrm{T}} \mathbf{A}_i (\mathbf{x}_k - \mathbf{v}_i) \tag{3.3.7}$$

where the norm $\mathbf{A}_i$ is a positive definite symmetric matrix associated with cluster i. For estimation of Euclidean distance between $\boldsymbol{x}_k$ and $\boldsymbol{v}_i$, $\mathbf{A}_i = \mathbf{I}\ \forall i$, where $\mathbf{I}$ is a unit matrix.

The first constraint Eq. (3.3.4) requires that the memberships of a chosen input feature vector over all the c fuzzy clusters should sum to 1.0. It is meaningful to assign very small membership values to a feature vector if it is representative of a catchment whose hydrologic response is quite dissimilar to hydrologic response of the other catchments considered for clustering. However, the first constraint does not permit the iterations of the FCM procedure to converge to a solution for which the memberships of the feature vector in all the c clusters do not sum to 1. Thus, in the solution of the FCM algorithm, there is a possibility that certain sites which do not fit in any of the identified regions would still have considerable membership values in all the clusters, such that they sum to one. This would, in turn, affect the homogeneity of the resulting clusters. To alleviate this problem, region adjustment procedures of Hosking and Wallis (1997), which are discussed in the first chapter, could be useful.

In the last decade, certain modifications have been proposed to the conventional FCM algorithm to overcome the aforementioned ill-effect of the first constraint

(Dave and Krishnapuram, 1997). However, these modified FCM clustering techniques are yet to find application in regional flood frequency analysis and investigating their advantage relative to conventional FCM in regionalization is still an open research issue.

The second constraint Eq. (3.3.5) ensures that the sum of membership degrees at a fuzzy cluster over the N feature vectors lies between 0 and N. If the sum of membership degrees at a fuzzy cluster is equal to zero, then it implies that the cluster does not contain any site. In contrast, the fuzzy cluster contains all the feature vectors if the sum equals N. Thus, each cluster has at least one feature vector in the optimal partition provided by the FCM algorithm.

The weight exponent μ in Eq. (3.3.3) determines the fuzziness of the clusters. It controls the extent of membership shared among fuzzy clusters. At $\mu = 1$, FCM converges in theory to the traditional k-means solution. In other words, the membership values u_{ik} tend to either 1 or 0 as the value of μ tends to 1. For $\mu \rightarrow \infty$, feature vectors tend to have equal membership in all the c clusters. Thus, in general, the degree of membership of the k-th rescaled feature vector $\boldsymbol{x}_k$ in the i-th fuzzy cluster, u_{ik} tends to $1/c$. Increase in the value of μ reduces the contribution to objective function from large values of $d^2(\boldsymbol{x}_k, \boldsymbol{v}_i)$. In other words, the sites whose characteristics are most dissimilar to the average characteristics of clusters (depicted by their centroids) are penalized less. As a consequence, the clusters tend to accommodate more sites.

The iterative procedure of FCM algorithm (Bezdek, 1981) is summarized below:

(i) Initialize fuzzy partition matrix $\boldsymbol{U}$ (or fuzzy cluster centroid matrix $\boldsymbol{V}$) using a random number generator.

(ii) If the FCM algorithm is initialised with fuzzy partition matrix $\boldsymbol{U}$, adjust the initial memberships u_{ik}^{init} of $\boldsymbol{x}_k$ belonging to cluster i using Eq. (3.3.8) so that Eq. (3.3.4) is satisfied.

$$u_{ik} = \frac{u_{ik}^{init}}{\sum_{i=1}^{c} u_{ik}^{init}} \quad \text{for} \quad 1 \leq i \leq c, \quad 1 \leq k \leq N \tag{3.3.8}$$

If the FCM algorithm is initialised with fuzzy cluster centroid matrix $\boldsymbol{V}$ (containing c fuzzy cluster centroids $\boldsymbol{v}_1^{init}, \ldots, \boldsymbol{v}_c^{init}$), determine memberships u_{ik} of $\boldsymbol{x}_k$ belonging to cluster i using Eq. (3.3.10) with $\boldsymbol{v}_i^{init}$ replacing $\boldsymbol{v}_i$.

(iii) Compute the fuzzy centroid $\boldsymbol{v}_i$ for $i = 1, 2, \ldots, c$ by Eq. (3.3.9)

$$\boldsymbol{v}_i = \frac{\sum_{k=1}^{N} (u_{ik})^{\mu} \boldsymbol{x}_k}{\sum_{k=1}^{N} (u_{ik})^{\mu}} \tag{3.3.9}$$

(iv) Update the fuzzy membership u_{ik} using

$$u_{ik} = \frac{\left(\frac{1}{d^2(\mathbf{x}_k,\mathbf{v}_i)}\right)^{1/(\mu-1)}}{\sum_{i=1}^{c}\left(\frac{1}{d^2(\mathbf{x}_k,\mathbf{v}_i)}\right)^{1/(\mu-1)}} \quad \text{for} \quad 1 \le i \le c, \quad 1 \le k \le N \tag{3.3.10}$$

Repeat steps (iii) and (iv) until change in the value of the memberships between two successive iterations becomes sufficiently small. At this point, the traditional methods of fuzzy cluster analysis recommend defuzzification of the fuzzy partition matrix, $\boldsymbol{U}$ (shown in Eq. (3.3.6)), to ultimately assign the feature vectors to clusters.

The fuzzy partition matrix can be defuzzified or hardened using the maximum-membership method or the nearest-center classifier (Ross, 1995, p. 398). In the maximum-membership method, the largest element in each column of $\boldsymbol{U}$ is assigned a membership value of unity and all the other elements in the column are assigned a membership value of zero Eq. (3.3.11). In other words, a feature vector is assigned to the cluster to which it has maximum resemblance. On the other hand, in the nearest-center classifier, each of the rescaled feature vectors, $\boldsymbol{x}_k$, is assigned to the cluster to whose centroid it is closest in terms of Euclidean distance Eq. (3.3.12).

$$u_{jk} = \max_{1\le i\le c}\{u_{ik}\} = 1; \quad u_{ik} = 0 \quad \text{for all} \quad i \ne j \tag{3.3.11}$$

$$\text{If } d_{jk} = \min_{1\le i\le c}\{d_{ik}\} = \min_{1\le i\le c}\|\mathbf{v}_i - \mathbf{x}_k\| \quad \text{then} \quad u_{jk} = 1; u_{ik} = 0 \quad \text{for all} \quad i \ne j \tag{3.3.12}$$

In hydrology, Hall and Minns (1999) used both Eqs. (3.3.11) and (3.3.12) to form hard clusters in fuzzy cluster analysis. The results of Rao and Srinivas (2006b) indicate that the effort needed to form homogeneous regions for RFFA is greatly reduced if fuzzy clusters are formed, rather than hard clusters by hardening the fuzzy partition matrix.

A fuzzy cluster is formed by assigning to it the sites whose memberships in the cluster exceed the specified threshold value. In general, the choice of a threshold value to form fuzzy clusters is subjective. In the fuzziest partition, the memberships of a feature vector in all the clusters would be equal to $1/c$. Therefore the value of $1/c$ is believed to be an acceptable choice for the threshold fuzzy membership.

The FCM algorithm may converge to a local minimum of the objective function. The optimal value of the objective function depends on initial guesses of number of clusters, cluster centers, and fuzzy memberships. These *a priori* assumptions are necessary but do not guarantee convergence to global minimum. Over the past two decades, researchers have been developing several heuristic cluster validity criteria to address the issue of convergence. This issue is discussed in Section 3.4.

3.3.2 Assignment of New Sites to Fuzzy Clusters

When a new site (gauged or ungauged) is considered for flood frequency analysis, its memberships in all the c clusters (determined by FCM algorithm), are computed by Eq. (3.3.10) with $\boldsymbol{x}_k$ replaced by a feature vector containing rescaled attributes of the new site, $\boldsymbol{x}_{\text{new}}$. In mathematical terms,

$$u_i^{\text{new}} = \frac{\left(\frac{1}{d^2(\boldsymbol{x}_{new}, \boldsymbol{v}_i)}\right)^{1/(\mu-1)}}{\sum_{i=1}^{c}\left(\frac{1}{d^2(\boldsymbol{x}_{new}, \boldsymbol{v}_i)}\right)^{1/(\mu-1)}} \quad \text{for} \quad 1 \le i \le c \tag{3.3.13}$$

where, u_i^{new} is the membership of the new site in i-th fuzzy cluster and $\boldsymbol{v}_i$ denotes the cluster centroid of the i-th fuzzy cluster. The attributes of the new site should be the same as those used to form the homogeneous regions and they should be rescaled using Eq. (3.3.1). If the new site is ungauged, it is assigned to all the cluster(s) in which it has membership greater than the specified sensible value of the threshold fuzzy membership (such as $1/c$). However, if the new site is gauged, it is assigned to those clusters in which it has membership greater than the threshold fuzzy membership after ensuring that the addition of the new site does not lead to significant increase in their statistical heterogeneity. As mentioned in Section 1.3.1, the at-site flood statistics must not be used as attributes to form regions for flood frequency analysis because they are used as the basis of an independent test of the homogeneity of the regions.

The desired flood quantile for a site that is common to two or more fuzzy regions, can be computed by using weighted average of the flood quantile values for the site estimated from the fuzzy regions. The weights may be assigned in proportion to the degree of membership of the site in the fuzzy clusters.

3.4 Fuzzy Cluster Validity Measures

Validity evaluation is a procedure to evaluate and compare clusters obtained from a clustering algorithm for different choices of parameters, or to compare clusters resulting from different clustering algorithms (Backer and Jain, 1981). In fuzzy cluster analysis the validity evaluation is carried out by using fuzzy cluster validity measures that are considered different from the objective function being optimized by the fuzzy clustering algorithm.

The criteria that are considered in cluster evaluation and selection include *compactness* and *separation* of clusters.

- *Compactness*: Optimal partition requires that the members of each cluster should be as close to each other as possible. A common measure of compactness is the variance, which should be minimized. If only compactness is considered as the validation criterion, then the best partition is obtained when each data point is considered as a separate cluster.

- *Separation*: Optimal partition requires that the clusters should be widely spaced. In other words, clusters should be far from each other. If only optimal separation is considered as the validation criterion, then the best partition is obtained when all the data points are included in a single cluster. For this case, the separation distance to nearest cluster is infinity.

Before describing various fuzzy cluster validity indices, definition of some terms is in order. Consider a fuzzy partition of the data set $\boldsymbol{X} = [\boldsymbol{x}_k; k = 1, \ldots, N]$ with $\boldsymbol{v}_i(i = 1, 2, \ldots, c)$ denoting the centroid of each cluster. Let u_{ik} $(i = 1, 2, \ldots, c; k = 1, 2, \ldots, N)$ denote the fuzzy membership of feature vector k in cluster i. Cardinality, variation, compactness and separation of a fuzzy cluster are defined as follows:

Cardinality of a fuzzy cluster:

Cardinality of a fuzzy cluster i, N_i^f, is equal to the sum of memberships of all the feature vectors in the cluster. In other words, it denotes fuzzy number of feature vectors in the cluster.

$$N_i^f = \sum_{k=1}^{N} u_{ik} \tag{3.4.1}$$

The fuzzy cardinality N_i^f need not be an integer. However, in the context of hard clustering, cardinality of a cluster is an integer. The sum of cardinalities of all the fuzzy clusters is equal to N.

$$\sum_{i=1}^{c} N_i^f = N \tag{3.4.2}$$

Variation of a fuzzy c-partition:

The fuzzy deviation of $\boldsymbol{x}_k$ from cluster i, d_{ik}, is defined as the distance between $\boldsymbol{x}_k$ and centroid of the cluster i, $\boldsymbol{v}_i$, weighted by u_{ik}:

$$d_{ik} = u_{ik} \, \|\boldsymbol{x}_k - \boldsymbol{v}_i\| \tag{3.4.3}$$

where $\|\cdot\|$ is the Euclidean norm. Instead, some other distance metric can also be used. Variation of fuzzy cluster i is defined as:

$$\sigma_i^f = \sum_{k=1}^{N} (d_{ik})^2 \tag{3.4.4}$$

The total variation of a data set (σ) with respect to the fuzzy c-partition is the summation of variations of all the clusters formed from the data set.

$$\sigma = \sum_{i=1}^{c} \sigma_i^f = \sum_{i=1}^{c}\sum_{k=1}^{N} (d_{ik})^2 = \sum_{i=1}^{c}\sum_{k=1}^{N} u_{ik}^2 \left\| \boldsymbol{x}_k - \boldsymbol{v}_i \right\|^2 \tag{3.4.5}$$

An optimal partition would result in smaller value for σ. Further, one may note that σ is the same as the objective function of the fuzzy c-means algorithm $J(\boldsymbol{U}, \boldsymbol{V}: \boldsymbol{X})$ for $\mu = 2$ in Eq. (3.3.3)

Compactness of a fuzzy c-partition:

The ratio of variation of a cluster to its cardinality is referred to as compactness of the cluster π_i.

$$\pi_i = \frac{\sigma_i^f}{N_i^f} \tag{3.4.6}$$

There are some alternate ways to define the compactness of a fuzzy c-partition, π. These include (i) the ratio of the total variation of a fuzzy c-partition to the size of the data set (σ/N), (ii) average compactness of clusters $(\sum_{i=1}^{c} \sigma_i^f / c)$, and (iii) compactness of a cluster that is largest.

$$\pi = \frac{\sigma}{N} \quad \text{or} \quad \frac{\sum_{i=1}^{c} \sigma_i^f}{N} \quad \text{or} \quad (max\ \pi_i, i = 1, \ldots, c) \tag{3.4.7}$$

Separation of a fuzzy c-partition (Xie and Beni, 1991):

Separation of a fuzzy c-partition may be defined as the minimum distance between cluster centroids $d_{\min}$.

$$d_{\min} = \min_{i \neq j} \left\| \boldsymbol{v}_i - \boldsymbol{v}_j \right\| \tag{3.4.8}$$

where $\boldsymbol{v}_i$ and $\boldsymbol{v}_j$ are the centroids of clusters i and j respectively ($1 \leq i \leq c, 1 \leq j \leq c$). A larger $d_{\min}$ indicates that the clusters are well separated.

In the following, various fuzzy cluster validity indices in vogue in literature are briefly described. Some cluster validity indices use only the membership values of a fuzzy partition of data. Examples include partition coefficient (Bezdek, 1974a,b), partition entropy (Bezdek, 1975), partition exponent (Windham, 1981), uniform data functional (Windham, 1982). These and similar indices may not be reliable because they have no connection to any property of the data.

(i) *Partition coefficient*: It was designed by Bezdek (1974a,b) to measure the amount of overlap between clusters.

$$V_{PC}(\mathbf{U}) = \frac{1}{N} \sum_{i=1}^{c}\sum_{k=1}^{N} (u_{ik})^2 \tag{3.4.9}$$

(ii) *Partition entropy (or classification entropy)*: Bezdek (1981) defines the classification entropy of a fuzzy c-partition as:

$$V_{PE}(\mathbf{U}) = -\frac{1}{N}\left[\sum_{i=1}^{c}\sum_{k=1}^{N} u_{ik}\ \log_a(u_{ik})\right] \tag{3.4.10}$$

where logarithmic base a $\in (1, \infty)$. Two other indices, namely fuzziness performance index V_{FPI}, and normalized classification entropy V_{NCE}, that are introduced by Roubens (1982), are determined using V_{PC} and V_{PE} as:

$$V_{FPI}(\mathbf{U}) = 1 - \frac{c \times V_{PC}(\mathbf{U}) - 1}{c - 1} \tag{3.4.11}$$

$$V_{NCE}(\mathbf{U}) = \frac{V_{PE}(\mathbf{U})}{\log_a(c)} \tag{3.4.12}$$

The optimal partition corresponds to a maximum value of V_{PC} (or minimum value of V_{PE}, V_{FPI} and V_{NCE}), which implies minimum overlap between clusters. The range of variation of V_{PC} is $[1/c, 1]$, while that of V_{PE} is $[0, \log_a(c)]$. On the other hand, the range of variation of V_{FPI} and V_{NCE} is [0, 1]. For a crisp (or hard) partition, V_{PC} is equal to 1, while V_{PE}, ,V_{FPI} and V_{NCE} are all equal to 0.

The V_{PC} takes the value $1/c$ and V_{PE} takes the value $\log_a(c)$ when the memberships of each feature vector in all the clusters are equal (i.e., $u_{ik} = 1/c\forall i, k$), which is the fuzziest c-partition.

The disadvantage of V_{PC}, V_{PE}, V_{FPI} and V_{NCE} is the lack of direct connection to any property of the data. In recent years, these validity indices have been used in hydrologic literature (Bargaoui et al., 1998; Hall and Minns, 1999; Güler and Thine, 2004)

While V_{PC} exhibits monotonic decreasing tendency with increase in the fuzzifier value, V_{PE}, V_{FPI} and V_{NCE} exhibit monotonic increasing tendency with increase in the value of this parameter. Furthermore, V_{PC} and V_{PE} are sensitive to the value of fuzzifier as $\mu \to 1$ and $\mu \to \infty$ (Halkidi et al., 2001, p. 138).

(iii) *Fukuyama and Sugeno index*: Fukuyama and Sugeno (1989) presented a validity measure, V_{FS}, by exploiting the compactness and the separation of clusters. Minimum value of V_{FS} implies an optimal partition, which corresponds to compact and well-separated clusters.

$$V_{FS}(\boldsymbol{U}, \boldsymbol{V} : \boldsymbol{X}) = \sum_{k=1}^{N}\sum_{i=1}^{c}(u_{ik})^{\mu}\ \|\boldsymbol{v}_i - \boldsymbol{x}_k\|_A^2 - \sum_{k=1}^{N}\sum_{i=1}^{c}(u_{ik})^{\mu}\ \|\boldsymbol{v}_i - \bar{\boldsymbol{v}}\|_A^2 \tag{3.4.13}$$

where $\|\cdot\|$ is the Euclidean norm, $\bar{\boldsymbol{v}}$ is the mean vector of $\boldsymbol{X}$ Eq. (3.4.14), $\boldsymbol{A}$ is a positive definite, symmetric matrix, and $\|\boldsymbol{X}\|_A = \sqrt{\boldsymbol{X}^T\boldsymbol{A}\boldsymbol{X}}$ is a inner product norm. When $\boldsymbol{A}$ is equal to unit matrix $\mathbf{I}$, the distance measure $\|\cdot\|_A^2$ in Eq. (3.4.13) becomes squared Euclidean distance.

$$\bar{\boldsymbol{v}} = \frac{1}{N}\sum_{k=1}^{N} \boldsymbol{x}_k \tag{3.4.14}$$

(iv) *Xie-Beni validity measure*: The index proposed by Xie and Beni (1991) is a function of the data set and the centroids of the clusters. It is defined as the ratio of overall compactness to the separation of a fuzzy c-partition (Xie and Beni, 1991, p. 842). The validity function is defined by

$$V_{XB}(\boldsymbol{U}, \boldsymbol{V} : \boldsymbol{X}) = \frac{\sum\limits_{i=1}^{c}\sum\limits_{k=1}^{N} (u_{ik})^2 \|\boldsymbol{v}_i - \boldsymbol{x}_k\|^2}{N \min\limits_{i \neq k} \|\boldsymbol{v}_i - \boldsymbol{v}_k\|^2} \tag{3.4.15}$$

where the term in the numerator is the sum of squares of fuzzy deviation of each feature vector $\boldsymbol{x}_k$ $(k = 1, \ldots, N)$ from the centroid of each fuzzy cluster $\boldsymbol{v}_i$ $(i = 1, \ldots, c)$. The magnitude of the term decreases with increase in compactness of the clusters. The denominator term, which measures the minimum separation between cluster centroids, has a larger value for clusters that are well separated. Minimum value of V_{XB} implies a good partition, which corresponds to compact and well-separated clusters. Xie and Beni (1991, p. 843) recommend substituting $(u_{ik})^\mu$ for $(u_{ik})^2$ in Eq. (3.4.15) when $\mu \neq 2$ in Eq. (3.3.3). Pal and Bezdek (1995, p. 374) refer to this as extended FCM Xie-Beni index $(V_{XB,m})$ which is given by

$$V_{XB,m}(\boldsymbol{U}, \boldsymbol{V} : \boldsymbol{X}) = \frac{\sum\limits_{i=1}^{c}\sum\limits_{k=1}^{N} (u_{ik})^\mu \|\boldsymbol{v}_i - \boldsymbol{x}_k\|^2}{N \min\limits_{i \neq k} \|\boldsymbol{v}_i - \boldsymbol{v}_k\|^2} \tag{3.4.16}$$

The value of V_{XB} monotonically decreases when the number of clusters gets large. To eliminate this problem, Kwon (1998) presented a new cluster validation measure V_K, which has a second term in numerator termed an *ad hoc* punishing function.

$$V_K(\boldsymbol{U}, \boldsymbol{V} : \boldsymbol{X}) = \frac{\sum\limits_{i=1}^{c}\sum\limits_{k=1}^{N} (u_{ik})^\mu \|\boldsymbol{v}_i - \boldsymbol{x}_k\|^2 + \frac{1}{c}\sum\limits_{i=1}^{c} \|\boldsymbol{v}_i - \bar{\boldsymbol{v}}\|^2}{\min\limits_{i \neq k} \|\boldsymbol{v}_i - \boldsymbol{v}_k\|^2} \tag{3.4.17}$$

Cluster validity has often been used to determine optimal number of clusters in a data set (e.g., Gath and Geva, 1989; Xie and Beni, 1991; Theodoridis and Koutroubas, 1999; Halkidi et al., 2001). The procedure requires fixing all the parameters of a clustering algorithm except number of clusters, c. Next, the parameter c is varied from 1 to a maximum value C_{max} in increments of 1. Values of a chosen cluster validity index that are computed for clusters obtained for each choice of c are analyzed to determine the optimal number of clusters in a given data set.

3.5 Example of Using Fuzzy C-Means Algorithm for Regionalization

3.5.1 Feature Extraction

The sensitivity of flood response of watersheds in Indiana to variation in the values of various physiographic, soil cover, land use, meteorological and geographical location attributes listed in Table 1.5.1 was discussed in Section 2.4.1 of the previous chapter. This lead to selection of the attributes mean annual precipitation, drainage area, slope of the main channel in the drainage basin, soil runoff coefficient, and storage. The geographic location attributes latitude and longitude are included in the feature vector to identify regions that are geographically contiguous.

Information pertaining to these attributes was available for the 245 stations. As before, only drainage area was transformed using logarithmic transformation. Then, each of the seven attributes was scaled so that their standard deviation is unity. Equal weight was assigned to all the attributes, implying equal importance to all the features.

3.5.2 Results from Fuzzy C-means Algorithm

Pal and Bezdek (1995, p. 370) mention that the FCM provides better performance for μ in the range 1.5–2.5. In this backdrop, the sensitivity of the results from the FCM algorithm to variation in the value of fuzzifier is examined by varying μ from 1.1 to 2.5 with an increment of 0.1. Variation in the value of objective function of FCM algorithm for c ranging from 2 to 10 and μ ranging from 1.1 to 2.0 is presented in Fig. 3.5.1.

It is evident from the figure that the value of objective function, in general, decreases with: (i) increase in the number of clusters for a specified value of fuzzifier μ and (ii) increase in the value of fuzzifier for a specified number of clusters. Theoretically, the objective function has a maximum value when all the sites (feature vectors) are grouped in a single cluster and a minimum value of zero when each cluster contains only one site.

The plausible hydrologic regions for flood frequency analysis are obtained by using two procedures. In the first procedure, the conventional defuzzification method

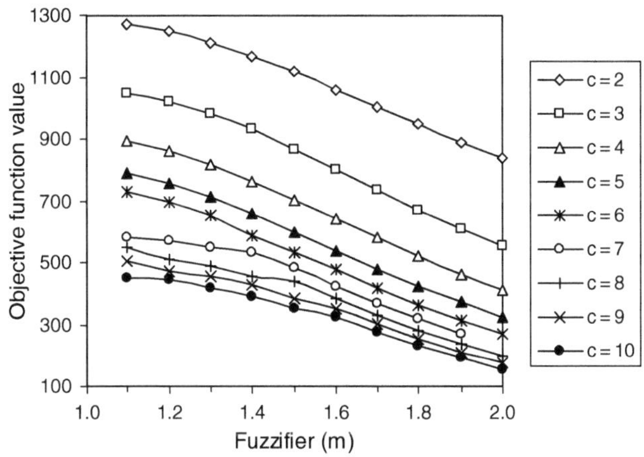

Fig. 3.5.1 Variation in the optimal value of objective function of the fuzzy clustering algorithm with change in the number of clusters (c) and the fuzzifier (μ) value

(Ross, 1995; Hall and Minns, 1999) given in Eq. (3.3.11) is used to harden the fuzzy partition matrices obtained for different values of c. Subsequently the resulting clusters are visually interpreted. Alternatively, in the second procedure, fuzzy cluster validity indices are used to determine optimal partition of watersheds and fuzzy clusters are formed by specifying a threshold value for fuzzy membership, as described in Section 3.3.1. The latter procedure retains fuzziness in the delineated regions.

3.5.2.1 Identification of Regions by Defuzzification and Visual Interpretation

To obtain plausible hydrologic regions in Indiana, the fuzzy partition matrix obtained from the FCM algorithm was defuzzified using the maximum-membership method described in Section 3.3.1. The resulting clusters were then visually interpreted. It is seen from Fig. 3.5.2 that for the choice of c equal to 2, the FCM algorithm provided two clusters with well-defined boundaries, one with watersheds in northern Indiana and the other with those in southern Indiana. It appears that the result is not sensitive to variation in the parameter μ for $c = 2$. However, it is evident from Fig. 3.5.3 that the two clusters tend to possess equal number of sites with increase in the value of fuzzifier μ. The change in the values of heterogeneity measures H_1 and H_2 with variation in μ is found to be insignificant and the cluster in the northern part of the state is highly heterogeneous (Fig. 3.5.4).

For the number of clusters equal to 3 and μ equal to 1.1, the FCM algorithm provided one well-defined cluster in the northern part of the state (shown in circles in Fig. 3.5.5), whereas the other two clusters that are shown in triangles and rectangles are vague in the sense that the boundaries between them are not well defined. Interestingly, the vagueness diminished and the clusters become well

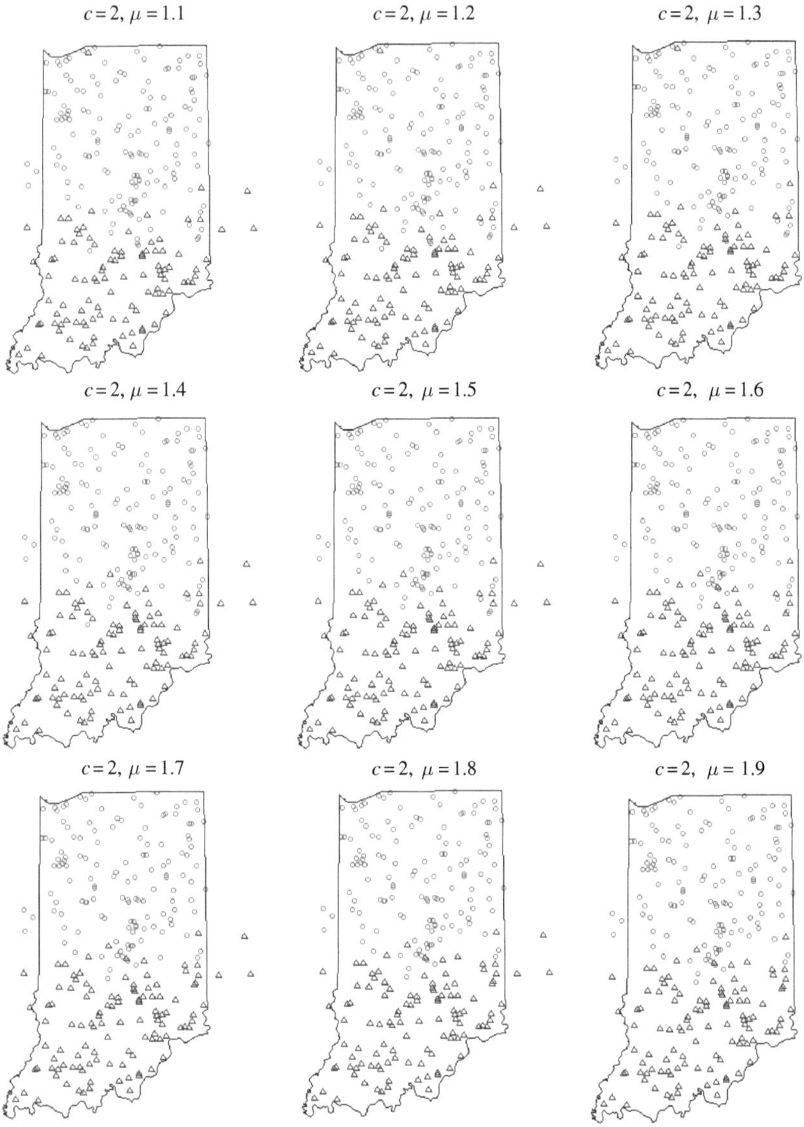

Fig. 3.5.2 Clusters of watersheds obtained in Indiana by hardening the fuzzy partition matrix obtained from FCM algorithm for $c = 2$. c denotes the number of clusters and μ refers to the value of fuzzifier. Each of the symbols in the diagram characterizes a site and different symbols denote different clusters

defined with increase in μ beyond 1.5. One of these clusters is predominantly in central Indiana, while the other cluster comprising of small drainage basins with steep slopes is identified in the karst area adjoining Kentucky in southern Indiana.

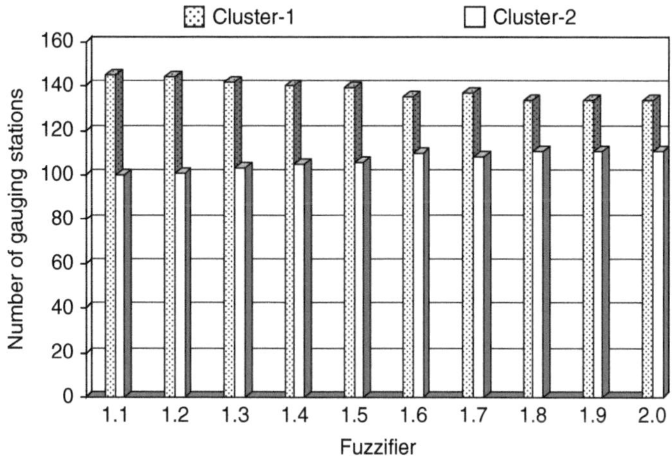

Fig. 3.5.3 Variation in size of clusters with increase in the value of fuzzifier – Results from FCM algorithm for $c = 2$

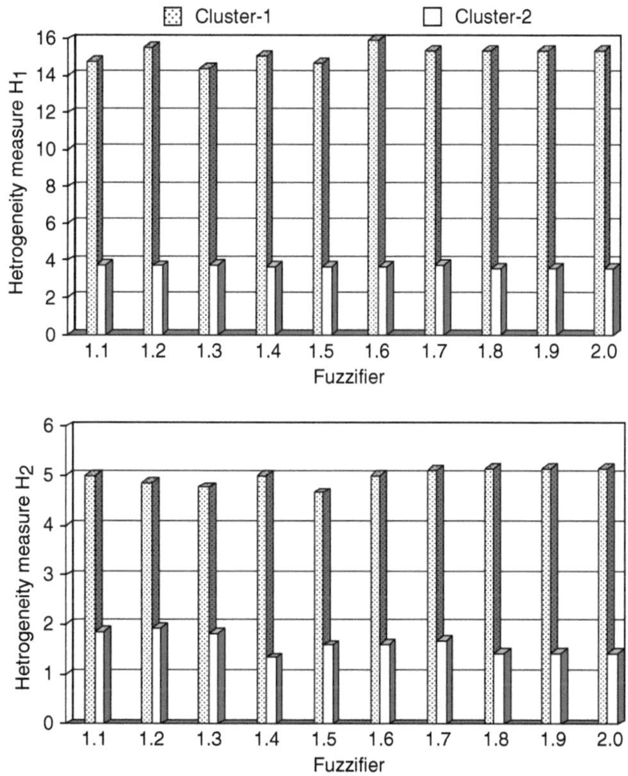

Fig. 3.5.4 Variation of heterogeneity measures with increase in the value of fuzzifier – Results from Fuzzy c-means algorithm for $c = 2$

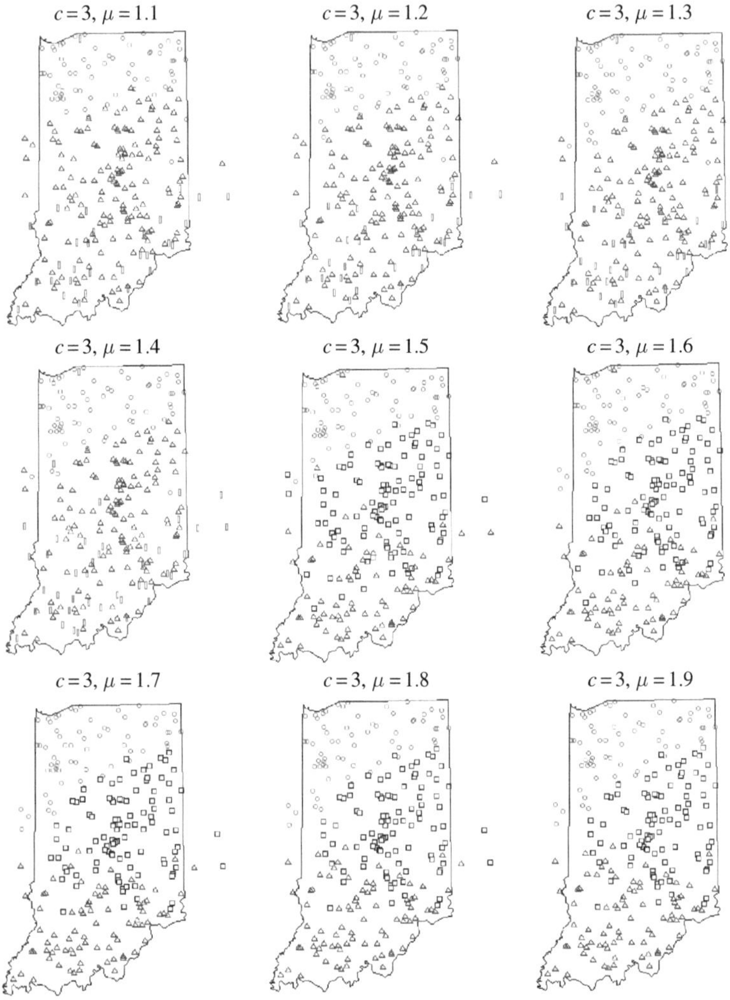

Fig. 3.5.5 Clusters of watersheds obtained in Indiana by hardening the fuzzy partition matrix obtained from FCM algorithm for $c = 3$. c denotes the number of clusters and μ refers to the value of fuzzifier. Each of the symbols in the diagram characterizes a site and different symbols denote different clusters

It is worth mentioning that the order of a cluster in the output of FCM algorithm might get altered with change in the value of fuzzifier. In other words, i-th cluster in the output of FCM algorithm may become j-th cluster with change in the value of fuzzifier μ.

It is seen from Fig. 3.5.6 that the three clusters tend to possess equal number of sites with increase in the value of fuzzifier μ. The cluster in the northern part of the state (shown in circles in Fig. 3.5.5) is highly heterogeneous and it consists of geographically neighboring drainage basins that have low runoff coefficient and

Table 3.5.1 Centers of the three clusters obtained from FCM algorithm for two typical values of fuzzifier (μ). In a few columns, the least or highest values of attributes are shown in bold font for each value of μ

μ	CN	A (miles)^2	Slope (ft/mile)	LAT	LONG	STOR (%)	P (in)	RC
1.1	1	45.75	8.34	41.20	86.26	**1.429**	36.70	**0.49**
	2	**0.27**	**111.62**	38.89	86.24	0.315	42.10	0.79
	3	61.28	11.92	39.59	86.18	0.438	39.91	0.76
1.5	1	41.42	10.77	41.09	86.39	**1.014**	36.92	**0.49**
	2	95.84	11.37	39.87	86.00	0.473	39.15	0.74
	3	**3.88**	**52.88**	38.90	86.39	0.567	41.99	0.79

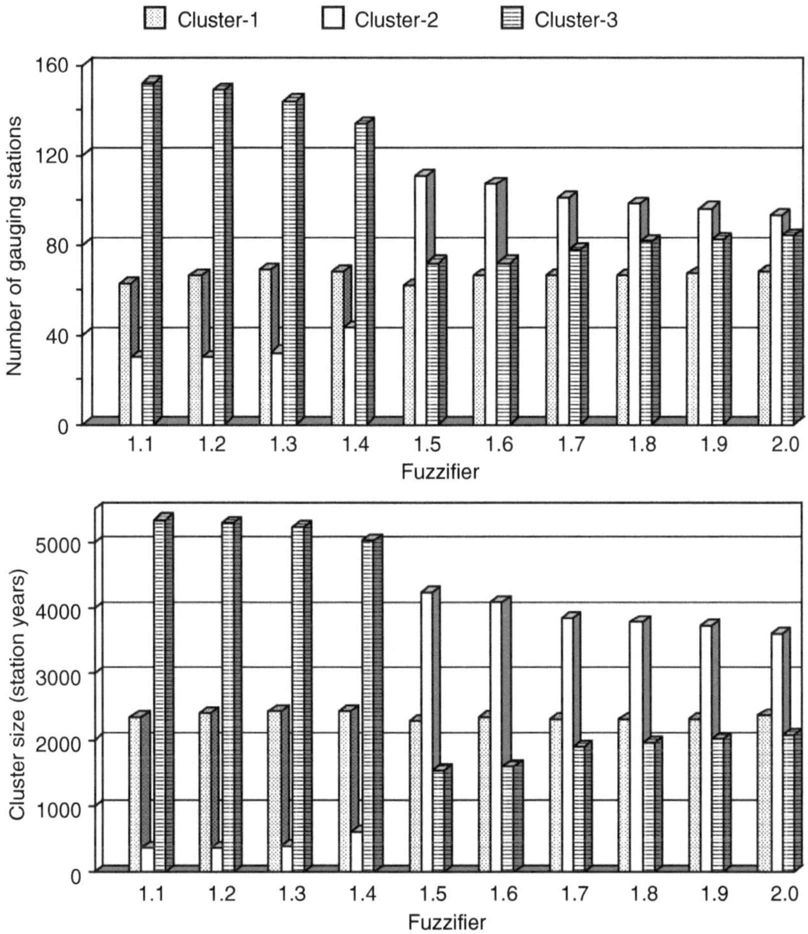

Fig. 3.5.6 Variation in size of clusters with increase in the value of fuzzifier – Results from FCM algorithm for $c = 3$

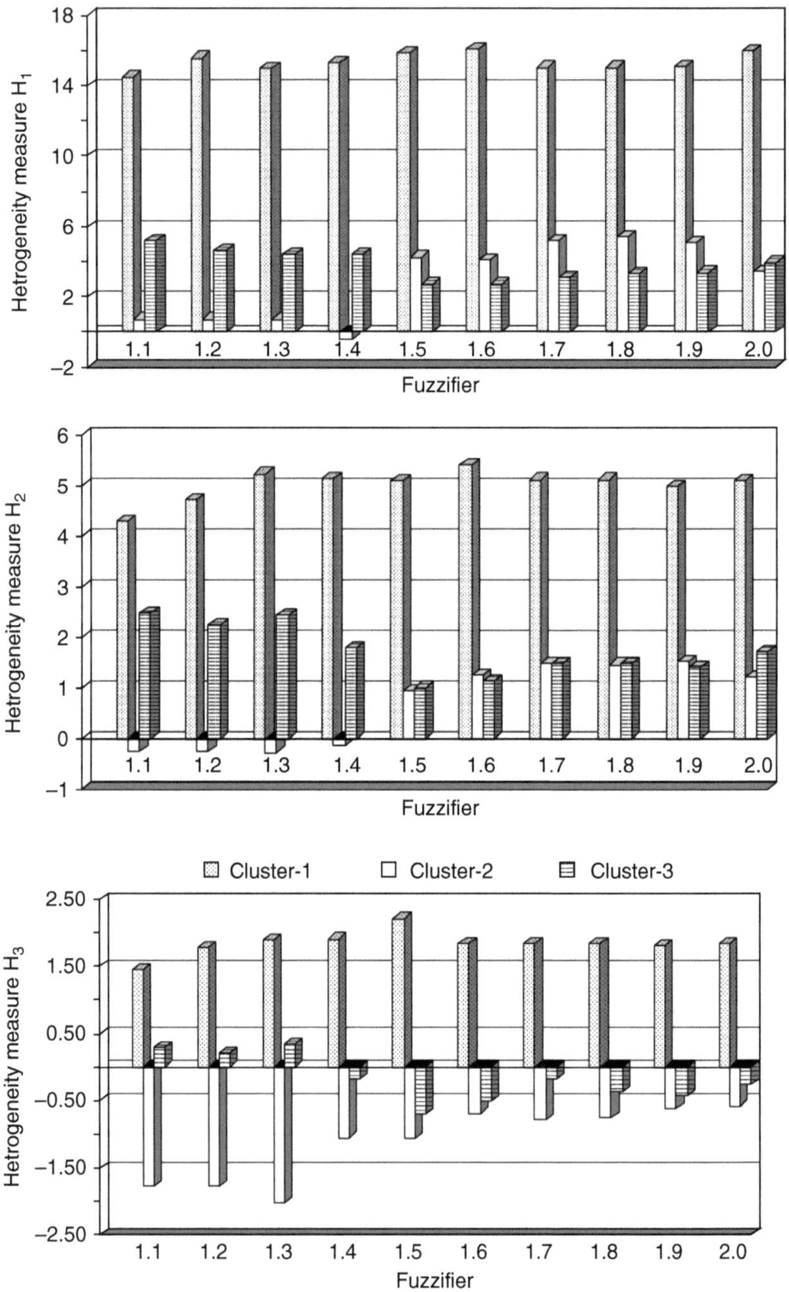

Fig. 3.5.7 Variation of heterogeneity measures with increase in the value of fuzzifier – Results from Fuzzy c-means algorithm for $c = 3$

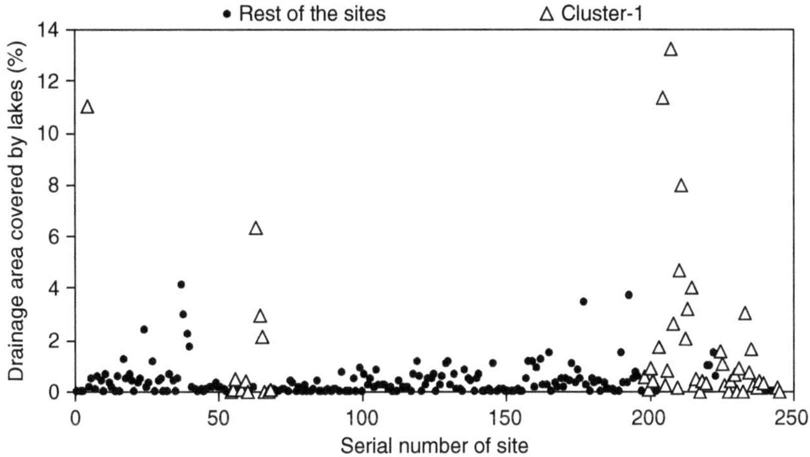

Fig. 3.5.8 Comparison of storage in drainage basins forming cluster-1 with storage at the other sites for $c = 4$ and $\mu = 1.1$

high surface storage (see cluster 1 in Table 3.5.1 and Fig. 3.5.7). Similar observations were made with regard to cluster-1 for the choice of c = 4 and $\mu = 1.1$ (Figs. 3.5.8 and 3.5.9). Change in the value of heterogeneity measure H_1 for the cluster in northern Indiana with variation in the value of μ is found to be insignificant (Fig. 3.5.7).

When c is 4 with μ equal to 1.1, the result showed two well-defined clusters (represented by circles and squares in Fig. 3.5.10), one in the northern part of the state

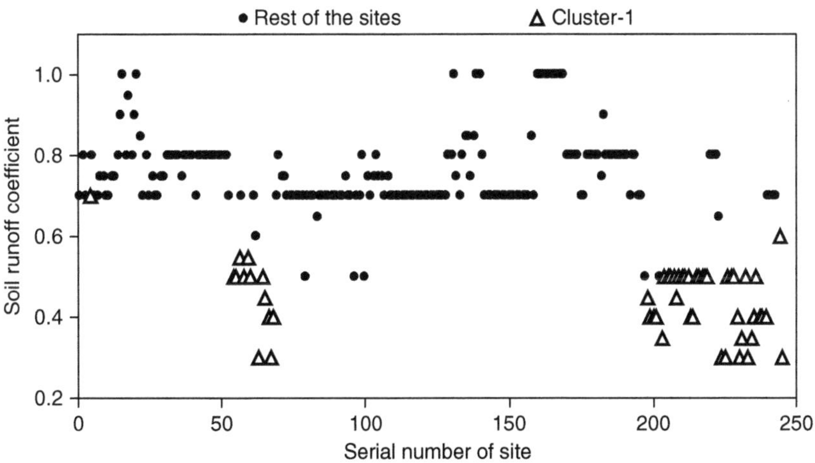

Fig. 3.5.9 Comparision of soil runoff coefficient values of drainage basins forming cluster-1 with those at the other sites for $c = 4$ and $\mu = 1.1$

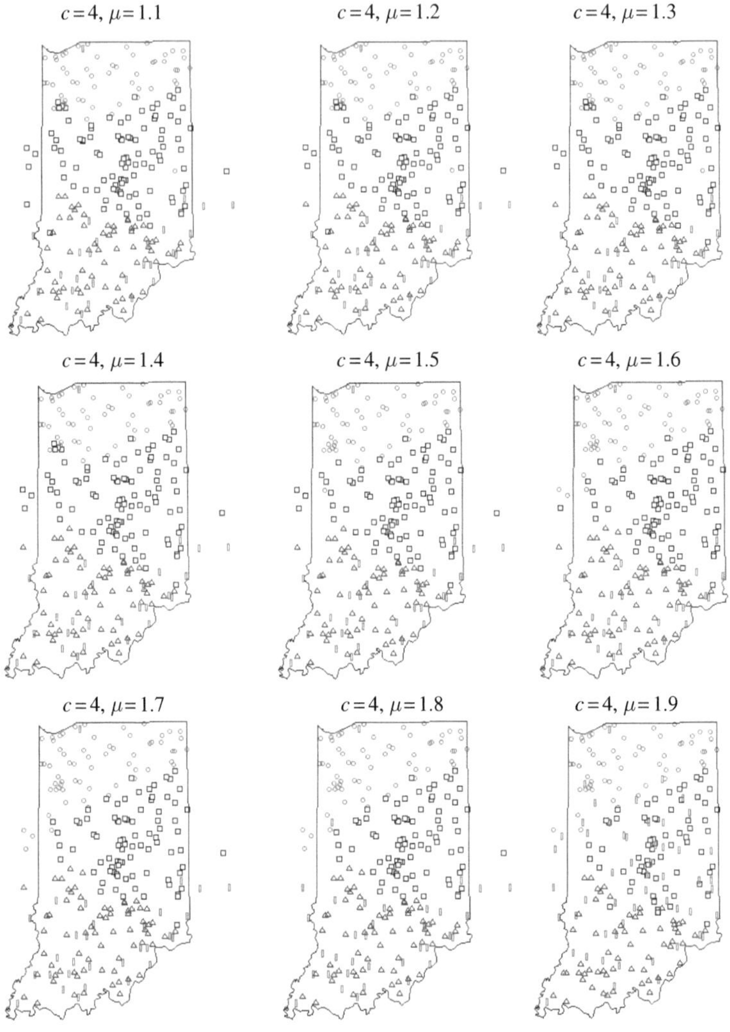

Fig. 3.5.10 Clusters of watersheds obtained in Indiana by hardening the fuzzy partition matrix obtained from FCM algorithm for $c = 4$

and the other in central Indiana. These clusters are very similar to those identified in the same parts of Indiana for c equal to 3. The other two clusters that are shown in Fig. 3.5.10, by triangles and rectangles are vague. High runoff coefficient and high precipitation is characteristic of catchments in these clusters, which are spread along the karst area in the southern part of the Indiana.

The cluster shown in triangles has sites with large drainage areas and mild slopes, whereas the cluster-4 shown in rectangles consists of small drainage basins with steep slopes (Table 3.5.2, Figs. 3.5.11 and 3.5.12).

Table 3.5.2 Centers of the four clusters obtained from FCM algorithm for two typical values of μ. In a few columns, the least or highest values of attributes are shown in bold font for each value of μ

μ	CN	A (miles)^2	Slope (ft/mile)	LAT	LONG	STOR (%)	P (in)	RC
1.1	1	54.36	7.45	41.29	86.27	**1.583**	36.69	**0.44**
	2	34.27	12.30	40.18	86.03	0.371	38.28	0.72
	3	**109.25**	12.20	38.88	86.44	0.590	41.87	0.80
	4	**0.26**	**116.20**	38.89	86.22	0.314	42.15	0.79
1.5	1	51.22	8.47	41.15	86.48	**1.009**	36.89	**0.47**
	2	50.59	11.43	40.17	85.86	0.463	38.35	0.72
	3	**87.22**	13.33	38.96	86.48	0.601	41.75	0.79
	4	**0.33**	**116.28**	38.91	86.24	0.485	42.15	0.80

CN is cluster number; A denotes drainage area; LAT and LONG refer to Latitude and Longitude in decimal degrees; STOR denotes drainage area covered by lakes; P stands for precipitation; RC is runoff coefficient.

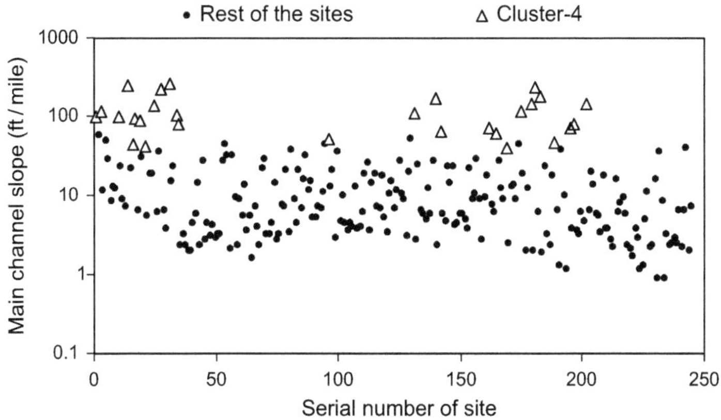

Fig. 3.5.11 Comparison of the main channel slopes of drainage basins forming cluster-4 with those at the other sites for $c = 4$ and $\mu = 1.1$ as well as $\mu = 1.5$

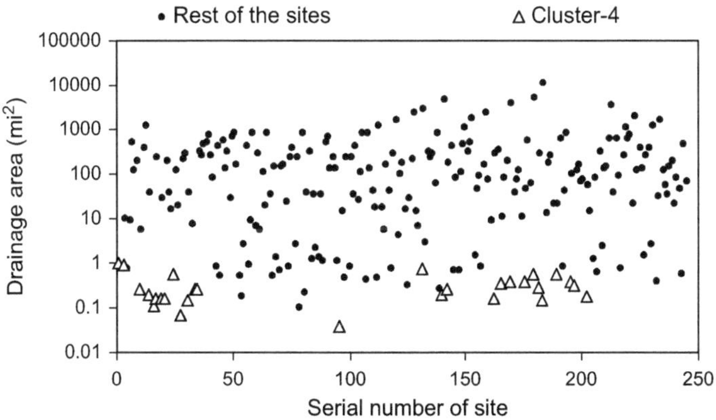

Fig. 3.5.12 Comparison of the drainage areas of basins forming cluster-4 with those at the other sites for $c = 4$ and $\mu = 1.1$ as well as $\mu = 1.5$

Comparison of the drainage areas at the sites in clusters 1, 2 and 3 show that drainage area at about 6 % of the sites is considerably high (Fig. 3.5.13). The site with serial number 184 (USGS station 3374000) has the largest drainage area equal to 11125 square miles. Movements of such sites with larger drainage area from one cluster to another affects the average drainage area of the concerned clusters dramatically. Thus, analyzing the cluster centers such as those shown in Tables 3.5.1 and 3.5.2 may not be sufficient to draw final conclusions. Figures such as 3.5.11 and 3.5.12 would be helpful in understanding the composition of clusters obtained from the FCM algorithm.

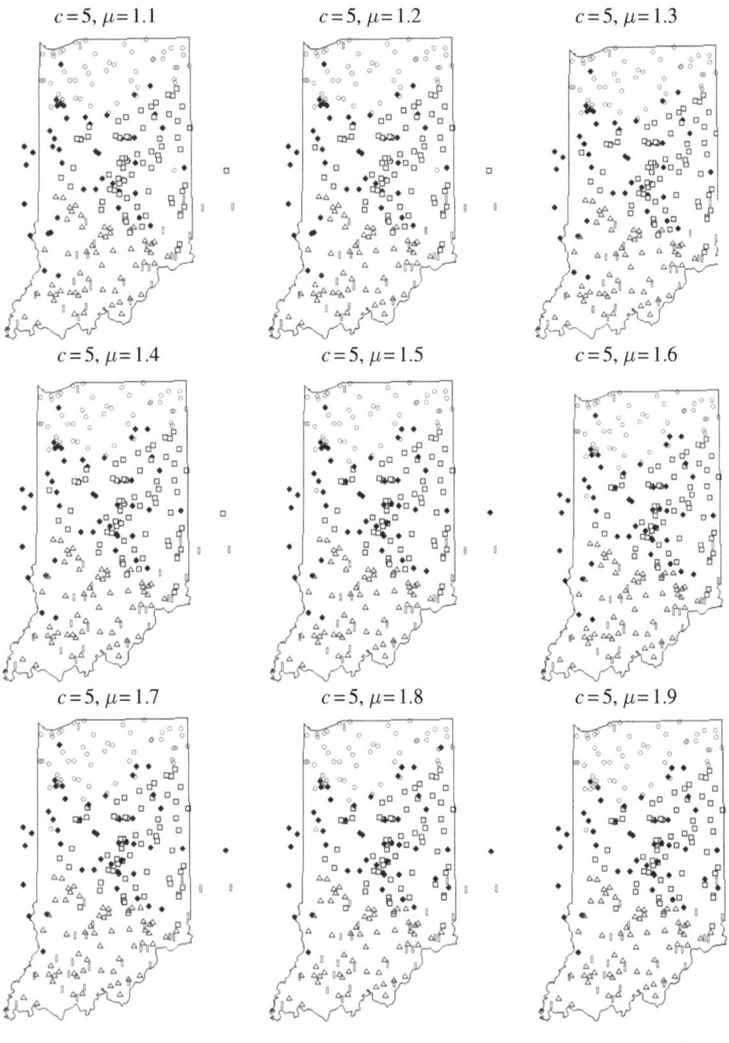

Fig. 3.5.13 Clusters of watersheds obtained in Indiana by hardening the fuzzy partition matrix obtained from FCM algorithm for $c = 5$

It is evident from Fig. 3.5.10 that the result for $c = 4$ is not sensitive to variation in fuzzifier value μ in the range 1.1–1.8. However, the degree of fuzziness in the result increased dramatically with increase in the value of μ beyond 1.8 which indicates that the clusters corresponding to μ greater than 1.8 are not suitable for further analysis.

The cluster observed in central Indiana for $c = 4$ split up into two vague clusters when c is increased to 5 (Fig. 3.5.13). These clusters, shown in squares and darkened diamonds, in central Indiana consist of geographically neighboring drainage basins that have similar soil runoff characteristics and mean annual precipitation. They are however, considerably distinct in their drainage areas (clusters 2 and 5 in Fig. 3.5.14 and Table 3.5.3). A significant difference is also noted in the slope of main streams draining these basins and the percentage of drainage areas covered by water bodies (Fig. 3.5.15 and Table 3.5.3).

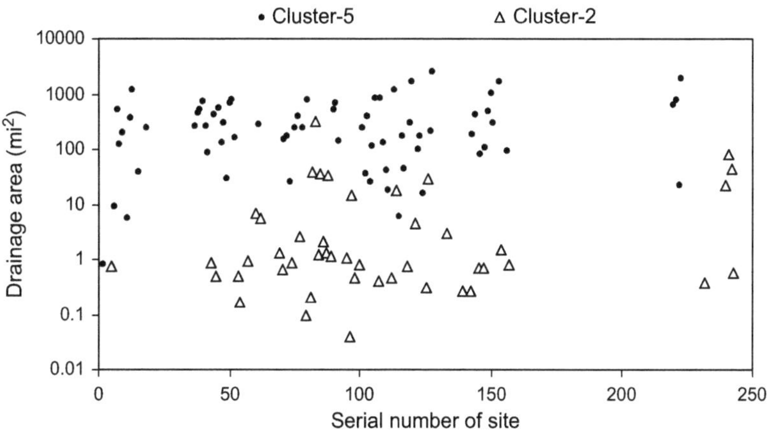

Fig. 3.5.14 Comparison of drainage areas of basins in clusters 2 and 5 for $c = 5$ and $\mu = 1.1$

Table 3.5.3 Centers of the five clusters obtained from FCM algorithm for two typical values of μ

μ	CN	A $(\text{miles})^2$	Slope (ft/mile)	LAT	LONG	STOR (%)	P (in)	RC
1.1	1	74.32	5.99	41.31	86.27	1.684	36.65	0.44
	2	1.67	23.87	40.13	86.65	0.128	38.40	0.68
	3	107.69	12.45	38.81	86.48	0.635	42.15	0.80
	4	0.27	124.90	38.90	86.04	0.348	42.28	0.81
	5	184.83	7.47	40.09	85.73	0.506	38.48	0.74
1.5	1	95.29	6.00	41.24	86.50	1.148	36.80	0.44
	2	177.65	7.61	40.13	85.75	0.528	38.46	0.73
	3	96.73	12.95	38.84	86.49	0.615	42.14	0.80
	4	0.30	133.81	38.83	86.24	0.456	42.53	0.81
	5	1.95	23.97	40.16	86.39	0.290	38.39	0.68

CN is cluster number; A denotes drainage area; LAT and LONG refer to Latitude and Longitude in decimal degrees; STOR denotes drainage area covered by lakes; P stands for precipitation; RC is runoff coefficient.

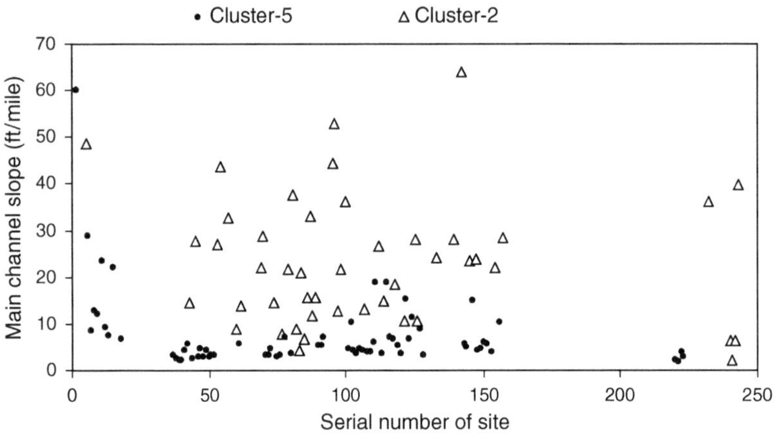

Fig. 3.5.15 Comparision of main channel slope at drainage basins in clusters 2 and 5 for $c = 5$ and $\mu = 1.1$

Comparison of size of clusters resulting from the FCM algorithm for c equal to 4 and 5 shows that the clusters tend to possess equal number of gauging stations (or sites) with increase in the value of fuzzifier μ. Further, all the clusters have sufficient size in terms of data (Figs. 3.5.16 and 3.5.17). Cluster-1, which consists of drainage

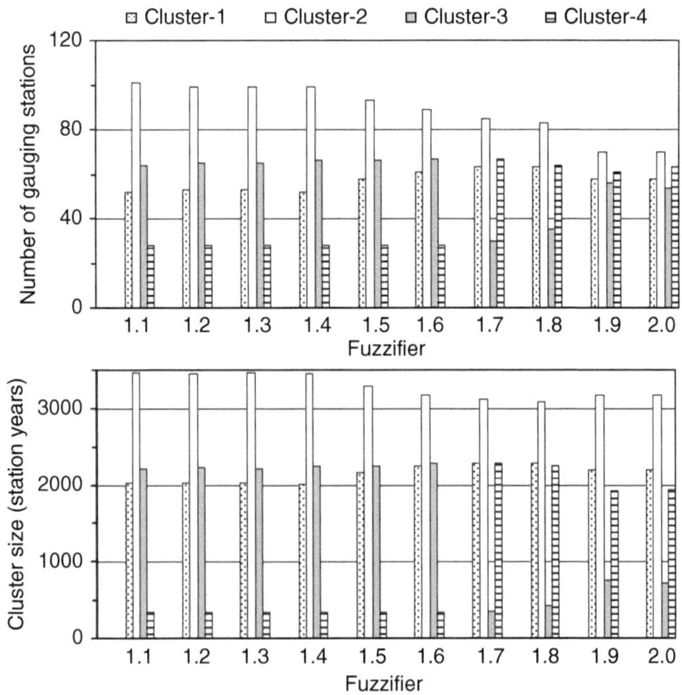

Fig. 3.5.16 Variation in size of clusters with increase in the value of fuzzifier – Results from FCM algorithm for $c = 4$

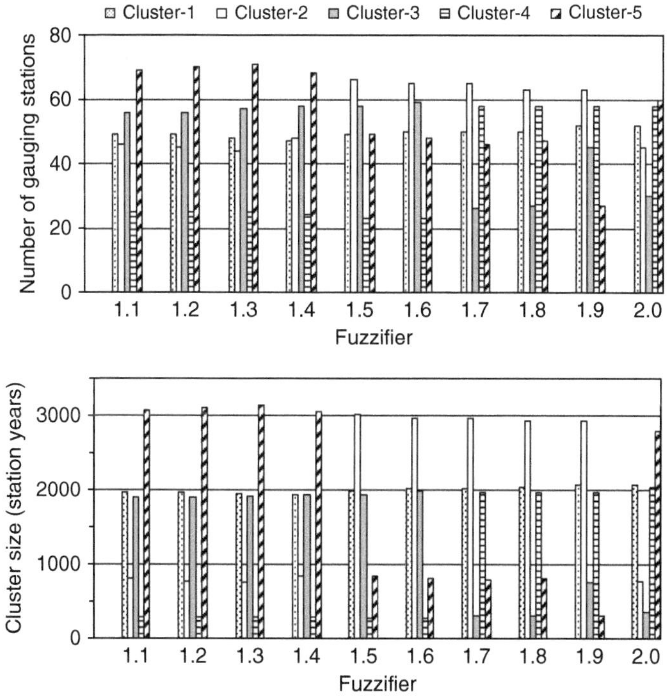

Fig. 3.5.17 Variation in size of clusters with increase in the value of fuzzifier – Results from FCM algorithm for $c = 5$

basins in the northern part of Indiana, is highly heterogeneous. Change in the values of heterogeneity measures with variation in μ is found to be insignificant for $c = 4$ and $c = 5$ (Figs. 3.5.18 and 3.5.19).

The partition achieved with c equal to 6 and μ equal to 1.1 consists of a cluster in the northern part of Indiana, which resembles the cluster that has been observed in northern Indiana for the choice of $c = 4$ and $c = 5$ (Figs. 3.5.20, 3.5.10 and 3.5.13). The location of sites forming this cluster are shown using circles in Fig. 3.5.20. In Figs. 3.5.21 and 3.5.22, this cluster refers to cluster-6 obtained from FCM algorithm for the choice of μ in the range 1.1–1.3 and it refers to cluster-1 obtained from the algorithm for the choice of μ in the range 1.4–2.0.

The composition of the cluster in northern Indiana was insensitive to variation in the value of the fuzzifier in the ranges 1.1–1.3 and 1.6–2.0. However, it splits up to form two clusters in northern Indiana for the choice of μ equal to 1.4 and 1.5 (see Fig. 3.5.20, and clusters 1 and 6 for $\mu = 1.4$ in Table 3.5.4). The larger of the two clusters consists of highly heterogeneous catchments in the Kankakee basin (Figs. 3.5.20, 3.5.21, and 3.5.22). It possesses predominantly medium size catchments with mildest slopes, moderate storage and least runoff coefficient values. On the other hand, the smaller cluster includes medium to large size catchments with milder slopes, high storage and low runoff coefficient values.

For $c = 6$ and the value of fuzzifier in the range 1.1–1.5, the clusters obtained in central Indiana and those obtained in southern Indiana are very similar to the

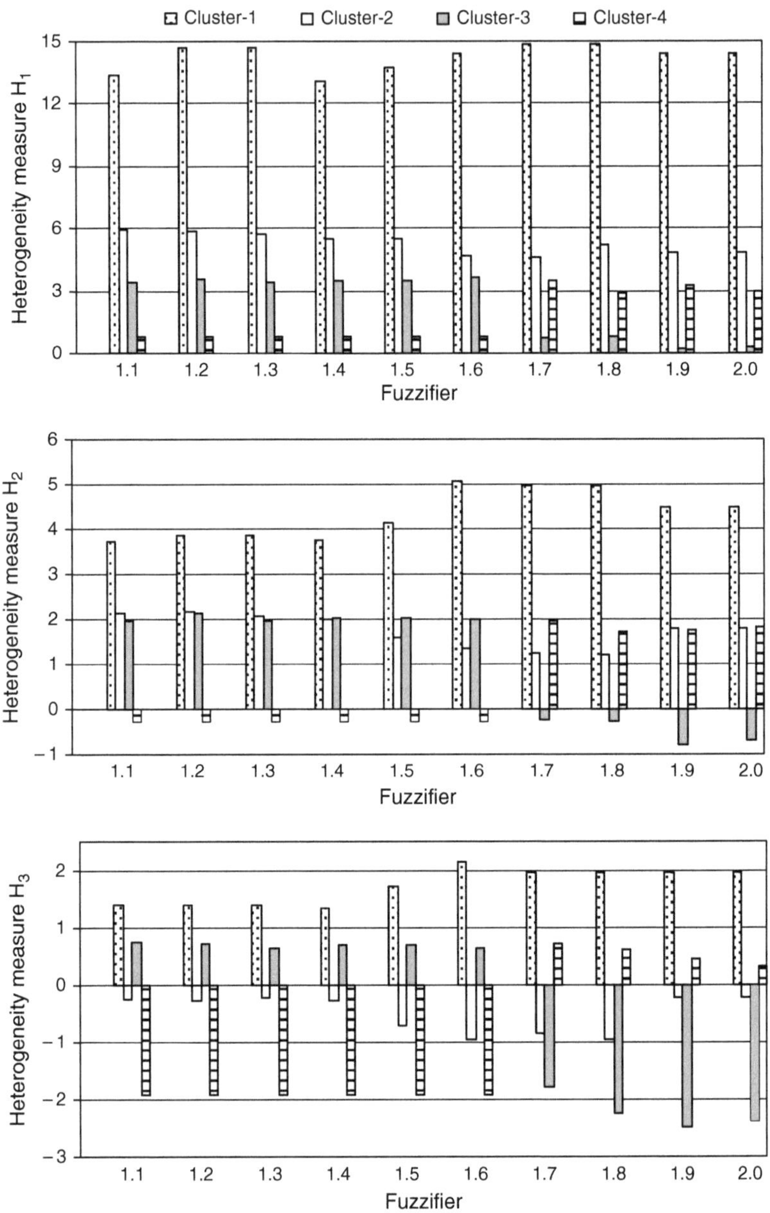

Fig. 3.5.18 Variation of heterogeneity measures with increase in the value of fuzzifier – Results from Fuzzy c-means algorithm for $c = 4$

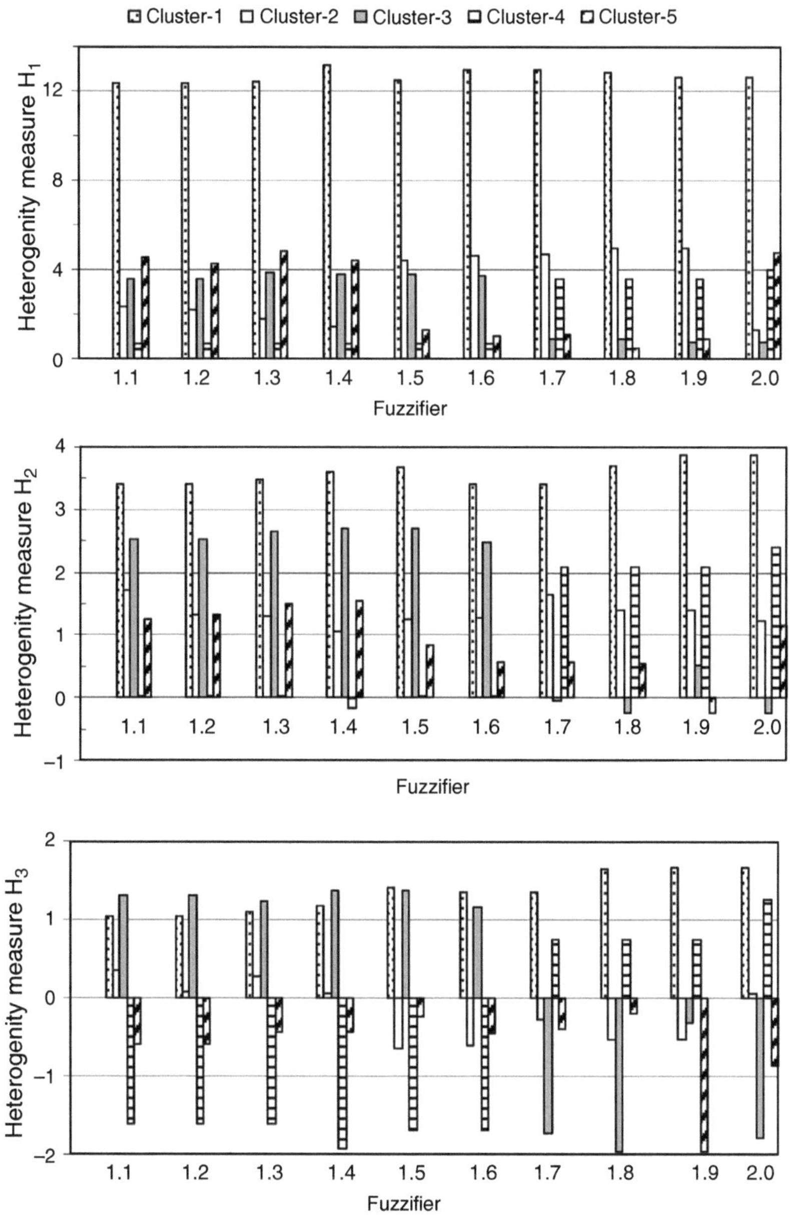

Fig. 3.5.19 Variation of heterogeneity measures with increase in the value of fuzzifier – Results from Fuzzy c-means algorithm for $c = 5$

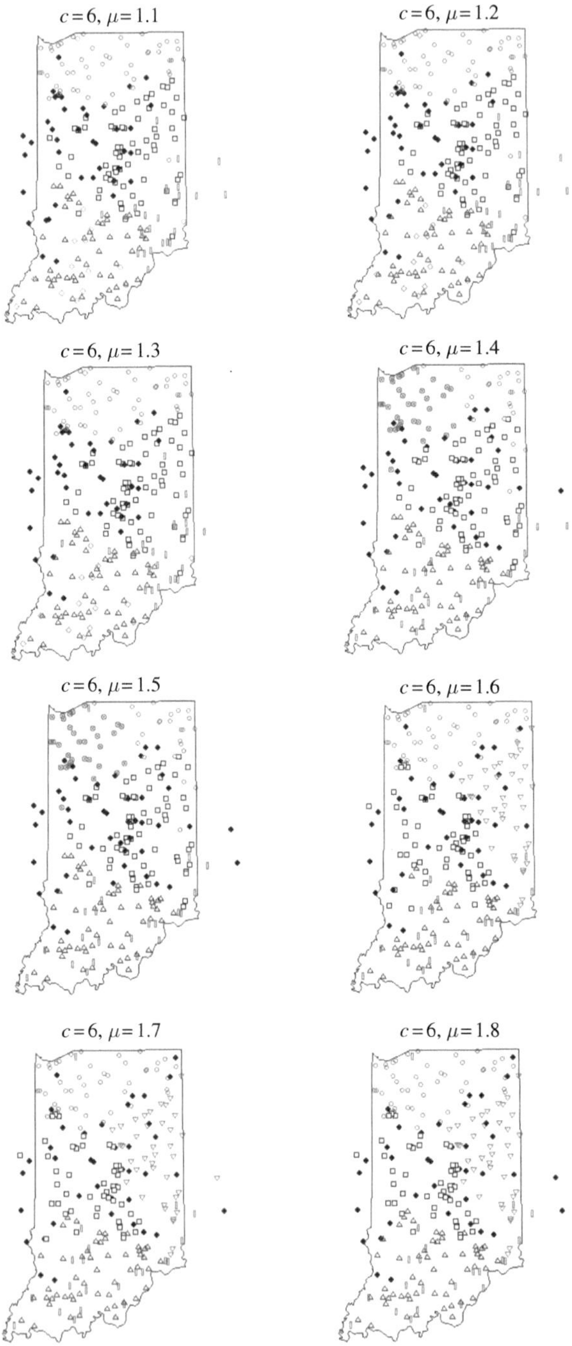

Fig. 3.5.20 Clusters of watersheds obtained in Indiana by hardening the fuzzy partition matrix obtained from FCM algorithm for $c = 6$

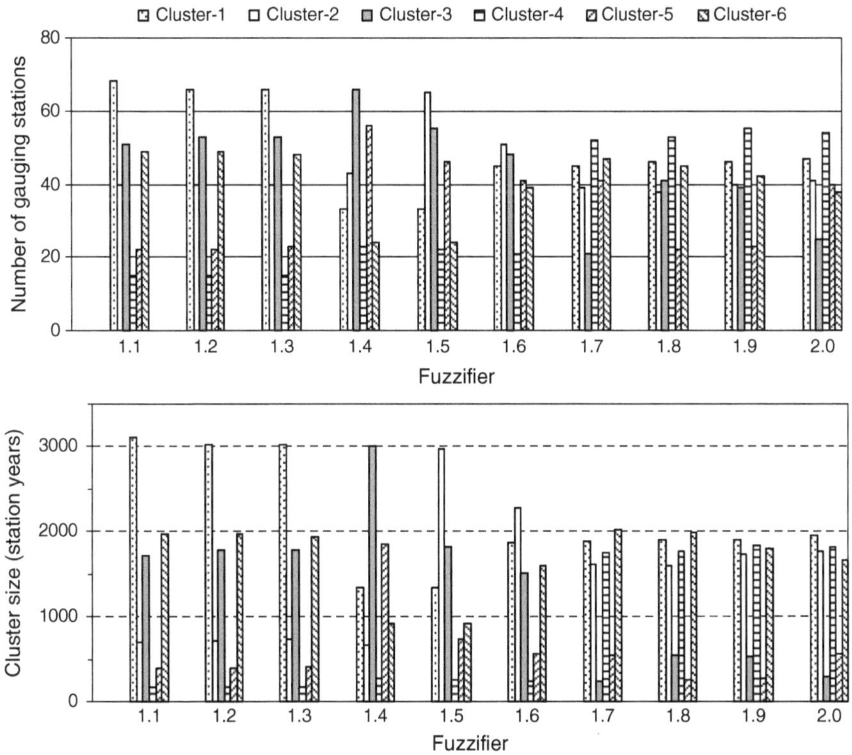

Fig. 3.5.21 Variation in size of clusters with increase in the value of fuzzifier – Results from Fuzzy c-means algorithm for $c = 6$

clusters in those parts of Indiana for $c = 5$. In Fig. 3.5.20, the squares in central Indiana denote a group of large catchments with milder slopes, moderate storage and moderate to high runoff coefficient values (cluster-1 for $\mu = 1.3$ in Table 3.5.4). The darkened diamonds, that are spread in central Indiana and along the Wabash river, represent smaller catchments with mild slope, small to moderate storage and moderate runoff coefficient values (see Cluster-2 for $\mu = 1.3$ in Table 3.5.4). The triangles in southern Indiana depict medium to large size catchments with mild slopes, high precipitation and high runoff coefficient values (see Cluster-3 for $\mu = 1.3$ in Table 3.5.4). The diamond symbols in southern Indiana denote small catchments with steepest slopes, moderate storage and high runoff coefficient values (see Cluster-4 for $\mu = 1.3$ in Table 3.5.4).

It is also evident from Fig. 3.5.20 that a new cluster emerged in the south-eastern part of the state for the choice of μ in the range 1.1–1.3 with $c = 6$. This cluster (shown in rectangles) comprises of small catchments with moderate slopes, moderate storage and highest runoff coefficient values (see Cluster-5 for $\mu = 1.3$ in Table 3.5.4). When the value of μ was increased beyond 1.5, a new cluster which includes predominantly the sites in central part of the state emerged.

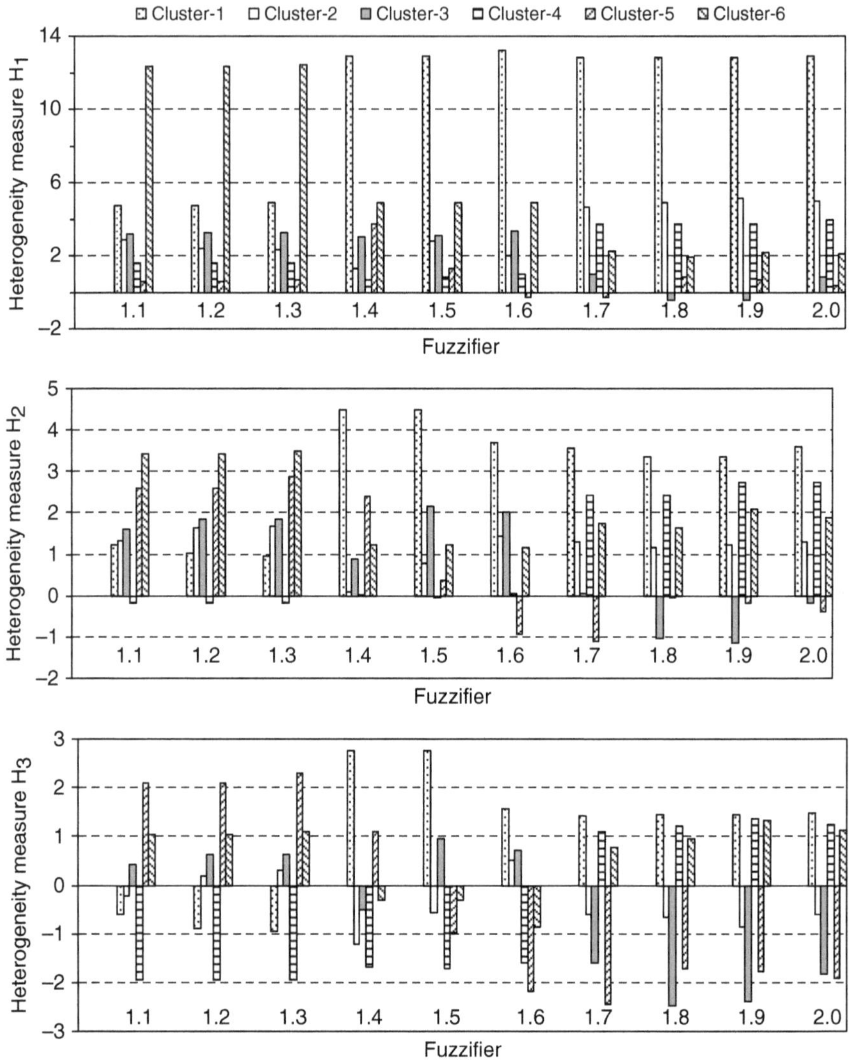

Fig. 3.5.22 Variation of heterogeneity measures with increase in the value of fuzzifier – Results from Fuzzy c-means algorithm for $c = 6$

It is characterized by large catchments with mild slopes and moderate storage (see Cluster-2 for $\mu = 1.6$ in Table 3.5.4). However, for μ greater than 1.5 the highly heterogeneous cluster identified in Kankakee basin for the choice of μ equal to 1.4 and 1.5 got merged with the cluster adjoining it in the northern part of Indiana. Further, the clusters in central and southern parts of Indiana appeared more vague (Fig. 3.5.20), suggesting that the solution provided by FCM algorithm for the value of μ greater than 1.5 is too fuzzy to identify the hydrologic regions.

Table 3.5.4 Centers of six clusters obtained from FCM algorithm for three typical values of μ

μ	CN	A (miles)^2	Slope (ft/mile)	LAT	LONG	STOR (%)	P (in)	RC
1.3	1	213.14	6.13	40.15	85.77	0.526	38.38	0.73
	2	1.89	22.81	40.23	86.70	0.205	38.22	0.66
	3	130.87	10.97	38.82	86.61	0.644	42.20	0.79
	4	0.26	166.51	38.76	86.51	0.291	43.04	0.79
	5	1.39	50.01	39.26	85.38	0.600	40.81	0.84
	6	97.28	5.58	41.30	86.36	1.352	36.58	0.44
1.4	1	73.57	6.06	41.16	86.88	0.650	37.44	0.44
	2	1.36	26.34	40.07	86.43	0.205	38.57	0.69
	3	181.72	7.66	40.03	85.81	0.402	38.71	0.74
	4	0.27	138.60	38.81	86.21	0.402	42.57	0.81
	5	95.02	13.00	38.80	86.51	0.611	42.27	0.80
	6	103.24	7.17	41.32	85.35	2.657	35.23	0.52
1.6	1	93.16	5.91	41.22	86.66	0.995	36.98	0.43
	2	155.99	9.14	39.88	86.36	0.431	38.98	0.71
	3	73.04	14.45	38.73	86.45	0.602	42.56	0.80
	4	0.30	144.14	38.79	86.35	0.406	42.76	0.80
	5	1.37	25.49	40.15	86.28	0.298	38.40	0.69
	6	132.03	8.66	40.31	85.45	0.695	38.13	0.74

CN is cluster number; A denotes drainage area; LAT and LONG refer to Latitude and Longitude in decimal degrees; STOR denotes drainage area covered by lakes; P stands for precipitation; RC is runoff coefficient.

When c was increased to 7, the clusters obtained showed remarkable resemblance to those obtained for the choice of $c = 6$. As per the 5T rule (Reed et al. in FEH 1999, p. 28, Vol. 3) the grouped stations should collectively supply five times as many station years of record as the target return period. Several of the clusters obtained for the choice of $c = 9$ and above were quite small in size. Such small clusters are not suitable for regional flood frequency analysis. Therefore the results obtained for $c = 9$ and above are not considered to be suitable for forming hydrologic regions. As noted earlier, it is evident that clusters tend to possess equal number of sites with increase in the value of fuzzifier μ (Figs. 3.5.23 and 3.5.25) and majority of clusters obtained from FCM algorithm tend to be closer to homogeneous with increase in the number of clusters (Figs. 3.5.24 and 3.5.26). Also, change in the value of heterogeneity measure H_1 with variation in the value of μ was found to be insignificant.

From the foregoing analysis, the following conclusions are drawn:

(i) The two clusters noted in northern part of Indiana for the choice of $c = 7$ and μ in the range 1.4–1.8 (Fig. 3.5.27) closely resemble the clusters obtained in the same region of Indiana by partitioning of the state into six clusters ($c = 6$ and $\mu = 1.4$–1.5 in Fig. 3.5.20). Similar observation is made for $c = 8$ (Fig. 3.5.28). Since the location of these two clusters is well defined, they can be considered as plausible hydrologic regions for flood frequency analysis. It may be recalled from the previous discussion that the cluster in northwestern Indiana consists of catchments in the Kankakee basin that are highly

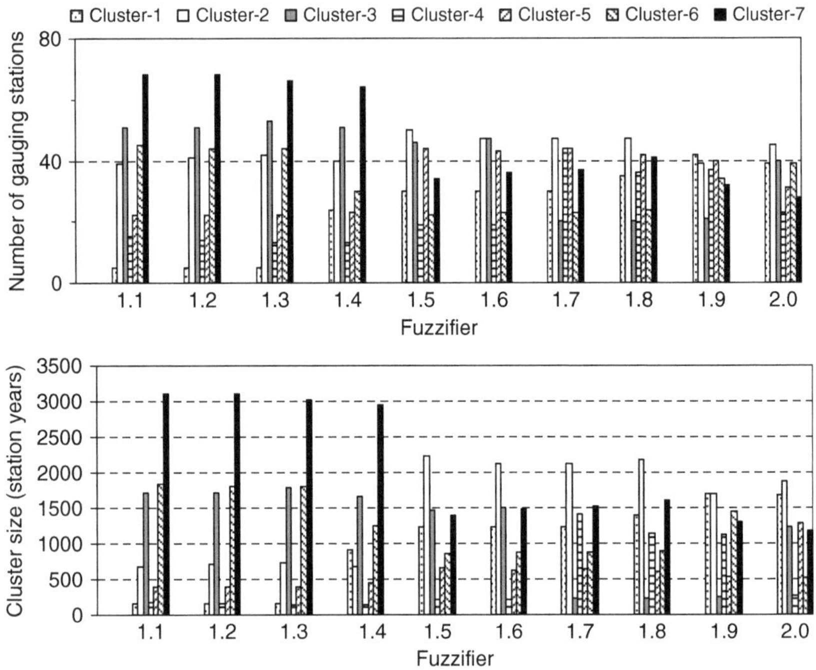

Fig. 3.5.23 Variation in size of clusters with increase in the value of fuzzifier – Results from Fuzzy c-means algorithm for $c = 7$

heterogeneous. It possesses predominantly medium size catchments with mild slopes, moderate storage, small runoff coefficient values and low precipitation. On the other hand, the cluster in northeastern Indiana consists of medium to large size catchments with milder slopes, high storage, low runoff coefficient values and low precipitation.

(ii) The cluster identified in southeastern part of Indiana for the choice of $c = 7$ and μ in the range 1.1–1.4 (or for the choice of $c = 8$ and μ in the range 1.1–1.8) closely resembles the cluster observed in the region for the choice of μ in the range 1.1–1.3 and $c = 6$. The catchments in this cluster are characterized by high runoff coefficient values. Since the location of this cluster is well defined, it may be considered as a plausible hydrologic region.

(iii) The clusters in southern Indiana adjoining Kentucky state, whose sites are shown as diamonds and triangles, retain their identity despite increase in the number of partitions from $c = 4$ to $c = 8$ (Figs. 3.5.10, 3.5.13, 3.5.20, 3.5.27 and 3.5.28). In fact, the sites comprising these two clusters were members of a single cluster for the choice of c equal to 2 and 3 (Figs. 3.5.2 and 3.5.5). Moreover, the cluster represented by diamonds is much smaller than that represented by triangles (Fig. 3.5.29). Consequently merging these two clusters is justified if the objective of regionalization is to identify geographically contiguous hydrologic regions. High precipitation and high runoff coefficient values

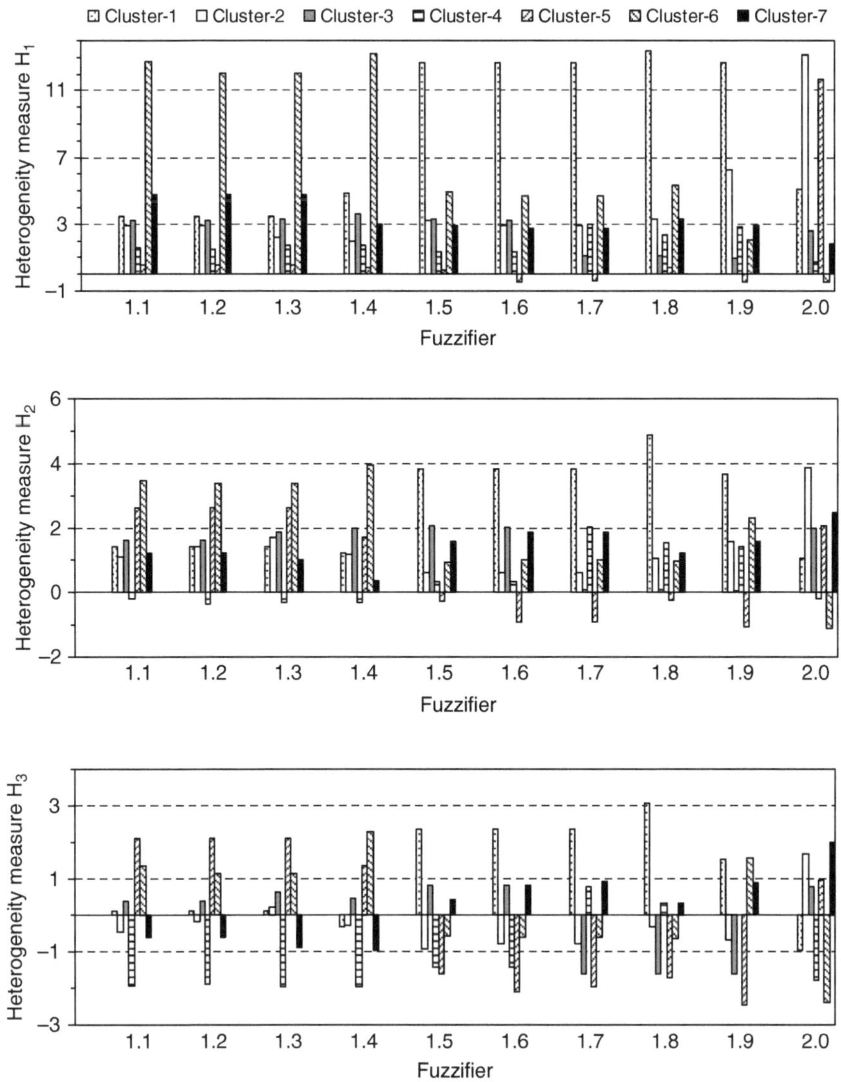

Fig. 3.5.24 Variation of heterogeneity measures with increase in the value of fuzzifier – Results from Fuzzy c-means algorithm for $c = 7$

characterize the catchments comprising these clusters. However, there is a significant difference in their catchment areas, basin storage and slopes. From the previous discussion it is known that the triangles in southern Indiana depict medium to large size catchments with mild slopes and moderate basin storage, whereas the diamonds in southern Indiana denote small catchments with steep slopes (Fig. 3.5.30) and low basin storage.

(iv) The clusters in central Indiana, whose members are shown as darkened diamonds and squares are characterized by mild to milder slopes and moderate

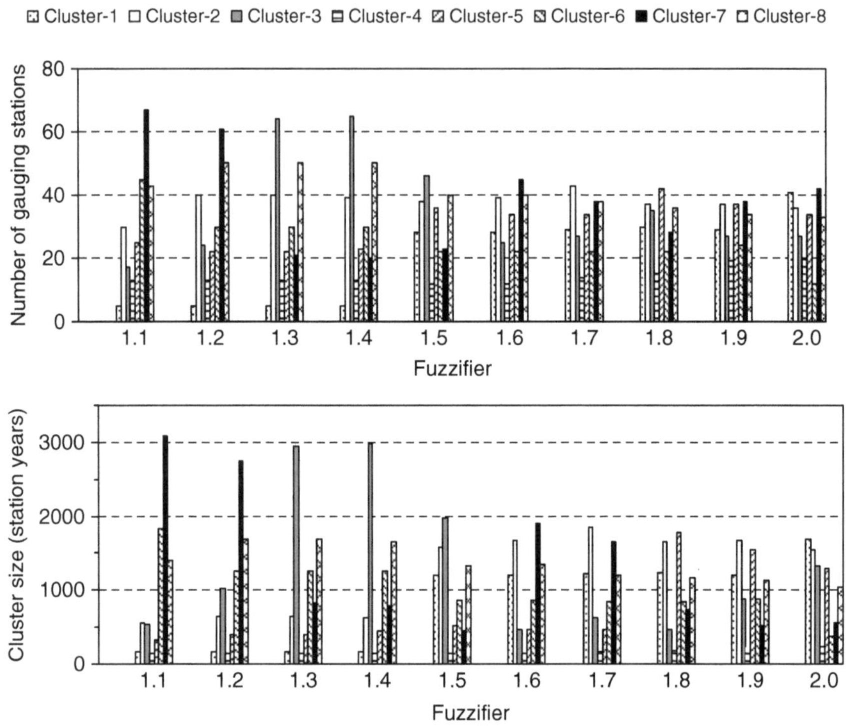

Fig. 3.5.25 Variation in size of clusters with increase in the value of fuzzifier – Results from Fuzzy c-means algorithm for $c = 8$

precipitation. The differences in their slope, basin storage and runoff coefficient are marginal. However, the difference in their drainage areas is considerable. Further, the darkened diamonds depict gauges with very short record lengths and small drainage areas (see cluster-2 in Fig. 3.5.31). For low values of μ, the darkened diamonds spread predominantly along the course of Wabash river and its tributaries, whereas for high values of μ they appear to be scattered in the entire central Indiana. This may be attributed to a few sites which keep shifting from one cluster to another in central Indiana with increase in fuzziness. The squares that depict larger drainage areas in central Indiana with moderate to high record lengths might be considered as a plausible hydrologic region. The effort needed to adjust the identified regions to improve their statistical homogeneity is found to be significant (Srinivas and Rao, 2003).

3.5.2.2 Identification of Regions by Fuzzy Cluster Validity Indices

The sites which are tightly linked to each cluster are identified by specifying a threshold fuzzy membership value equal to $1/c$. A fuzzy cluster is formed by

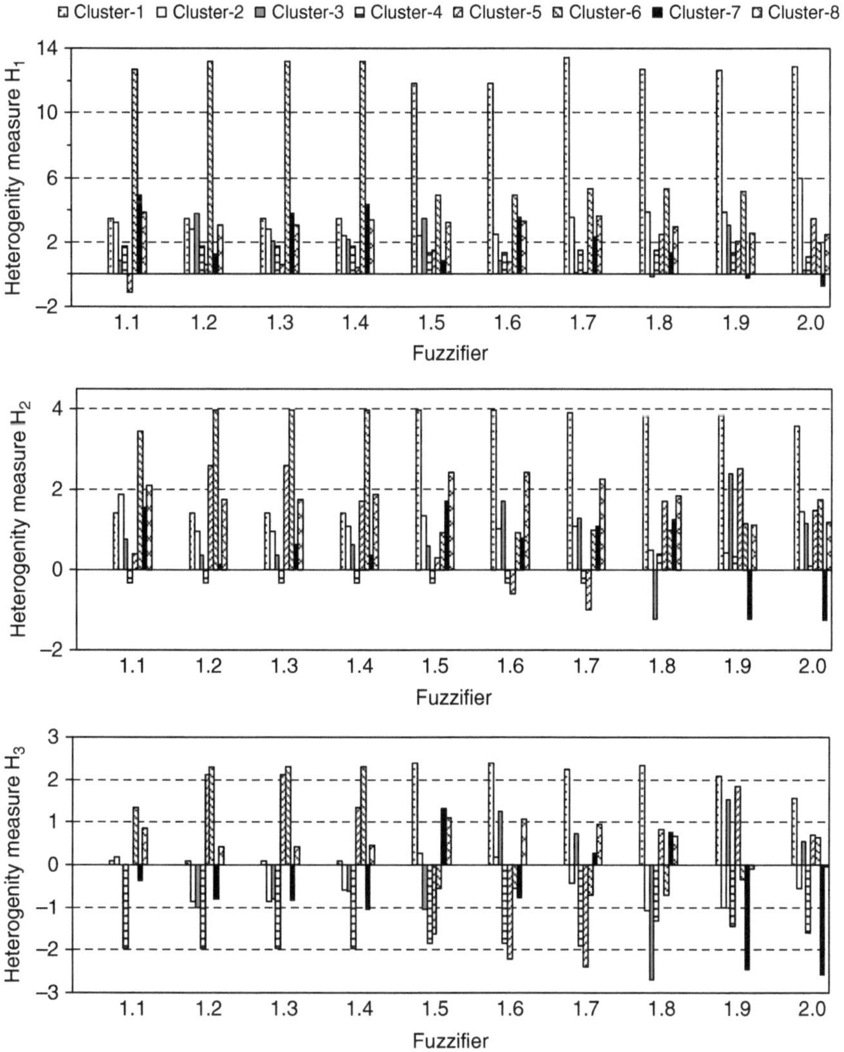

Fig. 3.5.26 Variation of heterogeneity measures with increase in the value of fuzzifier – Results from Fuzzy c-means algorithm for $c = 8$

assigning to it the sites whose memberships in the cluster exceed the specified threshold value. The upper limit on the number of clusters C_{max} has been fixed keeping in view the findings of Reed et al. (1999) on the minimum data requirement for a region.

The optimal number of clusters in the Indiana data set is identified by using seven fuzzy cluster validity measures: partition coefficient (V_{PC}), partition entropy (V_{PE}), fuzziness performance index (V_{FPI}); normalized classification entropy (V_{NCE}); extended Xie-Beni index ($V_{XB,m}$), Fukuyama-Sugeno index (V_{FS}), and Kwon's Index

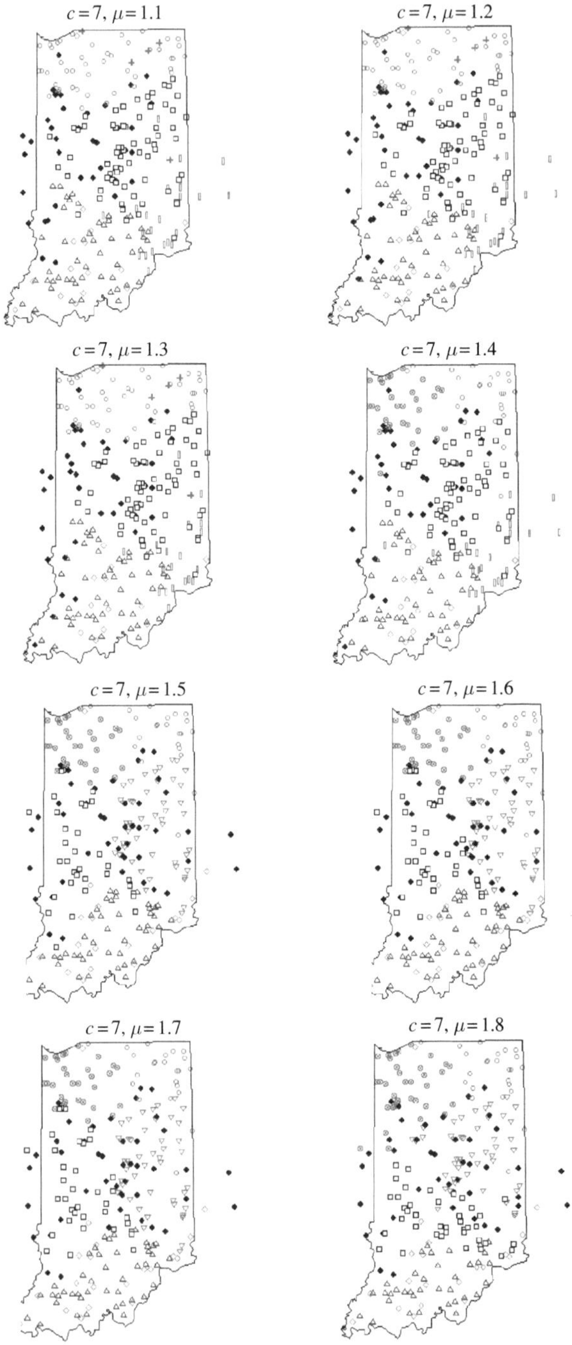

Fig. 3.5.27 Clusters of watersheds obtained in Indiana by hardening the fuzzy partition matrix obtained from FCM algorithm for $c = 7$

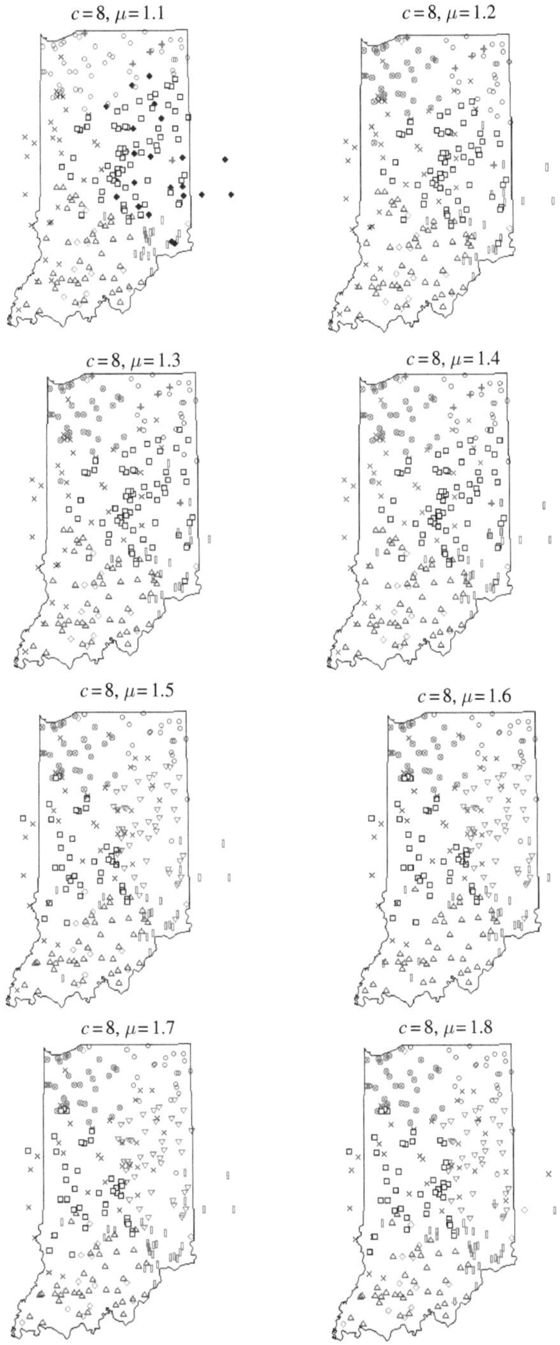

Fig. 3.5.28 Clusters of watersheds obtained in Indiana by hardening the fuzzy partition matrix obtained from FCM algorithm for $c = 8$

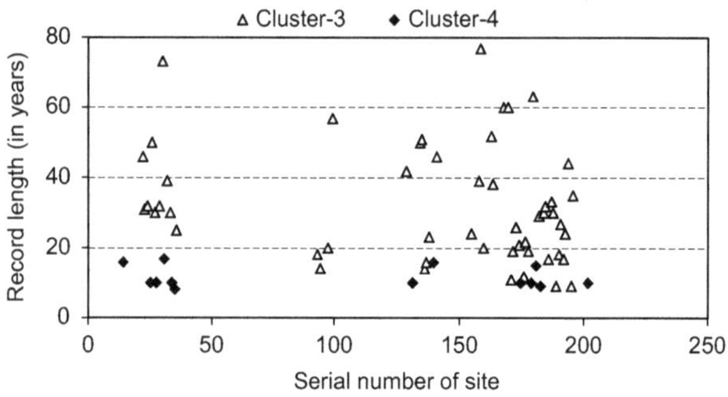

Fig. 3.5.29 Comparison of record lengths of stations in clusters in southern Indiana – Result from choice of $c = 7$ and $\mu = 1.4$. The triangles denote catchments of cluster-3, whereas the diamonds represent catchments of cluster-4

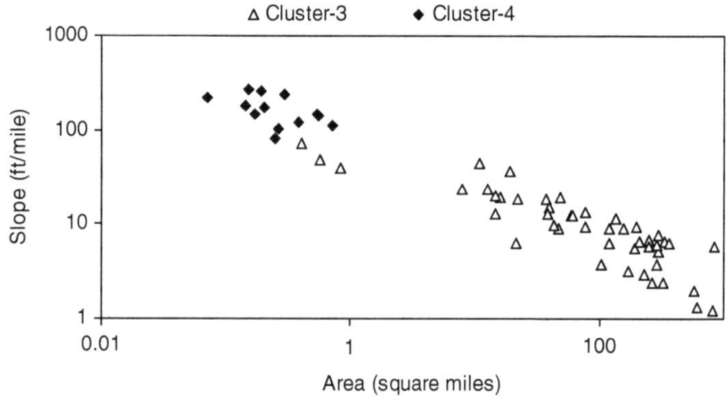

Fig. 3.5.30 Comparison of drainage area and slope of stations in clusters in southern Indiana – Result from choice of $c = 7$ and $\mu = 1.4$. The triangles denote catchments of cluster-3, whereas the diamonds represent catchments of cluster-4

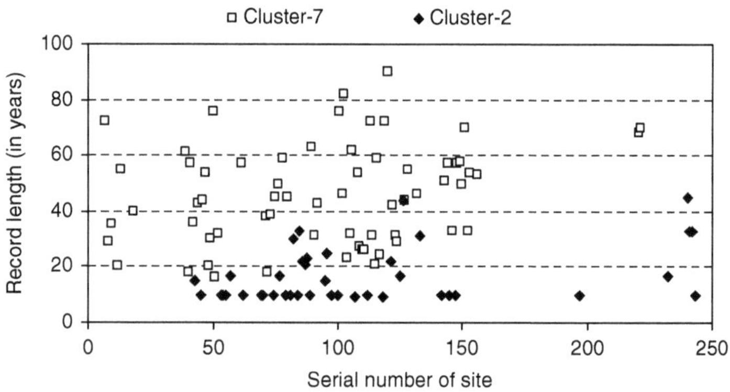

Fig. 3.5.31 Comparison of record lengths of stations in clusters in central Indiana – Result from choice of $c = 7$ and $\mu = 1.4$. The squares denote catchments of the cluster in predominantly the central Indiana region, whereas the darkened diamonds represent catchments of the cluster that is spread along the course of Wabash river and its tributaries

(V_K). Description of these measures is provided in Section 3.4. In general, the $V_{XB,m}$ and V_K indicate $c = 7$ as the optimal number of clusters for the value of μ in the range 1.1–1.5, while they identify $c = 4$ as the optimal partition for μ in the range 1.6–2.5 (Table 3.5.5).

Fuzziness in the results increases for μ greater than 1.5 (Fig. 3.5.32). This is reflected in the values of Xie-Beni and Kwon's Indices, which increase significantly beyond μ equal to 1.5 (Table 3.5.5), indicating a drop in the quality of resulting clusters. The optimal values of Xie-Beni and Kwon's Indices suggest that the best choice for μ is in the interval [1.3, 1.5], whose midpoint is 1.4.

The partition coefficient (V_{PC}) and partition entropy (V_{PE}) which always suggest $c = 2$ as the best partition are inefficient. In general, V_{PC} is maximum and V_{PE} is minimum at $c = 2$, irrespective of the chosen μ value (Table 3.5.5). While V_{PC} exhibits monotonic decreasing tendency with increase in both c and μ, V_{PE} exhibits monotonic increasing tendency with increase in the values of the model parameters (Fig. 3.5.33). As a result, both V_{PC} and V_{PE} often suggest $c = 2$ as optimal partition. Bargaoui et al. (1998), and Hall and Minns (1999) also make the same observation. For the Indiana data, the monotonic tendency is evident for the values of μ greater than 1.4 (Table 3.5.5)

Similar monotonic tendency is seen in case of V_{FPI} and V_{NCE} with increase in the value of fuzzifier (Table 3.5.5 and Fig. 3.5.33). The disadvantage of these indices is the lack of direct connection to the geometrical property or structure of data set (Xie and Beni, 1991; Halkidi et al., 2001). A validity measure should not exhibit monotonic tendency to be effective as an index for identifying optimal partition.

The Fukuyama-Sugeno index (V_{FS}), which is one of the seven fuzzy validity indices computed, exhibits monotonic decreasing tendency with increase in number of clusters (Table 3.5.5) and is therefore not effective in identifying optimal partition for the catchments in the study region.

Based on the foregoing analysis, the clusters given by the FCM algorithm with $c = 7$ and $\mu = 1.4$, are selected as optimal partition. Moreover, the choice of seven regions enables comparison of the results obtained in this section with the previously defined seven regions in Indiana by Glatfelter (1984).This decision on choice of μ is supported by Fig. 3.5.34 which show that the best partition of data for the value of c in the range 7–10 is achieved with the value of μ in the range 1.4–1.5, at which the partition coefficient is maximized and the classification entropy is minimized. Figure 3.5.35 shows the location of the fuzzy clusters in Indiana.

The optimal partition identified by using the fuzzy cluster validity indices has clusters which are very similar to the plausible hydrologic regions identified by visual inspection of clusters of watersheds in the foregoing section. Although this may not happen always, it gives confidence in the regions that are determined for flood frequency analysis.

The homogeneity of regions (clusters) corresponding to optimal partition ($c = 7$ and $\mu = 1.4$), which are formed without defuzzification, is tested by using heterogeneity measures of Hosking and Wallis (1997). The regions are, in general, adjusted following the eight options described in Section 1.4.1 to improve their homogeneity. In fuzzy cluster analysis, the knowledge of distribution of membership

Table 3.5.5 Comparison of the cluster validity measures for the data set pertaining to watersheds in Indiana, USA. The values in bold font denote optimal values of the validity measures. c: number of clusters; μ: Fuzzifier; V_{PC}: partition coefficient; V_{PE}: partition entropy; V_{FPI}: fuzziness performance index; V_{NCE}: normalized classification entropy $V_{XB,m}$: extended Xie-Beni index; V_{FS}: Fukuyama-Sugeno index; V_K: Kwon's index

c	μ	V_{PC}	V_{PE}	V_{FPI}	V_{NCE}	$V_{XB,m}$	V_{FS}	V_K
2	1.1	0.960	**0.066**	0.080	0.219	0.73	852.36	164.71
3		0.957	0.071	0.065	0.149	0.60	405.40	134.19
4		0.955	0.075	0.060	0.125	0.66	128.94	151.66
5		**0.963**	0.067	**0.046**	0.096	0.74	−215.97	174.12
6		0.949	0.093	0.061	0.120	0.65	−223.75	137.97
7		0.958	0.074	0.049	0.088	**0.51**	−532.72	**118.80**
8		0.959	0.074	0.047	**0.082**	0.53	**−602.51**	125.11
2	1.2	**0.913**	**0.147**	0.174	0.488	0.73	846.17	156.65
3		0.903	0.170	0.146	0.356	0.62	403.67	126.93
4		0.89	0.203	0.147	0.337	0.87	103.38	170.65
5		0.877	0.228	0.154	0.326	0.75	−89.24	140.25
6		0.881	0.227	0.143	0.292	0.63	−213.46	115.59
7		0.896	0.200	**0.121**	**0.237**	**0.49**	−516.54	**105.29**
8		0.87	0.245	0.149	0.271	0.90	**−577.64**	185.17
2	1.3	**0.857**	**0.240**	0.286	0.797	0.75	839.38	152.63
3		0.834	0.297	0.249	0.622	0.63	403.06	120.87
4		0.811	0.352	0.252	0.585	0.84	110.94	152.50
5		0.802	0.381	0.248	0.545	0.75	−76.76	129.05
6		0.801	0.398	0.239	0.511	0.61	−196.16	104.84
7		0.814	0.371	0.217	0.439	0.50	−481.19	97.02
8		0.811	0.379	**0.216**	**0.420**	**0.45**	**−596.52**	**87.93**
2	1.4	**0.799**	**0.328**	0.402	1.090	0.77	831.66	152.72
3		0.750	0.445	0.375	0.933	0.63	423.52	114.69
4		0.734	0.500	0.355	0.830	0.80	121.89	139.49
5		0.723	0.547	0.346	0.783	0.74	−54.64	123.13
6		0.675	0.654	0.390	0.840	1.05	−117.99	161.81
7		0.714	0.599	**0.334**	**0.709**	**0.53**	−309.16	**85.72**
8		0.699	0.653	0.344	0.723	0.58	**−358.91**	90.63
2	1.5	**0.746**	**0.404**	0.508	1.342	0.79	821.27	156.97
3		0.668	0.581	0.498	1.218	0.78	461.68	137.10
4		0.661	0.640	0.452	1.063	0.77	137.28	131.62
5		0.645	0.708	0.444	1.013	0.75	−27.35	122.35
6		0.635	0.756	**0.438**	0.972	0.65	−157.68	103.97
7		0.622	0.809	0.441	**0.957**	**0.52**	−255.76	**81.87**
8		0.600	0.883	0.457	0.978	0.61	**−298.24**	92.80
2	1.6	**0.699**	**0.466**	0.602	1.548	0.82	807.19	165.39
3		0.608	0.682	0.588	1.429	0.80	460.88	145.26
4		0.593	0.768	0.543	1.276	**0.73**	157.32	128.44
5		0.572	0.856	**0.535**	1.225	0.75	1.45	**126.25**
6		0.514	1.003	0.583	1.289	1.20	−55.56	188.39
7		0.513	1.044	0.568	1.235	0.91	−164.30	141.62
8		0.502	1.097	0.569	**1.215**	0.94	**−244.84**	144.56

Table 3.5.5 (continued)

c	μ	V_{PC}	V_{PE}	V_{FPI}	V_{NCE}	$V_{XB,m}$	V_{FS}	V_K
2	1.7	**0.660**	**0.515**	0.680	1.711	0.85	789.21	178.06
3		0.556	0.765	0.666	1.603	0.81	453.32	155.28
4		0.530	0.885	0.627	1.470	**0.70**	180.81	**129.85**
5		0.508	0.986	**0.615**	**1.411**	0.76	30.26	134.08
6		0.449	1.142	0.661	1.468	1.32	−25.44	222.88
7		0.414	1.264	0.684	1.496	1.11	−72.10	181.42
8		0.424	1.283	0.658	1.421	0.95	**−177.72**	156.11
2	1.8	**0.628**	**0.554**	0.744	1.840	0.89	767.65	195.23
3		0.512	0.834	0.732	1.748	0.82	440.82	167.84
4		0.472	0.990	0.704	1.644	**0.68**	207.02	**135.83**
5		0.450	1.100	**0.688**	**1.574**	0.75	58.22	146.24
6		0.395	1.260	0.726	1.619	1.51	2.75	283.95
7		0.355	1.398	0.753	1.654	1.34	−30.99	244.49
8		0.356	1.446	0.736	1.601	1.03	**−112.46**	187.92
2	1.9	**0.602**	**0.584**	0.796	1.940	0.94	743.05	217.37
3		0.475	0.891	0.788	1.867	0.82	425.22	183.88
4		0.414	1.097	0.781	1.822	**0.68**	257.39	**148.03**
5		0.399	1.200	**0.751**	**1.717**	0.76	84.41	164.97
6		0.348	1.361	0.782	1.749	1.85	28.06	395.98
7		0.310	1.503	0.805	1.778	1.70	−3.53	358.22
8		0.278	1.635	0.825	1.810	1.83	**−23.13**	382.14
2	2.0	**0.505**	**0.688**	0.990	2.285	14.49	844.66	3549.60
3		0.444	0.937	0.834	1.964	0.87	407.41	213.89
4		0.379	1.158	0.828	1.923	**0.74**	248.45	**182.44**
5		0.354	1.290	**0.808**	**1.846**	0.79	110.35	196.35
6		0.309	1.446	0.829	1.858	2.56	49.80	634.98
7		0.273	1.592	0.848	1.884	2.29	18.25	570.84
8		0.244	1.723	0.864	1.908	3.09	**−0.75**	771.73

of a catchment among the fuzzy regions is found to be useful in this task. In particular, there is no need to devote special effort for the options (ii), (iv), (v), (vi) and (vii) of the eight options if the threshold fuzzy membership value is sensibly chosen to form the fuzzy clusters.

In this study, adjustment option (i) is implemented on clusters obtained from FCM algorithm to form hydrologic regions which are statistically homogeneous. The sites that are flagged discordant by the discordancy measure are first identified. Though Hosking and Wallis (1997) provide critical values for the discordancy measure to declare a site unusual, it is worth identifying all the sites with high discordancy values. Secondly, the heterogeneity measures (H_1, H_2 and H_3) of the region to be adjusted are examined as they change with exclusion of each site from the region. In this context one site is eliminated at a time with replacement. Thirdly, the discordant site, whose exclusion reduces the heterogeneity measures for the region by a significant amount, is identified and removed from the region after ensuring that the site discordancy is not due to sampling variability. This procedure of region

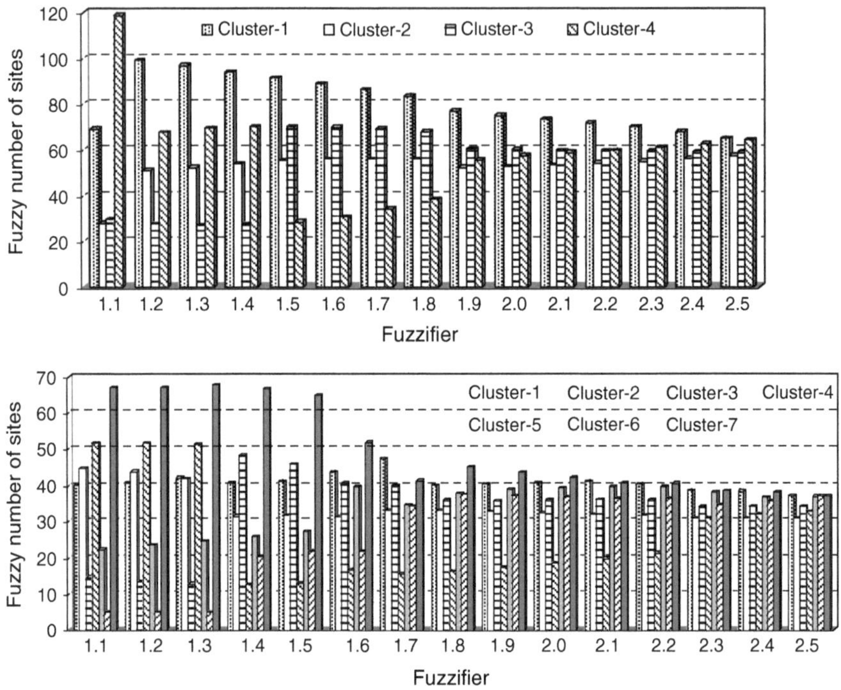

Fig. 3.5.32 Variation in cardinality of fuzzy clusters obtained from the fuzzy c-means algorithm for two typical values of c (4 and 7), with increase in the value of fuzzifier from 1.1 to 2.5 by an increment of 0.1

adjustment is described in Section 2.4.3. Hydrologic Regions 1, 2, 3, 4, 5 and 8 are formed by adjusting the clusters 1, 3, 5, 7, 6 and 4 respectively.

To adjust the highly heterogeneous cluster-2, several sites were eliminated from the region following the adjustment option (i). This amounts to splitting the highly heterogeneous cluster into two by using option (iii) for adjusting the regions. The first part containing a collection of highly heterogeneous sites formed hydrologic Region 6, whereas the second part with the homogeneous sites constituted hydrologic Region 7.

3.5.3 Testing the Regions for Robustness

The heterogeneity measures of Hosking and Wallis (1997) weigh information from each station in proportion to its record length. As a consequence, influence of stations with longer record length will be greater than that of stations with shorter record length. This may have adverse effects especially when some stations in a region have much longer record lengths than others. Therefore, the hydrologic regions are further examined for their robustness. By specifying various threshold values,

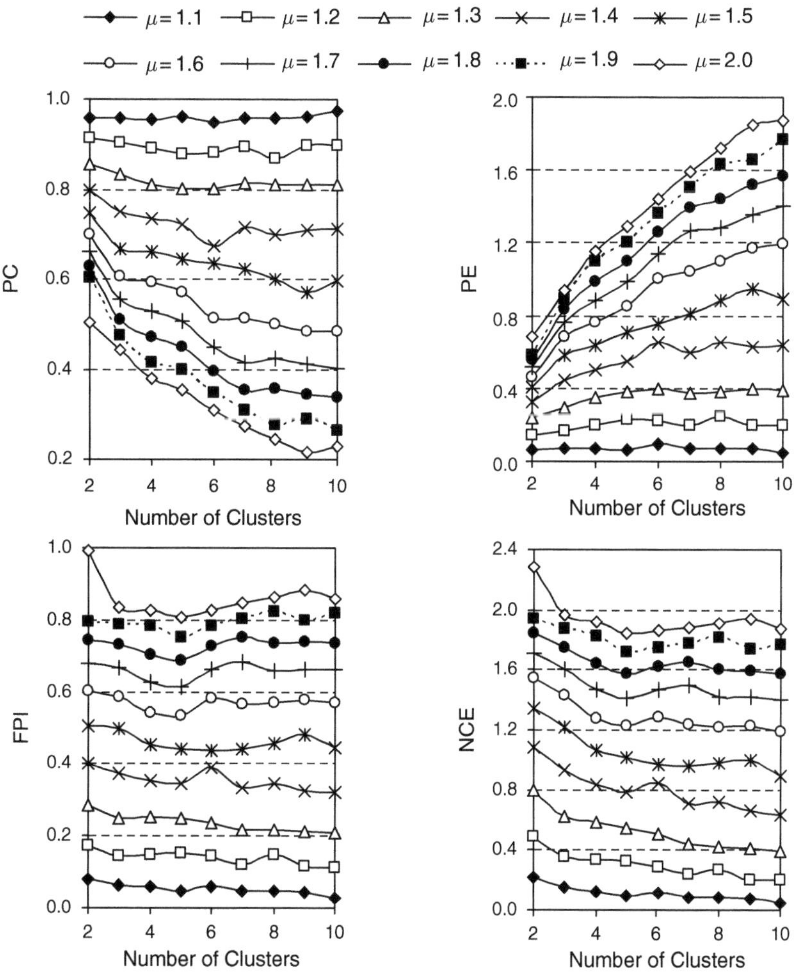

Fig. 3.5.33 Plots of fuzzy partition coefficient (PC), fuzzy partition entropy (PE), fuzziness performance index (FPI) and normalized classification entropy (NCE) values against the number of clusters for the Indiana watersheds obtained using FCM clustering. The optimal partition corresponds to a maximum value of PC (or minimum value of PE, FPI and NCE)

the stations with record lengths significantly different from those of the rest of the group are removed and the region with the remaining stations was examined for homogeneity. In this step, the stations that have an adverse effect on the homogeneity of the identified regions are excluded in an attempt to make the regions robust. The results of this analysis presented in Table 3.5.6 indicate that all the homogeneous regions identified are indeed robust.

Eleven sites, out of the 245 sites considered in this study, could not be allocated to any region. Three of these 11 sites were eliminated from the regions in the previous step to make the regions robust. Further, the remaining unallocated sites are those

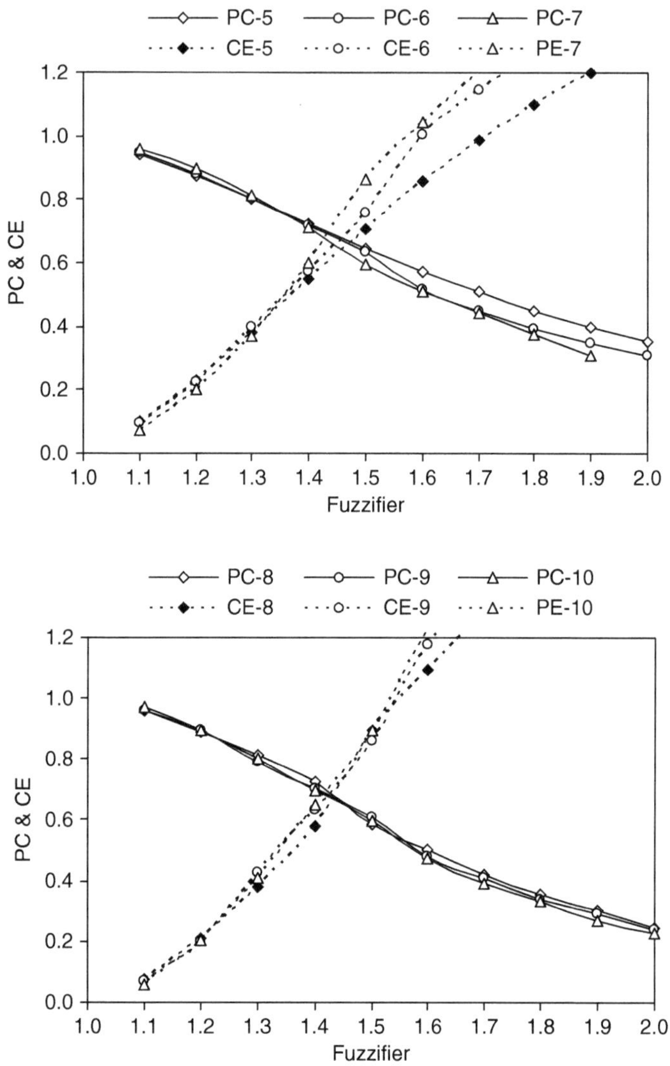

Fig. 3.5.34 Plot of Partition Coefficient (PC) and Classification Entropy (CE) for clusters obtained from FCM algorithm. The number that follows PC and CE denotes the number of clusters

which are highly discordant with sites in the clusters in which they have very strong membership.

It is evident from Figs. 3.5.36 and 3.5.37 that a few sites are eliminated from each fuzzy cluster to form the final regions. Further, the results presented in Table 3.5.7 indicate that except region 6, all the regions are acceptably homogeneous ($H_1 < 1$). The region-6 adjoining the Lake Michigan is highly heterogeneous and consists of 10 catchments in the Kankakee river basin of Indiana. Except for two of those 10

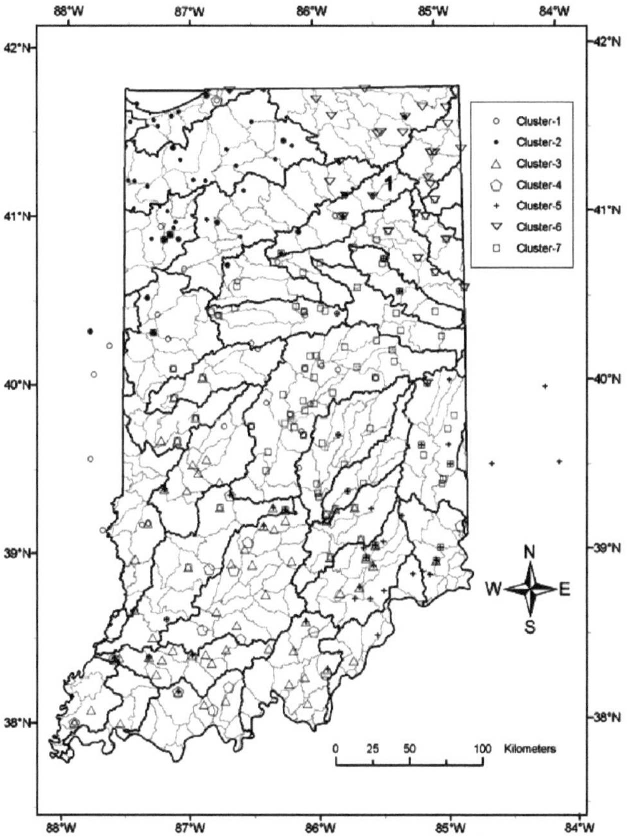

Fig. 3.5.35 Location of fuzzy clusters in Indiana obtained from fuzzy cluster analysis. The dark lines denote boundaries of eight digit watersheds, whereas the grey coloured lines are boundaries of 11 digit watersheds in Indiana

sites, all others have high membership in region 6 (Fig. 3.5.38). The average record length per station in the region 6 is 43-years, which is reasonably high.

The regions formed by Glatfelter (1984) using regional regression approach and those formed by applying FCM algorithm are presented in Figs. 3.5.39 and 3.5.40, respectively. It is evident from the figures that the regions identified differ significantly from those identified by Glatfelter (1984). Except region 8, all the homogeneous regions identified in this study have enough pooled data (Table 3.5.7). Following the region adjustment option (vi) mentioned earlier, the region-8 could be merged with region 2 which contains it geographically. However, it was decided not to merge these two regions because the sites of region 8 have low membership in region 2 (Table 3.5.8). It may also be noted that ten of the fourteen sites in the region 8 have high membership in the region 8 (Table 3.5.8 and Fig. 3.5.37).

The regions formed by using the FCM algorithm correlate well to regions obtained by hybrid clustering (Fig. 2.4.10) and also to geographical features and soil

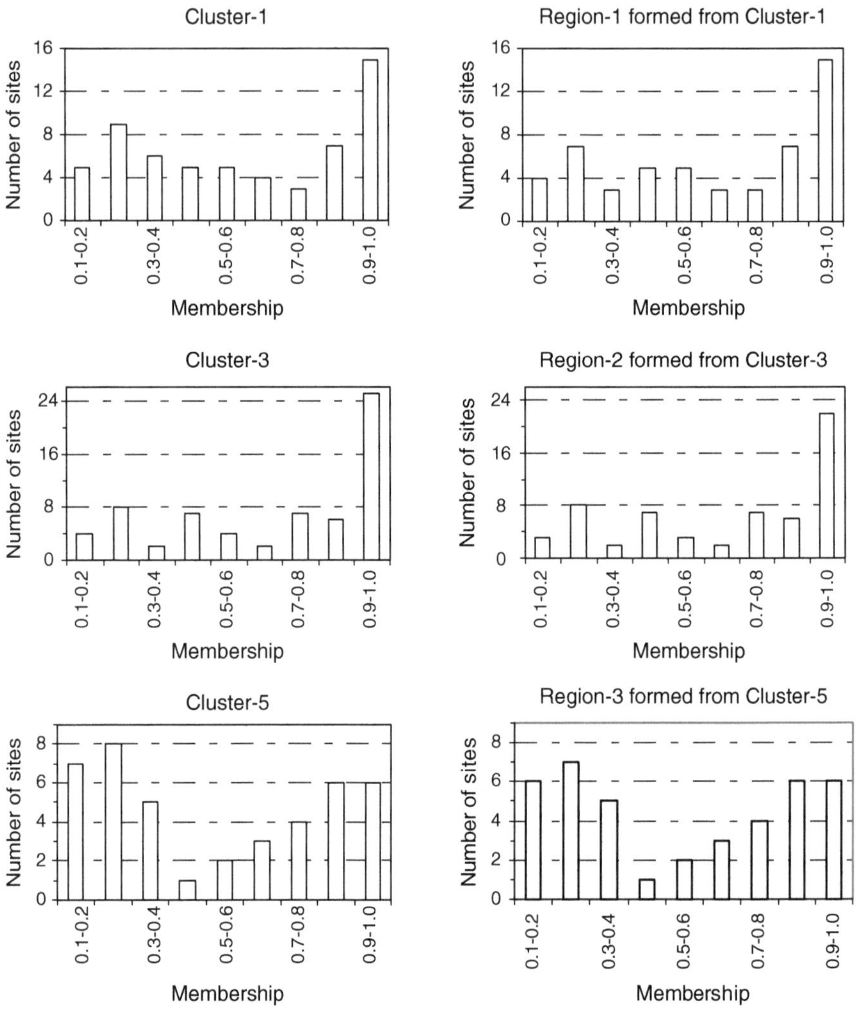

Fig. 3.5.36 Comparison of fuzzy clusters 1, 3 and 5 with the fuzzy regions formed by adjusting the same. The composition of a fuzzy region (cluster) is expressed using histograms of the fuzzy memberships of sites (feature vectors) belonging to the region (cluster). Sites with membership greater than $1/c$ in a cluster are considered to belong to the cluster

regions of Indiana (Figs. 3.5.41–3.5.43). This gives greater confidence in clustering techniques.

3.6 Concluding Comments

The fuzzy c-means (FCM) clustering algorithm is presented and its effectiveness in forming homogeneous regions for flood frequency analysis is illustrated through its application to watersheds in Indiana, USA.

Table 3.5.6 Results from the test of the regions for robustness

R	Condition	NS	RS	Heterogeneity measure			Region type
				H_1	H_2	H_3	
1	Entire region	52	911	0.63	0.83	−0.07	Homogeneous
	Sites with RL ≤ 10 are eliminated	27	664	1.03	1.09	0.58	Possibly Homogeneous
	Sites with RL≥30 are eliminated	45	665	0.63	0.15	−0.96	Homogeneous
2	Entire region	60	2010	0.89	0.92	−0.08	Homogeneous
	Sites with RL<20 are eliminated	44	1797	0.66	1.34	0.60	Homogeneous
	Sites with RL>50 are eliminated	48	1288	0.78	0.63	−0.23	Homogeneous
	Sites with RL≤10 and RL>50 are eliminated	42	1232	0.95	0.85	0.08	Homogeneous
3	Entire region	40	837	0.23	0.95	0.11	Homogeneous
	Sites with RL≤10 are eliminated	22	664	0.50	0.89	0.29	Homogeneous
	Sites with RL>40 are eliminated	35	579	0.61	1.13	−0.03	Homogeneous
4	Entire region	75	3274	0.79	0.80	−0.08	Homogeneous
	Sites with RL<30 are eliminated	58	2912	0.73	0.55	−0.30	Homogeneous
	Sites with RL>60 are eliminated	63	2401	0.78	0.88	−0.08	Homogeneous
	Sites with RL<30 and RL>60 are eliminated	46	2039	0.71	0.87	0.06	Homogeneous

R: Region; RL: record length; NS: Number of stations; RS: Region size in station years

Partition coefficient, partition entropy, fuzziness performance index, normalized classification entropy, extended Xie-Beni, Fukuyama-Sugeno, and Kwon's index are tested for their ability to identify optimal partition provided by the FCM clustering algorithm. Results obtained for Indiana watersheds suggest that extended Xie-Beni and Kwon's indices are effective in identifying optimal regions. The inability of other validity measures in identifying optimal regions could be attributed to shortcomings inherent in their formulation. Several of these indices are based only on fuzzy membership degrees and lack connection to structure of the data.

Furthermore, the optimal partition identified using the fuzzy cluster validity indices is seen to have clusters which are similar to the plausible hydrologic regions recognized by visual inspection of clusters formed by hardening fuzzy partition matrix. Although this may not happen always, it gives confidence in the regions that are determined for flood frequency analysis.

In general, groups of watersheds formed by using cluster analysis are statistically heterogeneous. Therefore it is necessary to adjust the clusters to improve their homogeneity. The Fuzzy memberships of sites in clusters are found to be useful in adjusting the regions. The effort needed to adjust a region is found to be significant

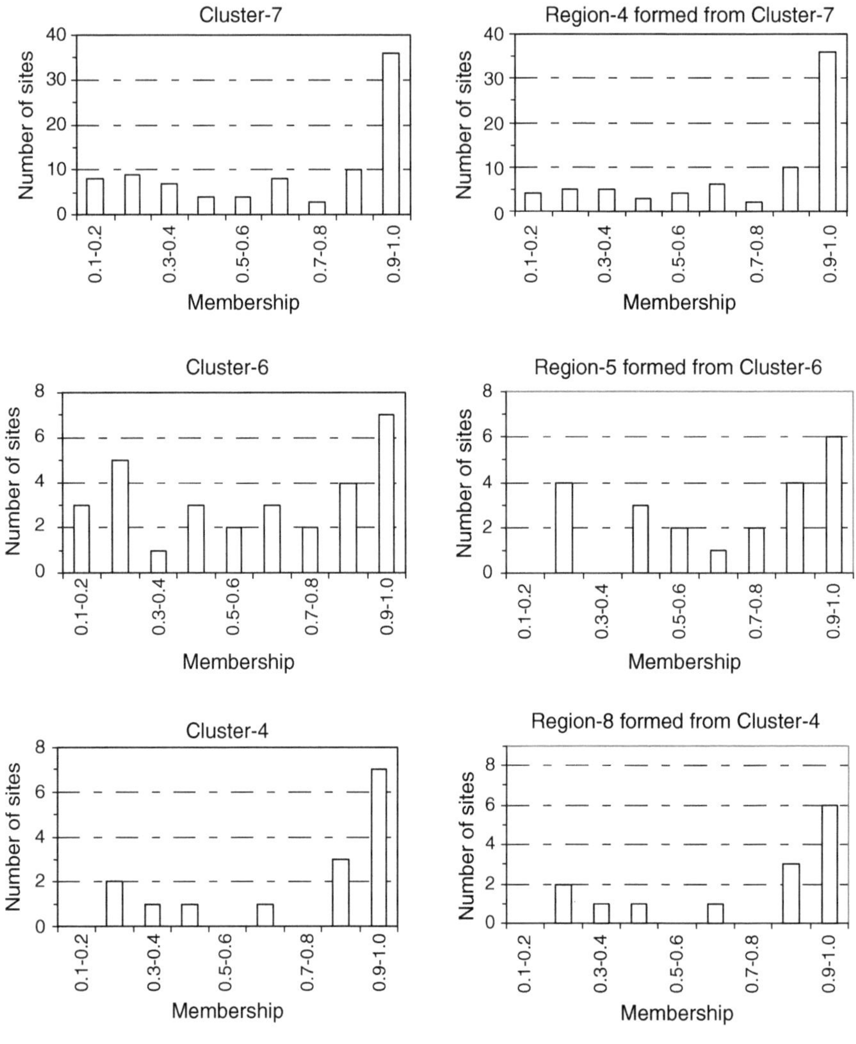

Fig. 3.5.37 Comparison of fuzzy clusters 7, 6 and 4 with the fuzzy regions formed by adjusting the same. The composition of a fuzzy region (cluster) is expressed using histograms of the fuzzy memberships of sites (feature vectors) belonging to the region (cluster). Sites with membership greater than $1/c$ in a cluster are considered to belong to the cluster

when hard clusters formed by defuzzification of result from fuzzy clustering algorithm are used to form hydrologic regions. In contrast, the effort needed to form homogeneous regions by adjusting fuzzy clusters derived from fuzzy clustering algorithm is found to be smaller.

The results shown in this chapter strongly support the use of fuzzy cluster analysis to derive homogeneous regions, which are effective for flood frequency analysis. The following conclusions are drawn:

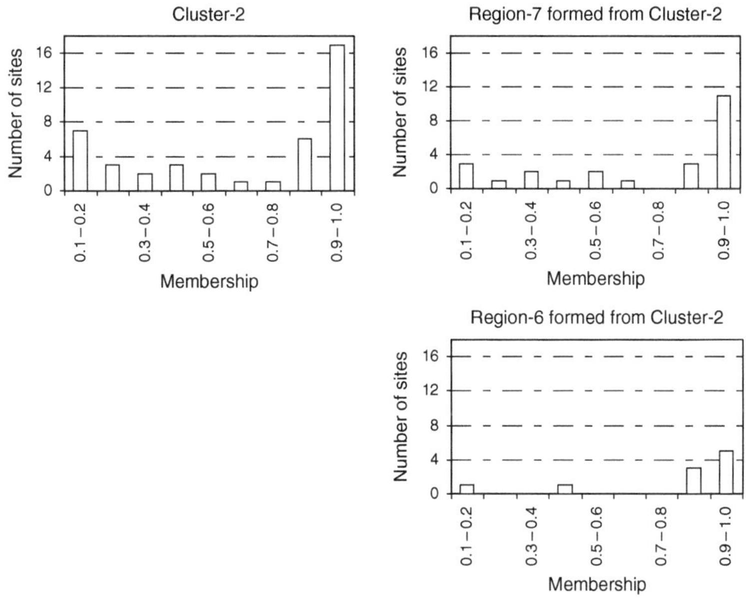

Fig. 3.5.38 Comparison of fuzzy cluster 2 with the fuzzy regions 6 and 7 formed by splitting the same

Table 3.5.7 Characteristics of the regions formed using FCM cluster analysis

Region number	NS	RS	Heterogeneity measure		
			H_1	H_2	H_3
1	52	911	0.63	0.83	−0.07
2	60	2010	0.89	0.92	−0.08
3	40	837	0.23	0.95	0.11
4	75	3274	0.79	0.80	−0.08
5	22	975	0.97	−0.13	−0.77
6	10	431	13.51	5.71	2.62
7	24	1012	0.48	0.03	0.79
8	14	160	0.99	−0.24	−1.70

(i) The optimal number of clusters (c) for regionalization has to be determined through detailed analysis of results from the fuzzy clustering algorithm and by using information from potential cluster validity measures.

(ii) The clusters formed by using FCM algorithm are sensitive to variation in the value of fuzzifier (μ) for a chosen number of clusters (c). Hence the optimal value of μ for regionalization has to be determined through careful investigation. It is not appropriate to use $\mu = 2$ as a default choice, which appears to be a common practice.

(iii) The fuzzy cluster validity measures such as partition coefficient, partition entropy, fuzziness performance index, and normalized classification entropy, which lack direct connection to structure in multi-dimensional space of feature

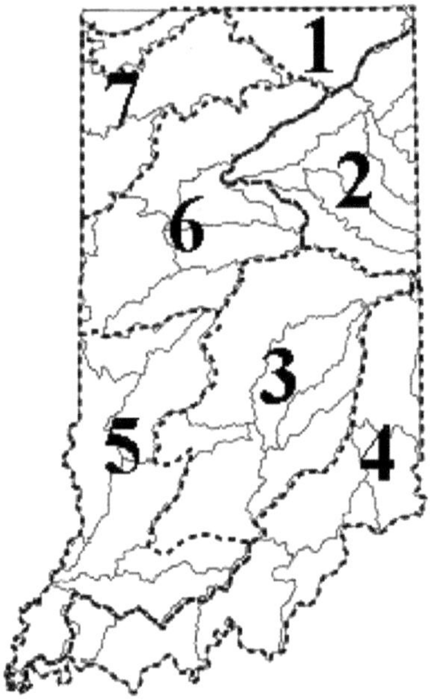

Fig. 3.5.39 The seven hydrological regions identified by Glatfelter (1984) for estimating the magnitude and frequency of floods on streams in Indiana

vectors prepared from watershed attributes, are inefficient in deriving hydrologically homogeneous regions.

(iv) Fuzzy cluster validity indices that simultaneously take into account the properties of the fuzzy membership degrees and the structure of the data (example, extended Xie-Beni Index, Kwon's Index), are effective in identifying optimal partition.

Table 3.5.8 Fuzzy memberships of the sites belonging to Region 8 in Regions 2 and 8

USGS site number	Membership of site in:	
	Region-2	Region-8
03373850	0.0003	0.9984
03373680	0.0015	0.9915
03302690	0.0014	0.9907
03360400	0.0027	0.9827
03303250	0.0059	0.9680
03302350	0.0041	0.9675
03372680	0.0167	0.8961
03276640	0.0162	0.8552
03303440	0.0463	0.8016
03356780	0.0525	0.6308
03303900	0.1574	0.4273
04095250	0.0509	0.3709
03378590	0.2065	0.2569
03376600	0.2580	0.2182

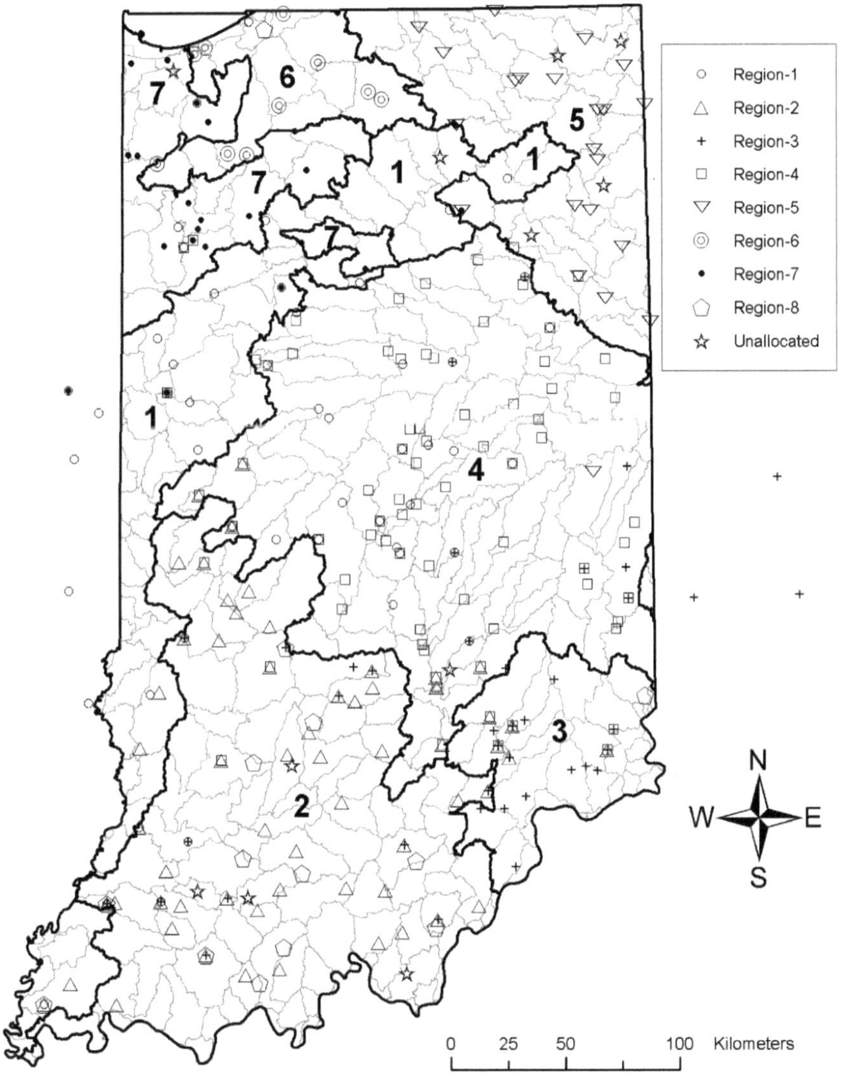

Fig. 3.5.40 Location of the regions defined by using the hybrid cluster analysis. The grey coloured lines within each region denote boundaries of 11 digit watersheds in Indiana, USA. Region 8, which is thoroughly mixed with Region 2, is not marked by soft boundary

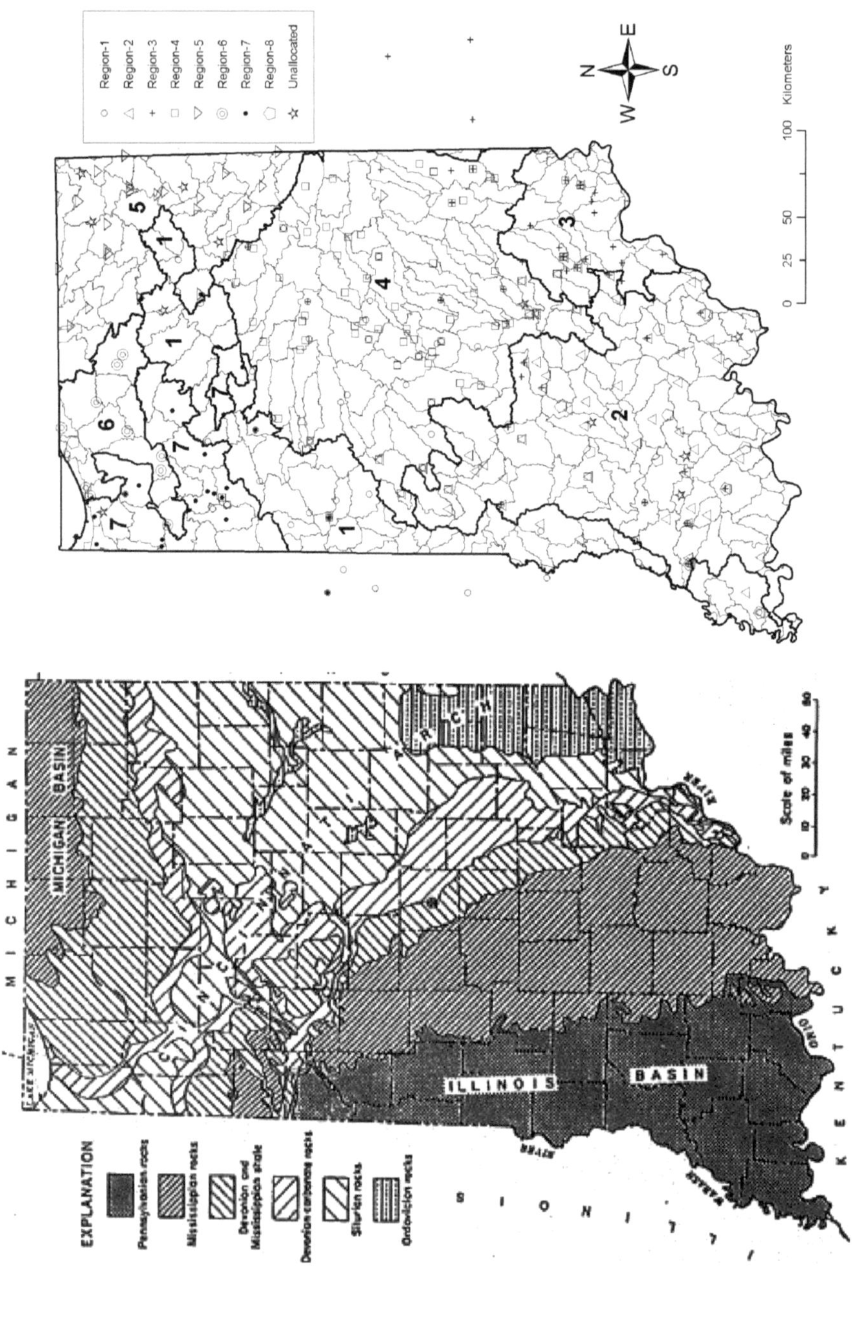

Fig. 3.5.41 Comparison of the hydrological regions identified in Indiana with geologic features of the state

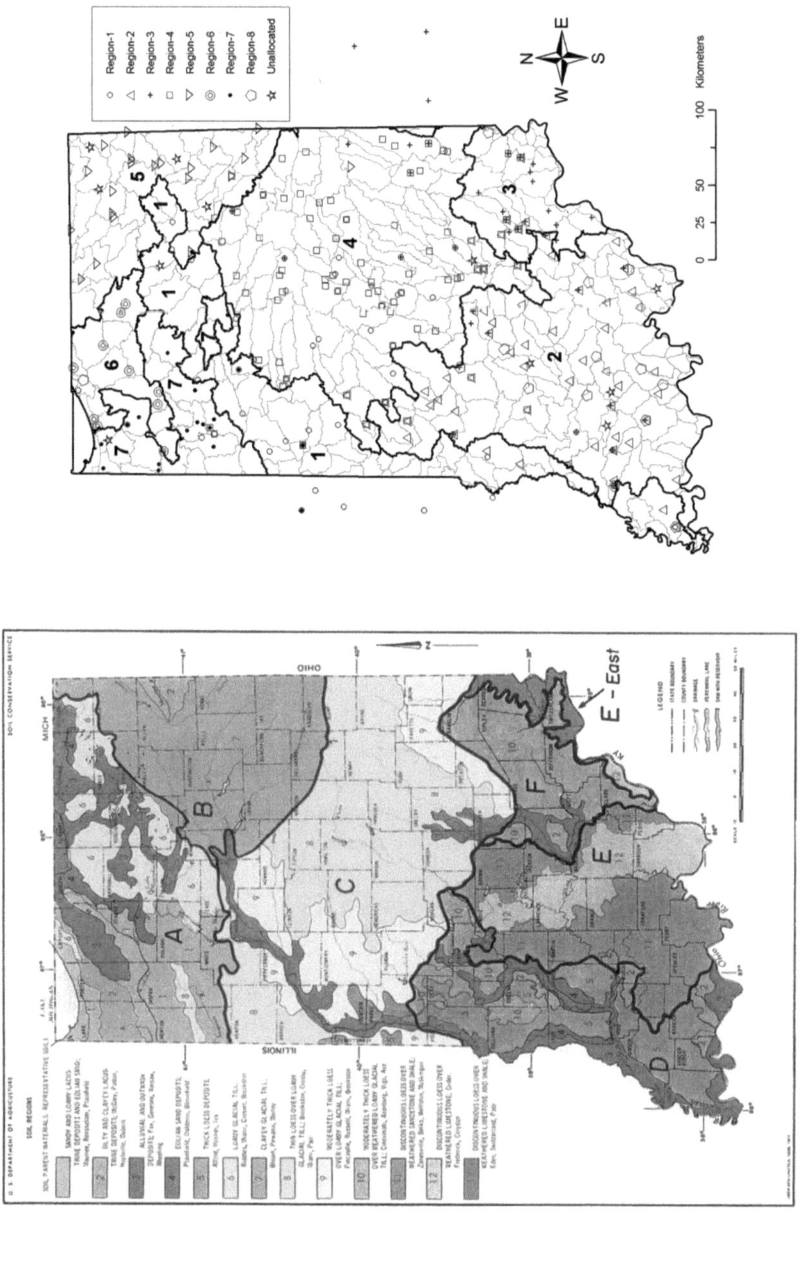

Fig. 3.5.42 Comparison of the hydrological regions identified in Indiana with soil regions in the state identified by soil conservation service, US Department of Agriculture

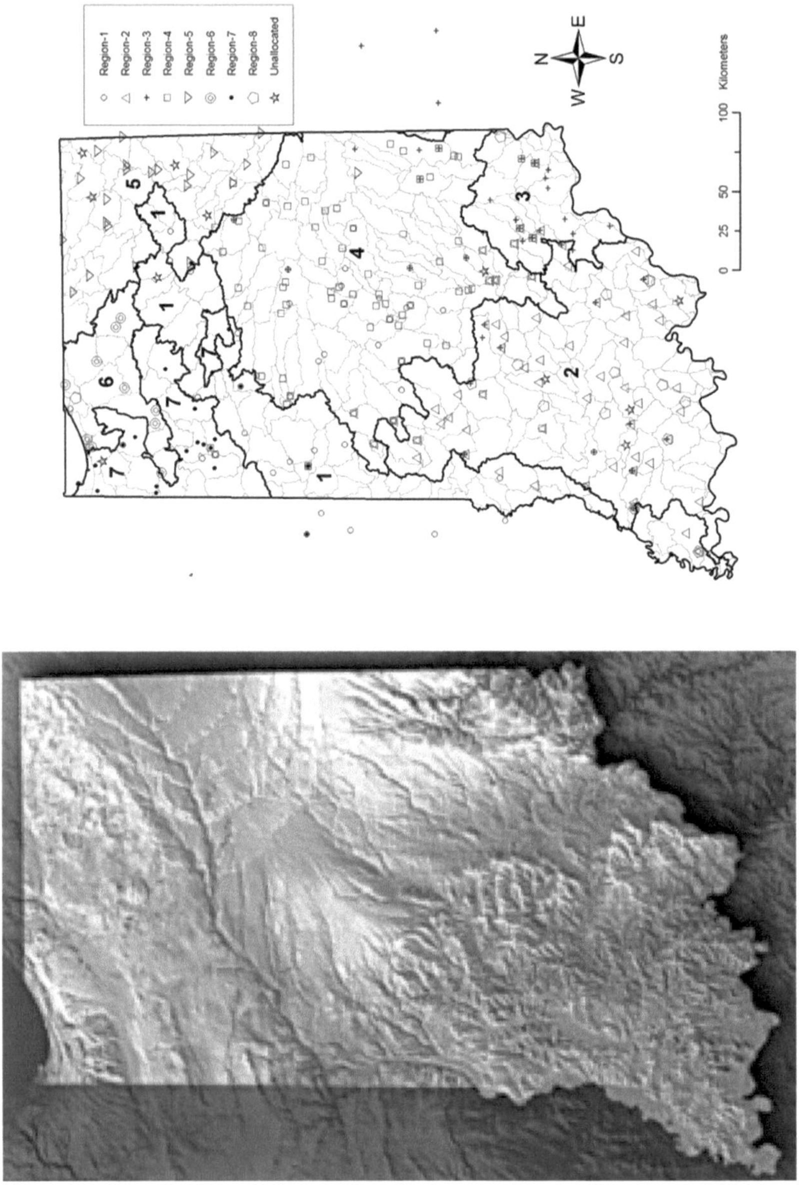

Fig. 3.5.43 Comparison of the hydrological regions identified in Indiana with tapestry produced by union of geology and topography of Indiana

(v) The effort needed to form homogeneous regions by adjusting the ‘hard clusters’ formed by defuzzification of FCM results could be considerable. Therefore, it is suggested that the final outcome of the fuzzy clustering algorithm be obtained as a set of ‘fuzzy clusters’, which could be adjusted with relatively smaller effort.

Chapter 4
Regionalization by Artificial Neural Networks

4.1 Introduction

Over the past two decades, artificial neural network (ANN) based models have been extensively developed and studied in an effort to simulate the behavior of the biological neurons in the human brain. The nonlinearity and flexibility embedded in ANNs makes them useful in a variety of physical science applications, including hydrology (Govindaraju and Rao, 2000; ASCE Task Committee on Artificial Neural Networks in Hydrology, 2000a,b). In this chapter, the efficacy of a special class of such networks called Kohonen Self-Organizing Feature Maps (SOFMs) for regionalization of watersheds is discussed.

4.2 Kohonen Self-Organizing Feature Maps (SOFMs)

The SOFMs, also called topology-preserving maps, belong to the category of competitive learning networks. They are based on unsupervised learning, which means that no target outputs are needed for classifying the given data. These networks attempt to find topological structure in the input data by mapping given data onto a feature map, also referred to as output layer or output space.

There are two different models of self-organizing neural networks (Su and Chang, 2000): Willshaw-Von Der Malsburg model (Willshaw and Malsburg, 1976) and Kohonen model (Kohonen, 1982). The former model is specialized to mappings where the dimension of input space is the same as that of output space, whereas the latter model is capable of generating mappings from high-dimensional input spaces to lower dimensional output space, known as Kohonen layer. An overview of self-organizing maps, including recent developments and their engineering applications are found in Kohonen et al. (1996), Obermayer and Sejnowski (2001), among others.

The SOFM (Kohonen, 1982) has been used in regionalization studies (Hall and Minns, 1998, 1999; Hall et al., 2002; Jingyi and Hall, 2004), with one-dimensional (1-D) Kohonen layer (also called linear Kohonen network). But the choice of number of nodes in the Kohonen layer remains subjective. Hall and Minns (1998, 1999) examined the utility of Kohonen network for regionalization by applying it to a sample of 101 gauged sites in southwest of England and Wales in the United Kingdom. Hall et al. (2002) applied the 1-D SOFM to three data sets from the south-west of England and Wales, Wales and Scotland and to the islands of Java and Sumatra in Indonesia. These studies did not report validation of the identified regions by using any heterogeneity measure. Jingyi and Hall (2004), however, use the heterogeneity measures of Hosking and Wallis (1997) to assess homogeneity of regions identified by using 1-D SOFM from 86 gauging stations in Jiangxi and Fujian provinces of China.

While learning, the SOFM allocates feature vectors depicting watershed characteristics to various output nodes. If well defined natural groupings are inherent in the data set, the feature vectors cluster around output nodes that are well separated from each other. It indicates that SOFMs select the optimal number of clusters automatically. This feature of SOFMs offers an advantage over hard and fuzzy clustering methods discussed in previous chapters, which can only partition given data set into specified number of clusters.

Nevertheless, in the absence of clearly distinguishable patterns in the given data, it is seldom possible to interpret clusters from the output of a SOFM, irrespective of its size and dimensionality. This is illustrated through a real world example in this chapter. In such situations, however, SOFMs may serve as a useful precursor to clustering algorithms. In a few recent attempts, clustering methodologies are devised to group the neighbouring nodes in the Kohonen layer to form clusters. A few algorithms which are based on this perspective are briefly reviewed in this chapter. Further, a novel clustering algorithm which is based on Fuzzy clustering of SOFM is presented for regionalization of watersheds for flood frequency analysis.

4.2.1 Algorithm of Kohonen Self-Organizing Feature Map

The SOFM (Kohonen, 1982) is one of the widely used artificial neural networks for identifying topological structure in given data. It has found applications in the areas of pattern recognition, biological modeling, data compression, signal processing and data mining (Kohonen, 1997). A schematic for the SOFM is presented in Fig. 4.2.1. The SOFM has an input layer and an output layer, each consisting of several nodes. The number of nodes in the input layer is equal to the number of watershed attributes considered for regionalization. Each node in the input layer is connected to each node in the output layer by synaptic links. Associated with each link is a connection strength or weight.

Let $\boldsymbol{y}_k$ denote the 'k-th' feature vector in n-dimensional space with coordinate axis labels $(y_1, \ldots, y_n)$, i.e., $\boldsymbol{y}_k = [y_{1k}, \ldots, y_{nk}] \in \Re^n$ and y_{ik} denotes the value of attribute i in the k-th n-dimensional feature vector $\boldsymbol{y}_k$.

The attributes of the feature vector $\boldsymbol{y}_k$ are rescaled as

$$x_{ik} = \frac{w_i}{\sigma_i}[f(y_{ik})] \quad \text{for} \quad 1 \le i \le n, 1 \le k \le N \tag{4.2.1}$$

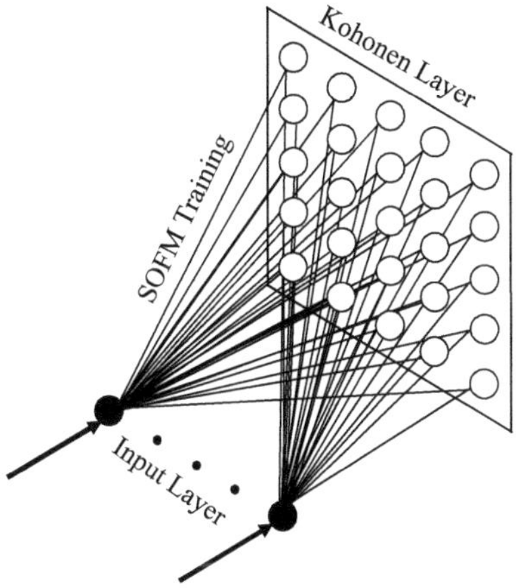

Fig. 4.2.1 A schematic of the Kohonen self organizing feature map. The input layer consists of rescaled watershed attributes

where x_{ik} denotes the rescaled value of y_{ik}; w_i is the weight associated with attribute i based on its relative importance; σ_i refers to the standard deviation of attribute i; $f(\cdot)$ represents the transformation function that is to be identified; and N represents the number of n-dimensional feature vectors. Rescaling the attributes is necessary because of the differences in their variance, relative magnitudes and importance.

In general, the k-th re-scaled feature vector is given by $\boldsymbol{x}_k = [x_{1k}, x_{2k}, \ldots, x_{nk}]$. The set of N rescaled feature vectors can be represented as $N \times n$ data matrix $\boldsymbol{X}$.

$$\boldsymbol{X} = \begin{bmatrix} x_{11} & \ldots & x_{1N} \\ \vdots & \ddots & \vdots \\ x_{n1} & \ldots & x_{nN} \end{bmatrix} \tag{4.2.2}$$

The number of nodes in the input layer of SOFM is equal to the dimension of the feature vector, n. The output layer, also referred to as competitive or Kohonen layer, has m nodes organised in a lattice that is usually one- or two-dimensional. The value of m can be chosen as maximum number of clusters to be formed (Fausett, 1994). In the context of regionalization in hydrology, the value of m is generally chosen to be much larger than the expected number of clusters, C_{Exp}. For a one-dimensional SOFM, Hall and Minns (1999) chose m to be at least $2C_{Exp}$, while Hall et al. (2002) considered the same to be at least $3C_{Exp}$.

The SOFM algorithm is as follows:

(i) Initialize weights $\{w_{ij}, i = 1, \ldots n, j = 1, \ldots m\}$ of the connections from the n input nodes to the m output nodes, randomly. These random weights are generally chosen from the same range of values as the components of the input vectors. Let $\boldsymbol{w}_j = \{w_{1j}, w_{2j}, \ldots, w_{nj}\}$ denote weight vector between the output node j and the nodes in the input layer. Set iteration $t = 0$.

(ii) Swap $\boldsymbol{X}$ to $\mathbf{X}'$.

(iii) Draw an input vector $\boldsymbol{x}_k$ from $\mathbf{X}'$ (randomly without replacement) and compute its distance from $\boldsymbol{w}_j$ by using Euclidean metric. Find the winning output node ω as

$$\omega = arg\min_j \left\| \boldsymbol{x}_k(t) - \boldsymbol{w}_j(t) \right\| \qquad j = 1, 2, \ldots, m \tag{4.2.3}$$

(iv) Update the weight vectors by

$$\boldsymbol{w}_j(t+1) = \boldsymbol{w}_j(t) + \eta(t) h_{j,\omega}(t) [\boldsymbol{x}_k(t) - \boldsymbol{w}_j(t)] \tag{4.2.4}$$

where $\eta(t)$ is the learning-rate parameter for iteration t, and $h_{j,\omega}(t)$ is called a neighborhood function. The quantity $\eta(t)$ is chosen to decrease monotonically with increase in t as

$$\eta(t) = \eta(0) \exp\left(-\frac{t}{\tau_1}\right) \tag{4.2.5}$$

where $\eta(0)$ is selected to have a value close to 0.1, τ_1 is a constant that is typically set equal to the maximum number of iterations $t_{\max}$ (e.g., 1000).

In Eq. (4.2.4), the neighborhood function $h_{j,\omega}(t)$, centered around the winning node ω, is given by

$$h_{j,\omega}(t) = \exp\left(-\frac{d_{\omega,j}^2}{2\sigma^2(t)}\right) \tag{4.2.6}$$

where $d_{\omega,j}$ is the topological distance between the winning node ω and its neighboring node j in the output layer,

$$d_{\omega,j} = \left\| \boldsymbol{r}_\omega - \boldsymbol{r}_j \right\| \tag{4.2.7}$$

where the discrete vector $\boldsymbol{r}_\omega$ defines the position of the winning node ω and the discrete vector $\boldsymbol{r}_j$ defines the position of its neighboring node j, both of which are measured in discrete output space.

In Eq. (4.2.6), the parameter $\sigma(t)$ is the effective width of the topological neighborhood $h_{j,\omega}(t)$ at time step t.

$$\sigma(t) = \sigma(0) \exp\left(-\frac{t}{\tau_2}\right) \tag{4.2.8}$$

where $\sigma(0)$ is set to be equal to the radius of the lattice in the output layer of SOFM (Fig. 4.2.2); τ_2 is a constant estimated by

$$\tau_2 = \frac{t_{\max}}{ln\sigma(0)} \tag{4.2.9}$$

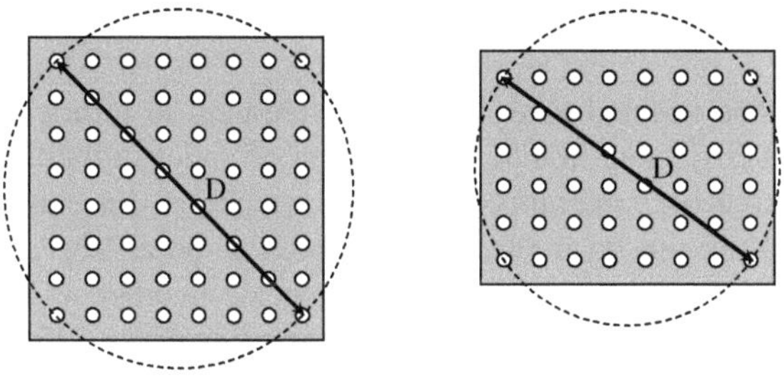

(a) (b)

Fig. 4.2.2 Schematic of the diameter (D) of the Kohonen lattice of SOFM – (**a**) square lattice and (**b**) rectangular lattice. The neighborhood of winning neuron at iteration $t = 0$ of SOFM is chosen as radius of the lattice, which is equal to D/2

From Eqs. (4.2.8) and (4.2.9) it follows that for $t = 0$, $\sigma(t) = \sigma(0)$, while for $t = t_{\max}$, $\sigma(t) = 1$. In other words, at $t = 0$ almost all nodes in the output layer centered on the winning node are updated, whereas at the end of iterations only a few neighboring nodes around a winning node are updated. Further, it can be inferred from Eq. (4.2.6) that $h_{j,\omega}(t)$ will shrink (or decay) exponentially with increase in t.

(v) If $\mathbf{X}'$ is empty go to step (vi), else go to step (iii).

(vi) If $t \geq t_{\max}$ go to (vii), else increment 't' to '$t + 1$' and continue with step (ii) until no noticeable changes in the feature map are observed.

(vii) Assign to each input vector $\boldsymbol{x}_k$ the label of its winner output node $j (j = 1, \ldots, m)$ using Eq. (4.2.3). The m' winning output nodes ($m' \leq m$) are referred to as prototypes. A detailed description of the SOFM algorithm is found in Haykin (2003).

4.3 Example of Using SOFMs for Regionalization

4.3.1 Features Used

The sensitivity of flood response of Indiana watersheds to variation in the values of watershed attributes is discussed in Section 2.4.1. As described before, this lead to selection of the attributes mean annual precipitation, drainage area, slope of the main channel in the drainage basin, soil runoff coefficient, and storage.

It is natural to expect that nearby watersheds exhibit similar extreme flow responses due to similarities in the causal precipitation events which form input to the flow generation process in the watersheds. Keeping this in view, the latitude and the longitude which reflect physical proximity of sites are chosen as attributes to identify regions that are geographically contiguous. Information pertaining to all these attributes was available for the 245 stations considered.

4.3.2 Results from SOFM

The selected seven attributes were used to form clusters of watersheds in Indiana, USA, using SOFM. The number of nodes in the input layer of the SOFM is equal to 7 (i.e., the number of attributes chosen for cluster analysis), while those in the Kohonen layer (KL) were chosen to be at least two times the expected number of clusters. In practice, the expected number of clusters in a study region is not known *a priori*, and was estimated as 12 ($\approx$245/20) assuming that regions with average size of 20 sites will be formed. Thus, the plausible grid size of KL was 24 ($= 2 \times 12$) nodes. Hosking and Wallis (1997, pp. 119–123) report that with an index-flood procedure, unless extreme quantiles with non-exceedance probability greater than 0.999 are to be estimated, there is little gain in the accuracy of flood quantile estimates by using regions larger than about 20 sites.

To arrive at an appropriate architecture for the KL, sensitivity in the spatial relationships of mapping produced by SOFM was analyzed for a few alternate KL sizes by visual inspection of several 1- and 2-D feature maps. Previous regionalization studies in hydrology (Hall and Minns, 1999; Hall et al., 2002; Jingyi and Hall, 2004) have considered 1-D feature maps. However, in view of the fact that regionalization of watersheds occurs in 2-D, it is expected that 2-D KL maps would be more revealing. Further, for the 2-D feature maps, a square grid was considered for the KL. Other possible shapes for the KL include rectangular and hexagonal. The grid size of 2-D KL was varied from 5×5 nodes to 11×11 nodes, with an increment of one node for each edge.

It is noted from the results of SOFM that irrespective of the architecture chosen for KL, same groups of feature vectors are always mapped onto nearby nodes on the KL. Further, increase in the size of grid resulted in increase in the count of nodes with zero (or no) mapping on the KL. It was observed that the feature maps generated did not reveal any information that would allow selection of an appropriate number of clusters. This could possibly be due to lack of well defined natural groups in the data. A typical example of result from linear (1-D) SOFM is shown in Fig. 4.3.1. It can be seen that the distribution of the number of hits at nodes on the lattice is fairly uniform with little indication of any preferential grouping. Figure 4.3.2 shows examples of 6×6 and 8×8 2-D Kohonen lattices. No specific patterns could be discerned with this classical application of SOFMs. While previous studies have utilized 1-D feature maps for regionalization (Hall and Minns, 1999; Hall et al., 2002; Jingyi and Hall, 2004), the results obtained herein suggest that additional steps are required for identification of homogeneous regions.

To form clusters using SOFM, two options are considered. In the first option, 1-D feature map is used with the number of nodes in its output layer equal to the number of clusters to be formed. As the second option, a novel procedure involving clustering of 2-D feature map formed with SOFM is considered (Srinivas et al., 2008). The results obtained from the first option are presented and discussed in the following Section 4.3.2.1, whereas the algorithm and results obtained for the second option are presented in the next section.

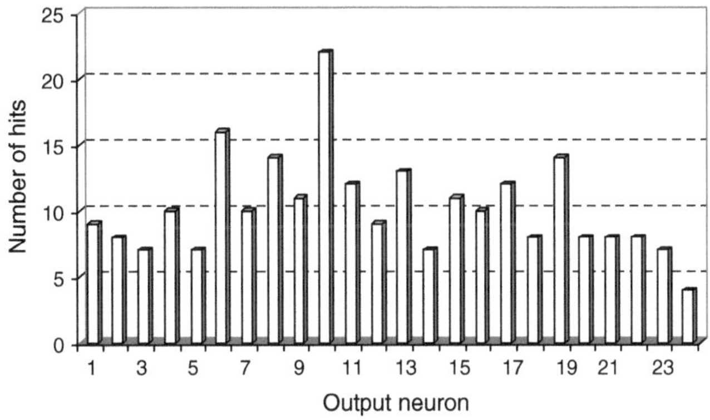

Fig. 4.3.1 Typical count map for one-dimensional Kohonen lattice obtained with classical SOFM developed for regionalization of watersheds in Indiana, USA, which illustrates the ambiguity in interpreting the number of clusters based on the number of hits at nodes (neurons) on the lattice

4.3.2.1 Identification of Plausible Regions Through 1-D SOFM

The 1-D feature map is used to form regions, by considering the number of nodes in the output layer to be equal to the number of clusters to be formed. To examine the sensitivity of results obtained from the feature map to number of nodes in the output layer of the SOFM, m is varied from 1 to 12. Further, three different scenarios were considered to form clusters of watersheds. In the first scenario, equal weight is assigned to all the seven attributes in Eq. (4.2.1). In the second scenario, weight assigned to drainage area is twice the weight assigned to other attributes. In the third scenario, feature maps are formed without geographic location attributes (latitude and longitude), with equal weight for all the other attributes. Comparison

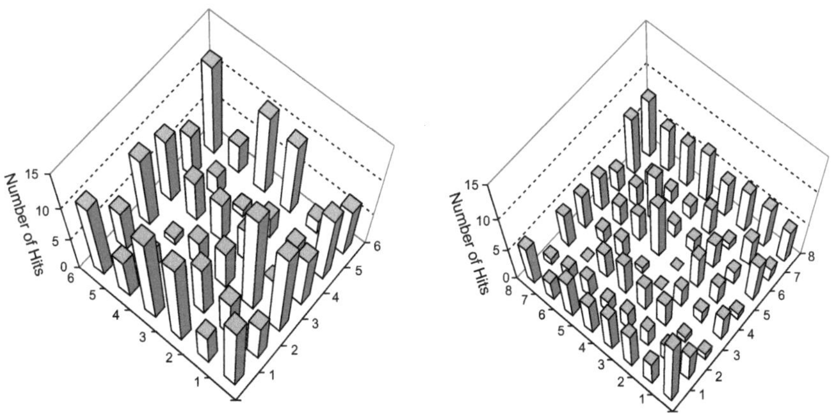

Fig. 4.3.2 Count maps for typical two-dimensional Kohonen lattices obtained with classical SOFM developed for regionalization of watersheds in Indiana, USA. Note that no clear picture is revealed for identifying the number of clusters for regionalization

of results from scenario-1 with those from scenario-3 helps in understanding the role of latitude and longitude in forming clusters that are geographically contiguous.

Further, the plausible hydrologic regions are identified by using two procedures. In the first procedure, the clusters provided by the feature map are visually interpreted. In the second procedure, hard cluster validity measures that have been described in Chapter 2 are adopted to determine optimal number of regions.

Identification of Regions Through Visual Interpretation of Clusters

The geographic location of clusters obtained from the 1-D feature map for the three scenarios is shown in Fig. 4.3.3(a–i), whereas the sizes of clusters and their homogeneity statistics are shown in Figs. 4.3.4–4.3.9.

The extent to which regional frequency analysis is preferable to at-site analysis depends on the number of sites in a region. In general, majority of identified clusters tend to be homogeneous with increase in m. However, increase in m provides several small clusters that are ineffective for regional flood frequency analysis. Hence, for

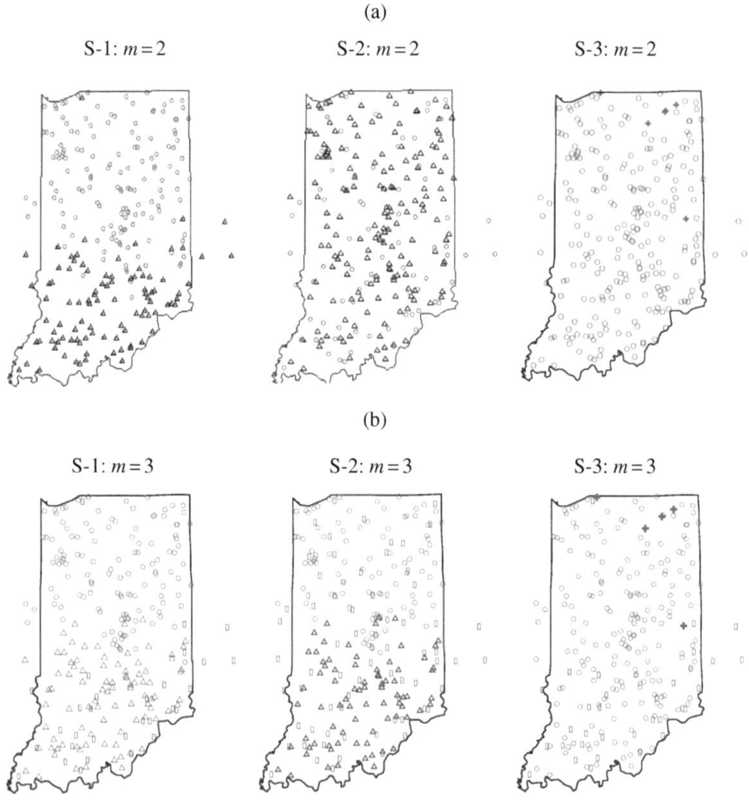

Fig. 4.3.3 Location of plausible hydrologic regions in Indiana obtained from 1-D SOFM. m denotes the number of output nodes of SOFM; S-1, S-2 and S-3 represent Scenario-1, Scenario-2 and Scenario-3 respectively. Each of the symbols in the diagram characterizes different clusters

Fig. 4.3.3 (continued)

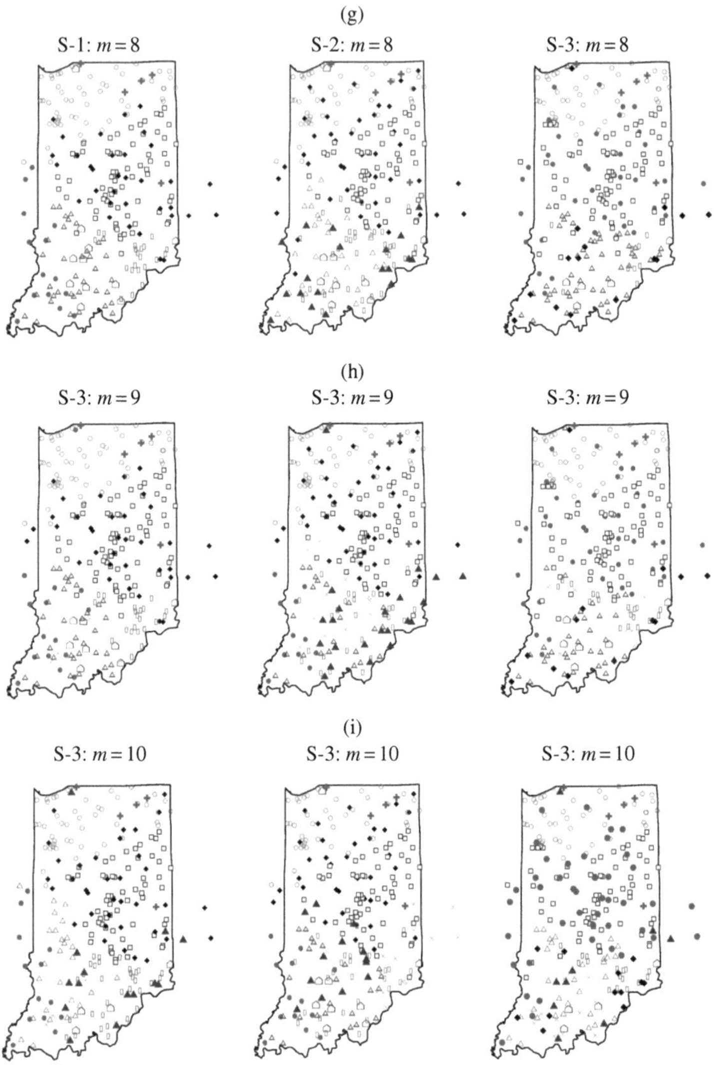

Fig. 4.3.3 (continued)

identifying plausible hydrologic regions that are effective for RFFA, optimal value of m is determined as a tradeoff between decrease in region size and increase in homogeneity of the region.

For $m = 2$ and scenario-1, the SOFM provides two clusters with well-defined boundaries, one consisting of watersheds in northern Indiana and the other consisting of those in southern Indiana. On the other hand, $m = 2$ and scenario-2 resulted in vague clusters in the sense that boundaries between them are not well defined. Interestingly, scenario-3 classified almost all the stations in Indiana as a

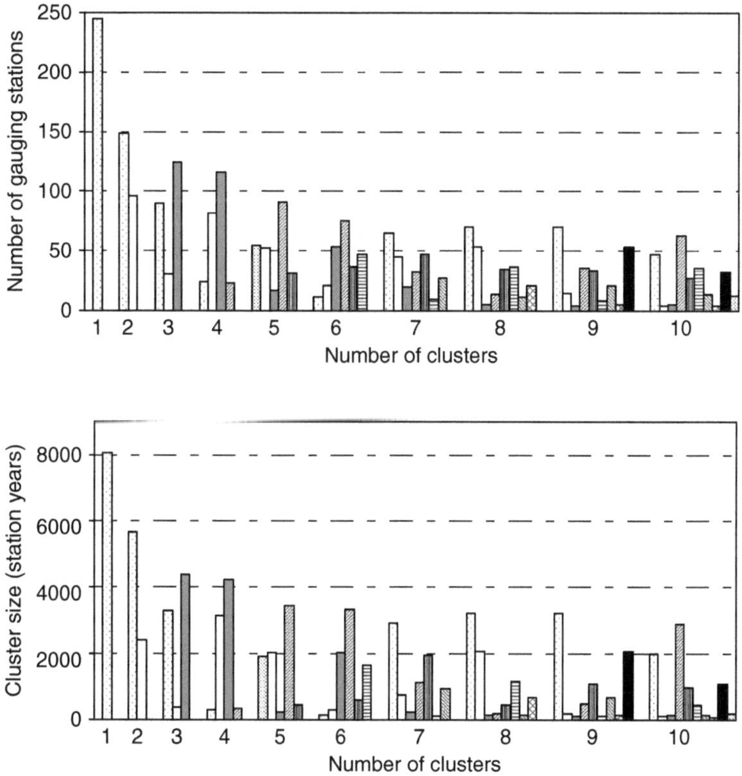

Fig. 4.3.4 Characteristics of clusters obtained from one-dimensional SOFM for the scenario-1 with equal weights to all the seven attributes

single cluster, except for a small group of 5 stations that are characterized by small drainage areas, milder slopes, large storage and low runoff coefficient values (cluster 2 in Table 4.3.1). It is seen from Figs. 4.3.4–4.3.9 that for all the three scenarios, majority of the sites belong to a larger cluster that is highly heterogeneous. Therefore $m = 2$ does not constitute a suitable classification for RFFA.

When m is increased to 3, scenario-1 provided a cluster in northern Indiana, whereas the other two clusters in the southern part of the state (shown in triangles and rectangles) are vague. The cluster shown in rectangles closely resembles a cluster identified in the same location of the state with scenario-3. Further, this cluster could also be viewed as a subset of the vague cluster that is spread throughout Indiana for scenario-2 (Fig. 4.3.3b). Possibly, this indicates a unique identity of the collection of those stations at m equal to 3. The first, second and third scenarios classify 51%, 39% and 85% of the sites as highly heterogeneous, respectively (Figs. 4.3.4–4.3.9). Therefore $m = 3$ does not constitute a suitable classification for RFFA.

With increase in m to 4, considerable improvement is evident in the results. The second scenario classified 40% of the stations into a highly heterogeneous cluster,

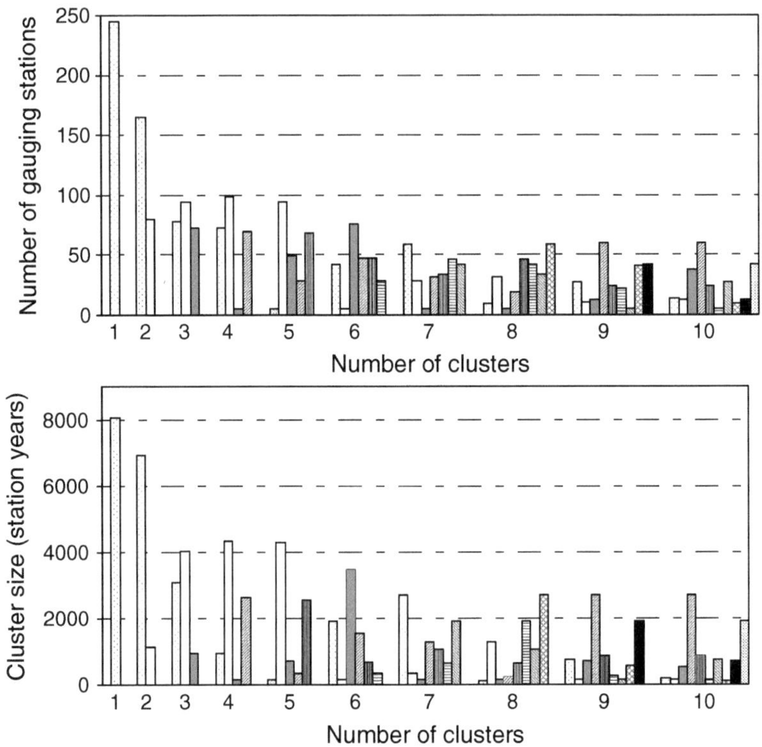

Fig. 4.3.5 Characteristics of clusters obtained from one-dimensional SOFM for the scenario-2 with weightage to drainage area twice that of all other attributes

while the first and third scenarios classified 47% and 37% of the sites into a heterogeneous cluster, respectively (Figs. 4.3.4–4.3.9).

From the results for all three scenarios shown in Table 4.3.2 (as cluster 1) it is evident that watersheds in southern-Indiana (shown as rectangles in Fig. 4.3.3(c)) consist of small drainage basins with steep slopes, low surface storage, high precipitation and moderate to high runoff coefficient values. Cluster 3 for scenario 2 and cluster 2 for scenario 3 (shown as + symbols in Fig. 4.3.3(c)) contain drainage basins with high storage and low runoff coefficients (Table 4.3.2). The cluster in northern Indiana consists of sites that are highly heterogeneous (cluster 3 for scenarios 1 and 3; cluster 2 for scenario 2; Figs. 4.3.7–4.3.9). In Fig. 4.3.3c, the triangles shown in south and south-central Indiana represent large drainage basins with milder slopes (cluster-2 for scenario-1; cluster-4 for scenarios 2 and 3 in Table 4.3.2). Since the fraction of stations classified into highly heterogeneous clusters was fairly high for the choice of m equal to 4, the clusters were not used to derive regions for flood frequency analysis.

For the choice of $m = 5$, the algorithm provided clusters shown in Fig. 4.3.3d. The triangles seen in south and south-central Indiana represent drainage basins with

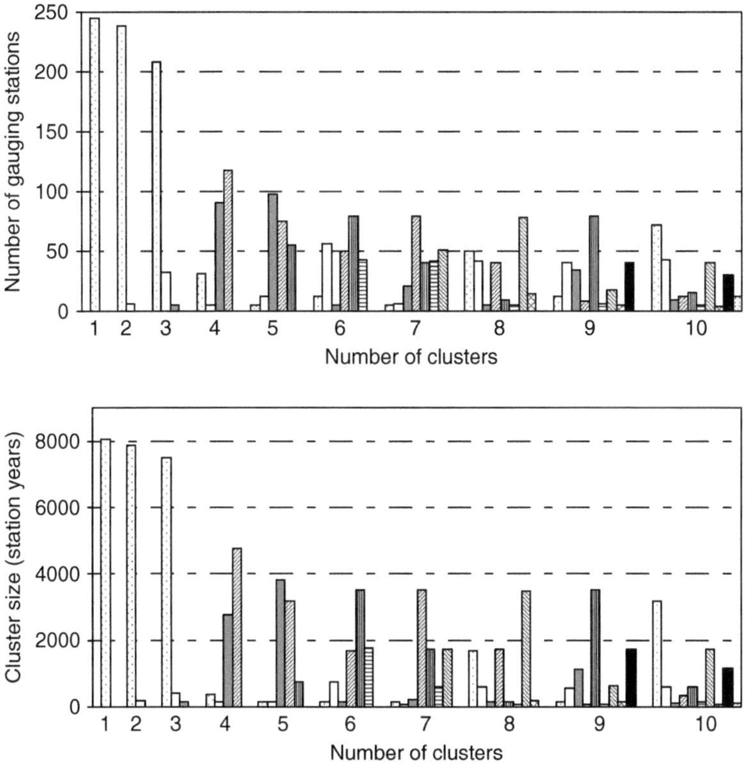

Fig. 4.3.6 Characteristics of clusters obtained from one-dimensional SOFM for the scenario-3 without latitude and longitude as attributes and with equal weights to all other five attributes

milder slopes and high runoff coefficient values (cluster-1 for scenario-1; cluster-5 for scenario 2; cluster-3 for scenario 3 in Table 4.3.3). Scenario-1 provided a cluster in central Indiana (represented by squares in Fig. 4.3.3d) which includes medium size basins that are characterized by milder slopes, low storage, moderate precipitation and runoff coefficient values.

The darkened circles include small drainage basins with short records, mild slopes and low storage (Fig. 4.3.3d, cluster-5 for scenarios 1 and 3 and cluster-3 for scenario-2 in Table 4.3.3). Average record length at the sites in these clusters is in the range 13–15-years. While 36% of the stations comprising the darkened circles in case of scenario-1 were found to have record lengths less than 10 years, the respective values for scenarios 2 and 3 were found to be 57–60%.

Characteristics of the cluster represented by '+' symbols in case of scenarios 2 and 3 are same as those reported for a similar cluster for the choice of m equal to 4. This cluster has just 5 stations and 157 station-years of data. Such small clusters are not suitable for regional flood frequency analysis. In case of scenario-3, this cluster is evident for all the values of m in the range 2–10, whereas in case of scenario-1 this cluster emerges at m equal to 8 and persists when m is increased beyond 8. For scenario-2, this cluster emerges for the choice of $m = 4$. In order to save space,

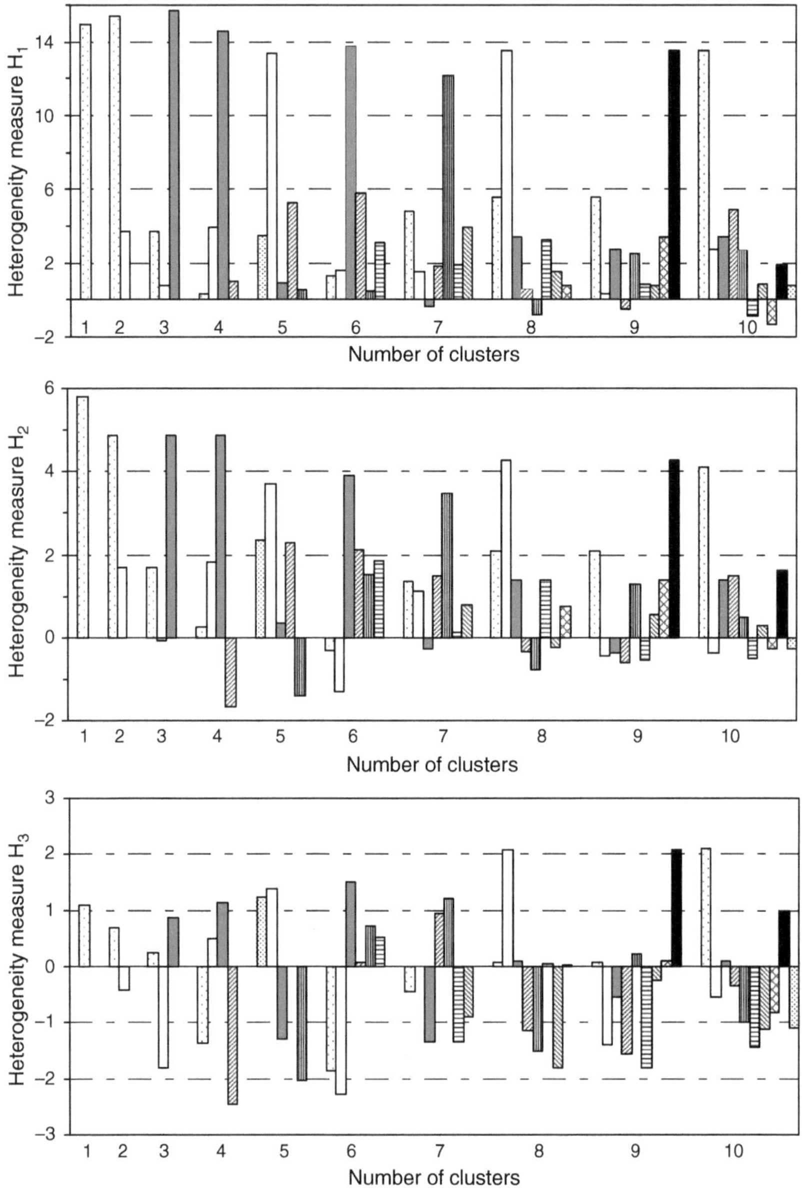

Fig. 4.3.7 Effect of increase in the number of clusters on heterogeneity measures – Results from one-dimensional SOFM for the scenario-1 with equal weights to all the seven attributes

characteristics of this cluster are not discussed repeatedly in the following part of this chapter.

In effect, for the choice of m equal to 5 we have three predominant plausible regions in case of scenario-1 (shown as circles, squares and triangles), while only two predominant plausible regions in case of scenarios 2 and 3 (shown as circles and

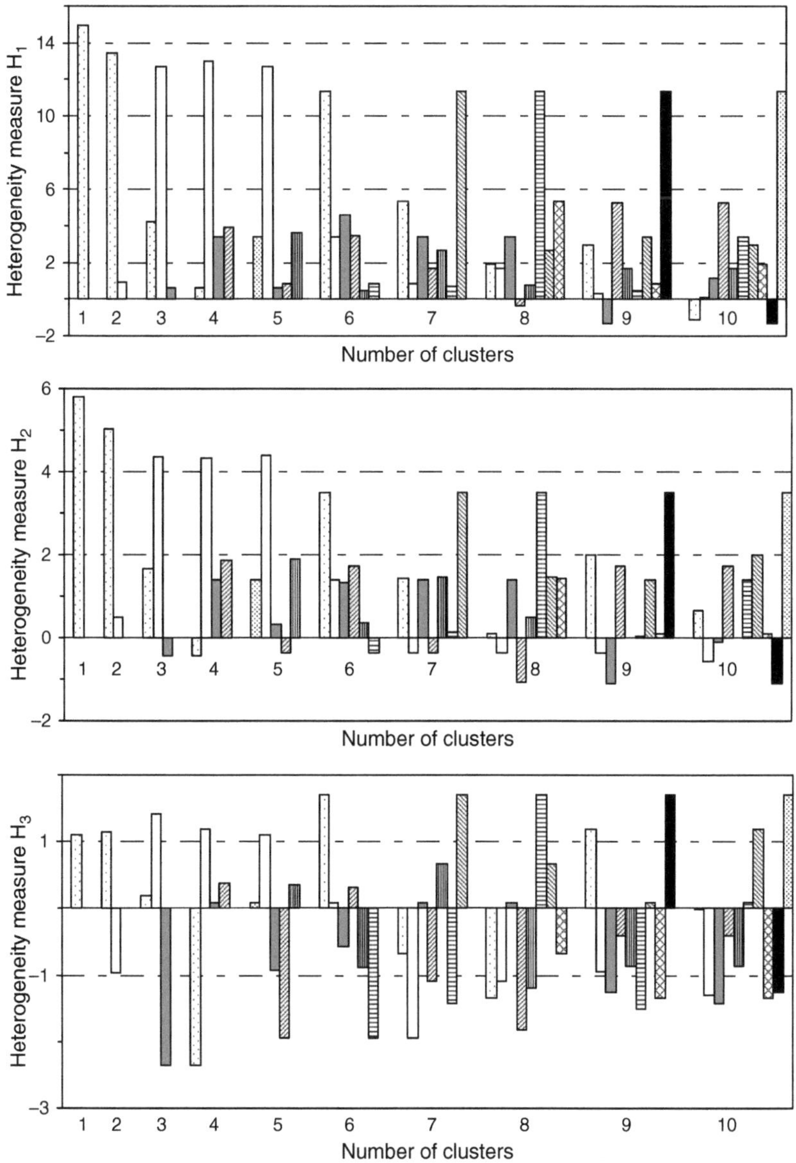

Fig. 4.3.8 Effect of increase in the number of clusters on heterogeneity measures –Results from one-dimensional SOFM for the scenario-2 with weightage to drainage area twice that of all other attributes

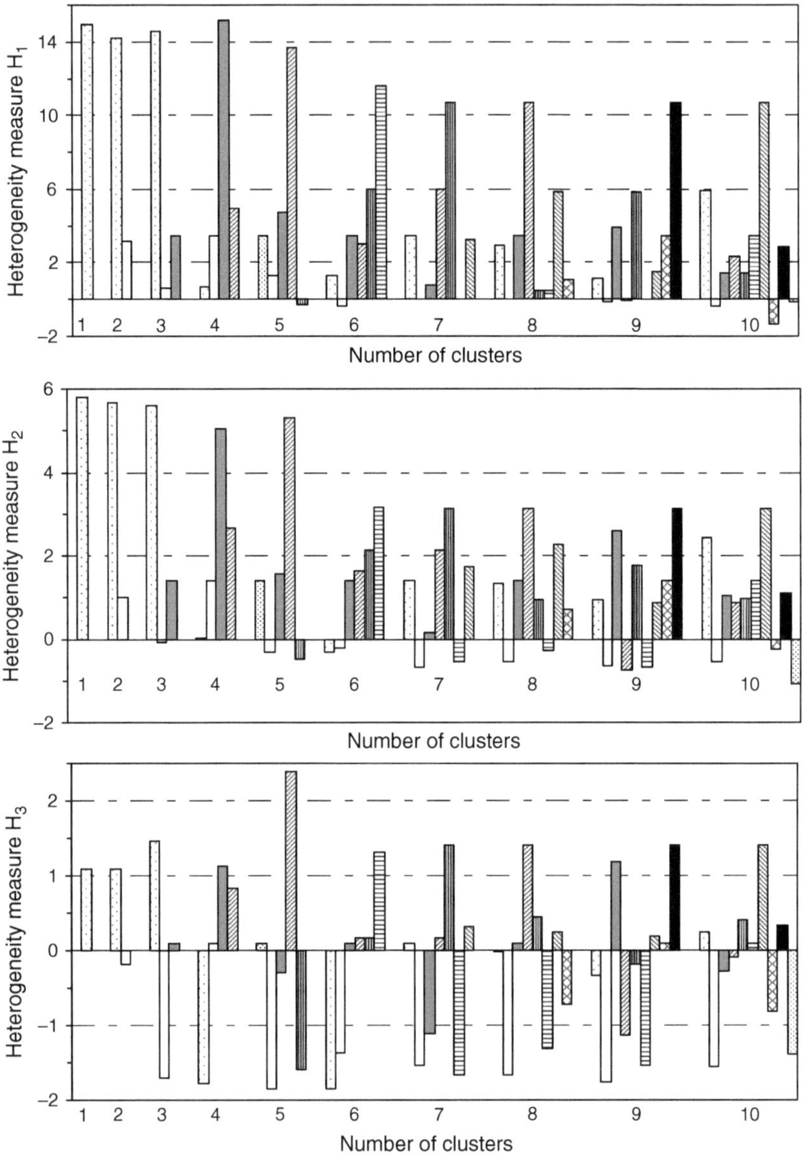

Fig. 4.3.9 Effect of increase in the number of clusters on heterogeneity measures – Results from one-dimensional SOFM for the scenario-3 without latitude and longitude as attributes and with equal weights to all other five attributes

triangles). All other clusters appear to be insignificant in terms of their information content.

Furthermore, it is noted that the first scenario classified only 21% of the stations into a highly heterogeneous cluster, whereas in the second and third scenarios

Table 4.3.1 Centers of the two clusters obtained from one dimensional SOFM for the three typical scenarios

S	CN	A (miles)^2	Slope (ft/mile)	LAT	LONG	STOR (%)	P (in)	RC
1	1	330.19	8.97	40.57	86.14	0.842	37.74	0.63
	2	348.66	45.37	38.91	86.34	0.442	41.84	0.80
2	1	500.18	6.65	40.04	86.22	0.836	39.12	0.68
	2	1.75	57.43	39.68	86.21	0.374	39.82	0.72
3	1	344.17	23.61	–	–	0.474	39.40	0.70
	2	69.15	8.15	–	–	9.105	37.17	0.50

S denotes Scenario; CN is cluster number; A denotes drainage area; LAT and LONG refer to Latitude and Longitude in decimal degrees; STOR denotes drainage area covered by lakes; P stands for precipitation; RC is runoff coefficient

39% and 31% of the sites belong to highly heterogeneous clusters, respectively (Figs. 4.3.4–4.3.9 and circles in Fig. 4.3.3d).

Among the six clusters obtained for the choice of m equal to 6, the sites in cluster-6 for scenario-2 and those in cluster-1 for scenarios 1 and 3 depict small drainage basins with short records, steep slopes, low surface storage, high precipitation and high runoff coefficient values (Table 4.3.4). These sites represented as rectangles are spread in southern Indiana (Fig. 4.3.3e). They closely resemble the clusters obtained in the same region of the state for $m = 5$.

Further, clusters represented by hollow circles in northern Indiana and those represented by triangles in southern Indiana closely resemble the clusters noted in the respective parts of Indiana for scenario-1 at $m = 5$ (Figs. 4.3.3d and 4.3.3e). The hollow circles represent a highly heterogeneous collection of medium size drainage basins with milder slopes, low precipitation and low runoff coefficient

Table 4.3.2 Centers of the four clusters obtained from ANN clustering algorithm for the three typical scenarios. In a few typical columns, the least and/or highest values of attributes are shown in bold font

S	CN	A (miles)^2	Slope (ft/mile)	LAT	LONG	STOR (%)	P (in)	RC
1	1	**0.37**	**124.25**	39.08	85.76	0.321	**41.98**	**0.82**
	2	**632.92**	10.81	39.08	86.24	0.495	41.32	0.78
	3	264.63	8.34	40.79	86.09	**1.007**	37.34	**0.60**
	4	2.83	37.23	39.42	87.20	0.121	39.70	0.69
2	1	**1.10**	**61.37**	39.69	86.24	**0.199**	39.82	0.72
	2	369.28	5.43	40.71	86.15	0.693	37.47	0.62
	3	56.18	9	41.23	85.66	**9.994**	37.80	**0.50**
	4	**663.07**	10.02	38.94	86.32	0.507	**41.67**	**0.79**
3	1	**0.38**	**109.51**	–	–	**0.285**	**41.98**	**0.81**
	2	56.18	9	–	–	**9.994**	37.80	**0.50**
	3	191.54	13.1	–	–	0.501	37.20	0.57
	4	**550.41**	**8.99**	–	–	0.538	40.38	0.77

S denotes Scenario; CN is cluster number; A denotes drainage area; LAT and LONG refer to Latitude and Longitude in decimal degrees; STOR denotes drainage area covered by lakes; P stands for precipitation; RC is runoff coefficient

Table 4.3.3 Centers of the five clusters obtained from ANN clustering algorithm for the three typical scenarios

S	CN	A (miles)^2	Slope (ft/mile)	LAT	LONG	STOR (%)	P (in)	RC
1	1	751.18	9.79	38.81	86.45	0.601	42.13	0.81
	2	283.86	7.60	41.29	86.27	1.75	36.76	0.45
	3	0.36	153.60	39.04	85.96	0.26	42.18	0.78
	4	299.87	9.87	40.18	85.91	0.35	38.30	0.73
	5	1.66	40.60	39.28	86.77	0.26	40.37	0.75
2	1	56.18	9.00	41.23	85.66	9.994	37.80	0.50
	2	386.65	5.10	40.70	86.16	0.714	37.49	0.62
	3	2.23	26.37	40.27	86.19	0.134	38.14	0.64
	4	0.33	113.86	38.85	86.30	0.311	42.32	0.81
	5	669.70	10.03	38.93	86.32	0.511	41.71	0.79
3	1	56.18	9.00	–	–	9.994	37.80	0.50
	2	0.31	175.54	–	–	0.206	43.08	0.79
	3	583.25	8.35	–	–	0.490	40.80	0.78
	4	335.34	6.04	–	–	0.741	36.84	0.55
	5	1.41	41.27	–	–	0.215	39.51	0.73

S denotes Scenario; CN is cluster number; A denotes drainage area; LAT and LONG refer to Latitude and Longitude in decimal degrees; STOR denotes drainage area covered by lakes; P stands for precipitation; RC is runoff coefficient

Table 4.3.4 Centers of the six clusters obtained from ANN clustering algorithm for the three typical scenarios

S	CN	A (miles)^2	Slope (ft/mile)	LAT	LONG	STOR (%)	P (in)	RC
1	1	0.31	175.54	38.94	86.45	0.206	43.08	0.79
	2	3.08	38.22	39.44	87.35	0.105	39.69	0.69
	3	278.61	7.72	41.27	86.27	1.717	36.79	0.45
	4	435.36	5.89	40.12	85.93	0.403	38.39	0.73
	5	3.93	40.49	39.61	85.55	0.324	39.78	0.79
	6	745.50	9.23	38.77	86.55	0.637	42.31	0.80
2	1	357.14	4.35	41.22	86.55	0.960	36.79	0.47
	2	56.18	9.00	41.27	85.66	9.994	37.80	0.50
	3	730.99	5.71	39.99	85.92	0.443	38.74	0.74
	4	249.75	13.06	38.77	86.42	0.644	42.29	0.82
	5	1.85	26.14	40.30	86.19	0.106	38.07	0.64
	6	0.33	113.86	38.85	86.30	0.311	42.32	0.81
3	1	0.31	175.54	–	–	0.206	43.08	0.79
	2	1.48	40.78	–	–	0.214	39.48	0.73
	3	56.18	9.00	–	–	9.994	37.80	0.50
	4	560.27	10.83	–	–	0.613	42.41	0.81
	5	510.30	5.74	–	–	0.428	38.42	0.73
	6	325.03	6.09	–	–	0.906	36.45	0.44

S denotes Scenario; CN is cluster number; A denotes drainage area; LAT and LONG refer to Latitude and Longitude in decimal degrees; STOR denotes drainage area covered by lakes; P stands for precipitation; RC is runoff coefficient values (cluster-3 for scenario-1; cluster-1 for scenario 2; cluster-6 for scenario-3 in Table 4.3.4). This cluster has 17–21% of the stations considered for cluster analysis (Figs. 4.3.4–4.3.6). The tendency of the SOFM algorithm to filter out highly heterogeneous collection of basins into a cluster is evident from the foregoing discussion of results for the choice of m in the range 2–6.

The triangles seen in south and south-central Indiana denote medium to large size drainage basins with milder slopes and high runoff coefficient values (Fig. 4.3.1e; cluster-6 for scenario-1, cluster-4 for scenarios 2 and 3 in Table 4.3.4). Also, the squares shown in central Indiana for the three scenarios depict medium to large size basins that are characterized by milder slopes, low storage, moderate precipitation and runoff coefficient values (cluster-4 for scenario-1; cluster-3 for scenario-2; cluster-5 for scenario-3 in Table 4.3.4).

For scenario-1, the sites represented as darkened circles for $m = 5$ split up into two clusters at $m = 6$ which are coded as darkened diamonds and darkened squares (Figs. 4.3.3d and 4.3.3e; clusters 2 and 5 in Table 4.3.4). A slight difference is evident in the runoff coefficient values and storage at the two resulting clusters. The collection of these stations closely resembles cluster-2 provided by scenario-3 that does not consider latitude and longitude as attributes for the cluster analysis. Thus, the split in case of scenario-1 seems to be primarily the effect of considering latitude and longitude as features for cluster analysis.

The collection of sites coded as darkened diamonds and darkened squares in case of scenario-1 and those represented as darkened circles for scenarios 2 and 3 consists of small drainage basins with short records, mild slopes and low storage (Fig. 4.3.3e; clusters 2 and 5 for scenario 1, cluster-5 for scenario 2, cluster-2 for scenario 3 in Table 4.3.4). These results are further corroborated by Figs. 4.3.10–4.3.12, which compare record lengths at sites in this cluster with those at other sites considered for cluster analysis.

When m was increased beyond 6, several of the clusters provided by the SOFM for all the three scenarios showed remarkable resemblance to those obtained for the choice of $m = 6$. Among those, clusters coded as hollow circles in northern Indiana, squares in central Indiana and triangles in southern Indiana are noteworthy. They can be considered as plausible hydrologic regions because they are all significant in terms of pooled information content.

Further, emergence of a cluster is noted in southeastern part of Indiana for $m \geq 7$. The catchments in this cluster are characterized by high runoff coefficient values. It may also be considered as one of the hydrologic regions. Several other clusters that are noted in southern Indiana for $m > 6$ are small (example, pentagons, x and darkened triangles in Figs. 4.3.3f–i).

Identification of Optimal Partition Through Cluster Validity Measures

The optimal number of clusters formed by SOFM for the Indiana dataset is identified by using four hard cluster validity measures, namely Dunn's index (V_{D}; Dunn, 1973), Davies-Bouldin index (V_{DB}; Davies and Bouldin, 1979), Calinski-Harabasz index (V_{CH}; Calinski and Harabasz, 1974), and Minimum description length (V_{MDL};

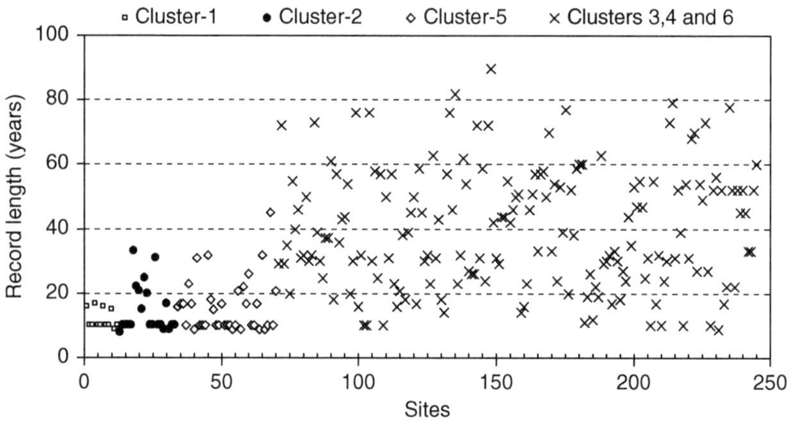

Fig. 4.3.10 Comparison of record lengths at sites comprising the clusters obtained for scenario-1 at $m = 6$

Bischof et al., 1999; Qin and Suganthan, 2004). These measures have been described in Section 2.3.4.

For optimal partition, the values of V_{DB} and V_{MDL} should be small, whereas the values of V_D and V_{CH} should be large. It is seen from Fig. 4.3.13 that V_{MDL} indicates 6 as the optimal value of m for all the three scenarios. The V_{DB} indicates 6 as the optimal value of m for scenario-1, and 8 as the optimal value of m for scenario-2. In addition, for scenario-3, the value of the index is seen to be low for $m = 2$ and $m = 3$. However, this is a false indicator of optimal partition because clusters obtained with the smaller values of m comprise excessively large regions, which are generally heterogeneous and hence not suitable for RFFA (Figs. 4.3.4–4.3.9). Similarly, it is noted that V_D also gives a false indication of optimal partition for

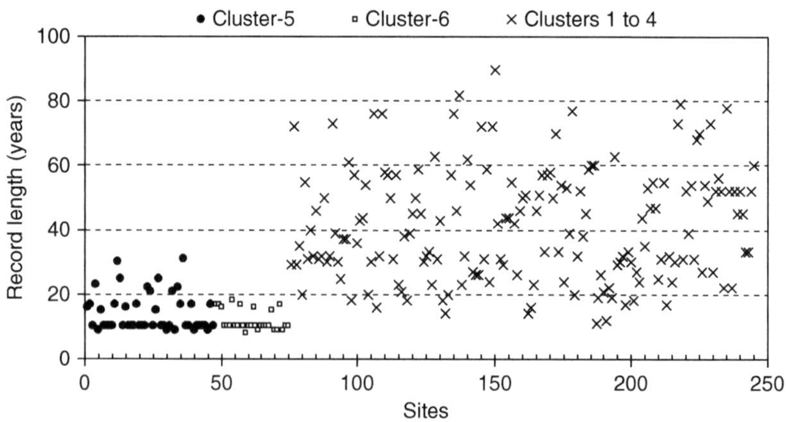

Fig. 4.3.11 Comparison of record lengths at sites comprising the clusters obtained for scenario-2 at $m = 6$

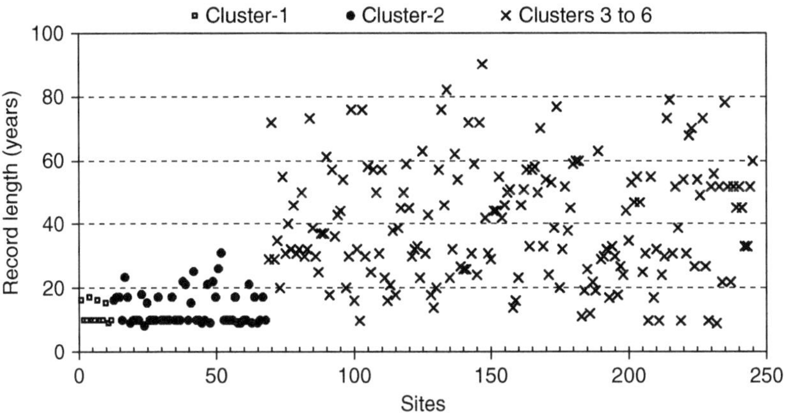

Fig. 4.3.12 Comparison of record lengths at sites comprising the clusters obtained for scenario-3 at $m = 6$

scenario-3, for m equal to 2 and 3. Interestingly, for m in the range 4–11, both V_{DB} and V_D show 6 as the optimal value of m for the scenario-3.

Furthermore, it is noted that the V_D shows 5 as the optimal value of m for scenario-1, and 8 as the optimal value of m for scenario-2. Conversely, the V_{CH}

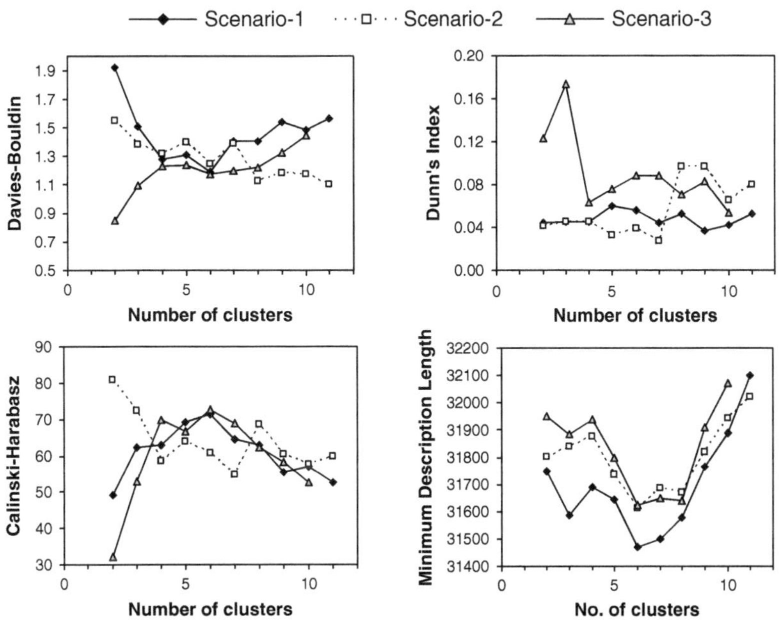

Fig. 4.3.13 Identification of optimal partition provided by the 1-D SOFM by using cluster validity measures. The partition with the maximum value for Dunn's index (or Calinski-Harabasz index) and the minimum value for Davies-Bouldin index (or minimum description length) is taken as the optimal partition

indicates 6 as the optimal value of m for scenarios 1 and 3, and 8 as the optimal value of m for scenario-2. The V_{CH} gives a false indication of optimal partition for the second scenario for $m = 2$ and $m = 3$.

In conclusion, majority of the cluster validity indices appear to indicate 6 as the optimal value of m for scenarios 1 and 3. On the contrary, for scenario-2 they appear to indicate 8 as the optimal value of m.

The optimal partition identified using the cluster validity indices for the scenarios 1 and 3 are found to be very similar to the plausible hydrologic regions recognized by visual inspection of clusters of watersheds in Indiana.

Comparison of the clusters obtained for $m = 6$ for the scenarios 1 and 3 show that four of the six clusters, which are coded as hollow circles in northern Indiana, squares in central Indiana, rectangles and triangles in southern Indiana are very similar (Fig. 4.4.3e). In addition, the sites which are shown as darkened circles for scenario-3 comprise the sites that are coded as darkened diamonds in eastern Indiana and darkened circles in western Indiana for scenario-1. The split up of the collection of sites in scenario-3 into eastern and western parts in scenario-1 is attributed to the use of location attributes (latitude and longitude) for regionalization in the scenario-1.

The reason underlying use of location attributes has already been mentioned in Section 4.3.1. In this setting, the plausible hydrologic regions (i.e., clusters) obtained from scenario-1 are adjusted following the procedure described in Section 2.4.3 and using the options listed in Section 1.4.1.

Considerable effort was required to adjust the clusters determined using SOFM. In the process, additional 28 gauging stations in Indiana having a minimum record length of 10 years were considered for inclusion in the regions based on geographical contiguity following option (viii) of Section 1.4.1. Hydrologic regions 1, 2, 3 and 4 shown in Fig. 4.3.15 resulted upon revising the clusters coded as darkened circles, triangles, darkened diamonds and squares, respectively, in Fig. 4.3.3(e). The highly heterogeneous cluster coded as hollow circles in northern Indiana was split into two parts. The part which had highly heterogeneous sites constituted Region 6 and the remaining sites were merged with sites eliminated while forming hydrologic region 4 to derive hydrologic region 5. Considerable effort was necessary for adjusting clusters to derive regions 3, 5 and 6. This could be because the set of plausible hydrologic regions (clusters) are identified using conventional SOFM, which has limitations in determining structure in hydrologic data.

In the region revision process, first the sites that are flagged discordant by the discordancy measure of Hosking and Wallis (1997) are identified. Secondly, the heterogeneity measures (H_1, H_2 and H_3) of the region to be adjusted are examined as they changed with exclusion of each site from the region. Thirdly, the discordant site, whose exclusion reduces the heterogeneity measures for the region by a significant amount, is identified and removed from the region after ensuring that the site discordancy is not due to sampling variability. In some instances, a site excluded from one region would fit in more than one region. Such a site is considered to be common to all the concerned regions.

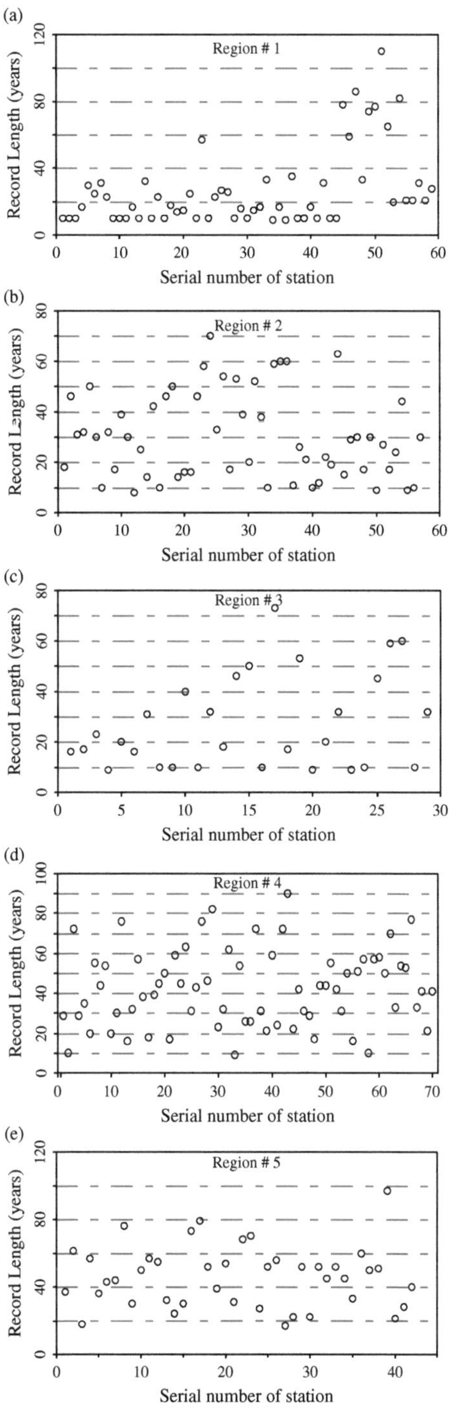

Fig. 4.3.14 Comparison of peak flow records for sites in the adjusted regions

4.3.3 Testing the Regions for Robustness

The Regions are tested for robustness following the procedure described in previous chapters. Peak flow records of sites in each region are examined (Fig. 4.3.14). Regions 1 to 4 have sites with large variation in their record length. Hence they are tested for robustness. By specifying various threshold values, the stations with record lengths significantly different from that of the rest of the group are removed and the region with the remaining stations was examined for homogeneity. The results presented in Table 4.3.5 show that the Regions 1 to 4 are robust.

The study resulted in delineation of Indiana into five homogeneous regions, one heterogeneous region and an unallocated residue of 23 stations of which 21 are located in Indiana (Fig. 4.3.15). These residual stations have a collective record of 506 station-years. The values of heterogeneity measures shown in Table 4.3.6 indicate that regions 1 to 5 are all acceptably homogeneous, while region-6 adjoining Lake Michigan is highly heterogeneous. All the homogeneous regions identified have sufficient pooled information to be effective for flood frequency analysis. Interestingly, the delineated regions bear remarkable resemblance with geological features and soil regions of Indiana (Figs. 4.3.16–4.3.18).

Table 4.3.5 Results from testing the regions for robustness. R denotes Region number, and NS represents number of stations

R	Condition	NS	Heterogeneity measure			Region type
			H_1	H_2	H_3	
1	Entire region	59	0.86	−0.12	−0.94	Homogeneous
	Sites with RL≤10 are eliminated	41	1.08	0.07	−0.73	Possibly Homogeneous
	Sites with RL<20 are eliminated	31	0.88	0.40	−0.24	Homogeneous
	Sites with RL≥35 are eliminated	51	0.39	−0.24	−0.75	Homogeneous
2	Entire region	58	0.85	0.43	−0.65	Homogeneous
	Sites with RL<20 are eliminated	36	0.58	1.16	0.56	Homogeneous
	Sites with RL>50 are eliminated	49	0.88	0.48	−0.76	Homogeneous
	Sites with RL≤10 and RL>50 are eliminated	40	0.85	0.87	−0.17	Homogeneous
3	Entire region	30	−0.46	0.66	0.28	Homogeneous
	Sites with RL≤10 are eliminated	21	−0.28	0.40	0.20	Homogeneous
	Sites with RL<20 are eliminated	15	−0.41	0.01	−0.10	Homogeneous
4	Entire region	73	0.48	−0.40	−1.78	Homogeneous
	Sites with RL<20 are eliminated	63	0.81	0.12	−1.03	Homogeneous

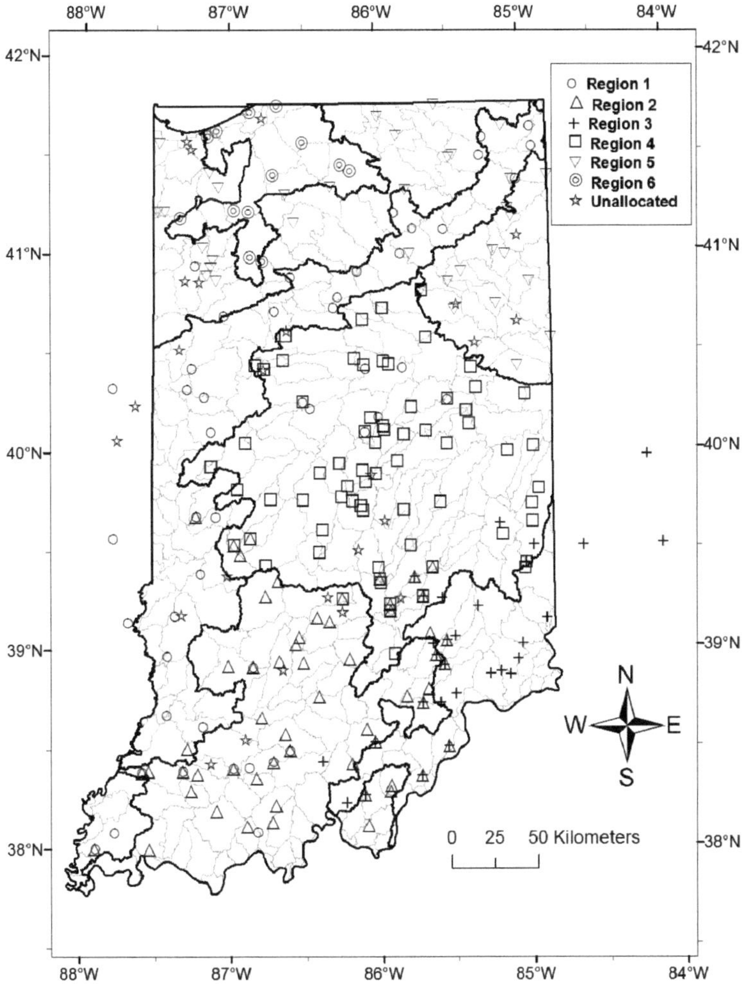

Fig. 4.3.15 Location of the regions defined using the 1-D SOFM

Table 4.3.6 Characteristics of the regions formed using SOFM

Region number	N	RS	Heterogeneity measure		
			H_1	H_2	H_3
1	62	1689	0.86	−0.12	−0.94
2	58	1730	0.85	0.43	−0.65
3	30	804	−0.46	0.66	0.28
4	73	3039	0.48	−0.40	−1.78
5	42	1938	0.04	−0.91	−0.85
6	14	519	13.69	6.33	2.94

N: Number of stations; RS: Region size in station years

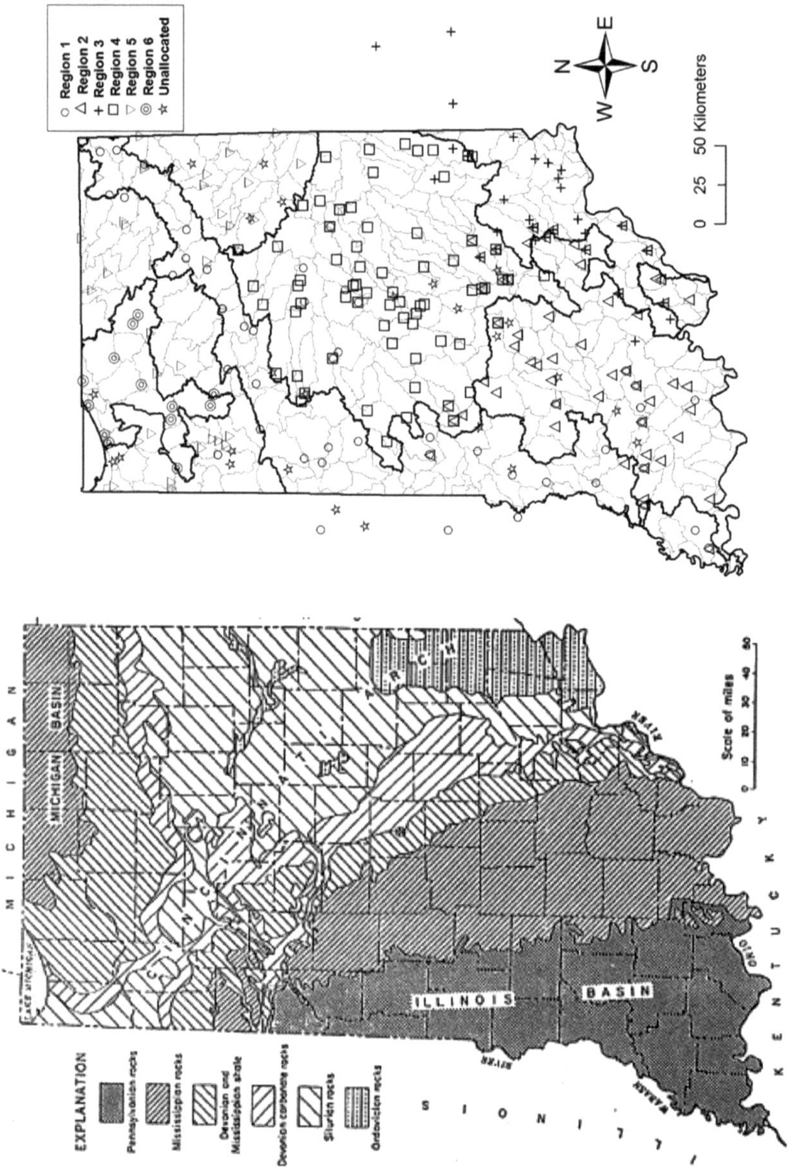

Fig. 4.3.16 Comparison of the hydrological regions identified in Indiana using SOFM with geologic features of the state

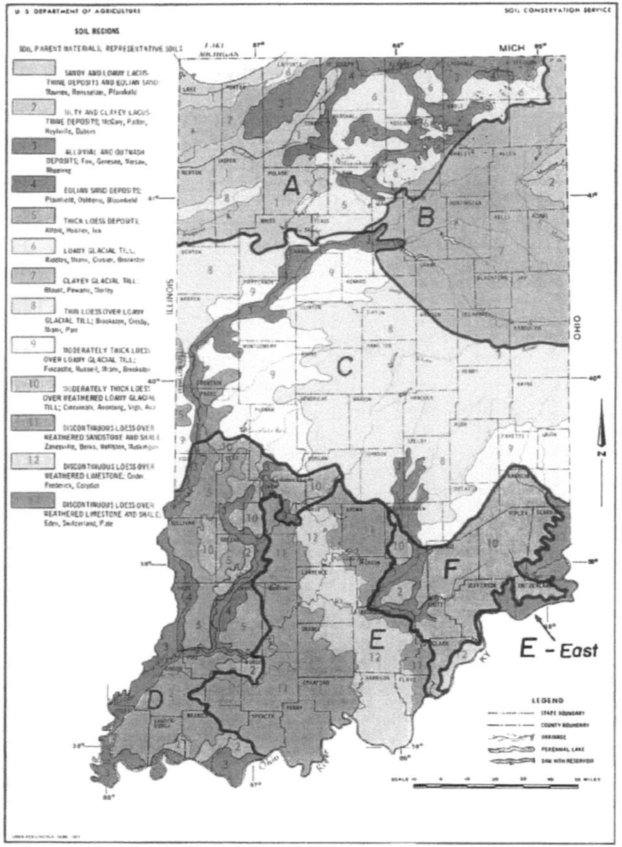

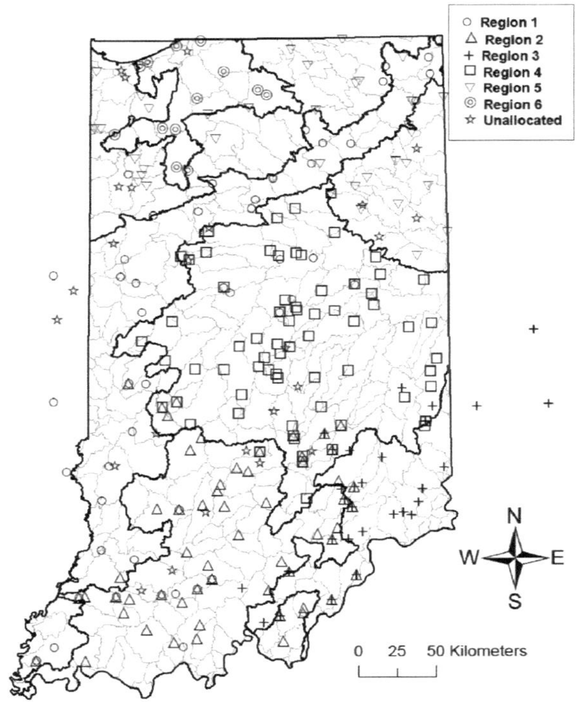

Fig. 4.3.17 Comparison of the hydrological regions identified in Indiana using SOFM with soil regions in the state identified by soil conservation service, US Department of Agriculture

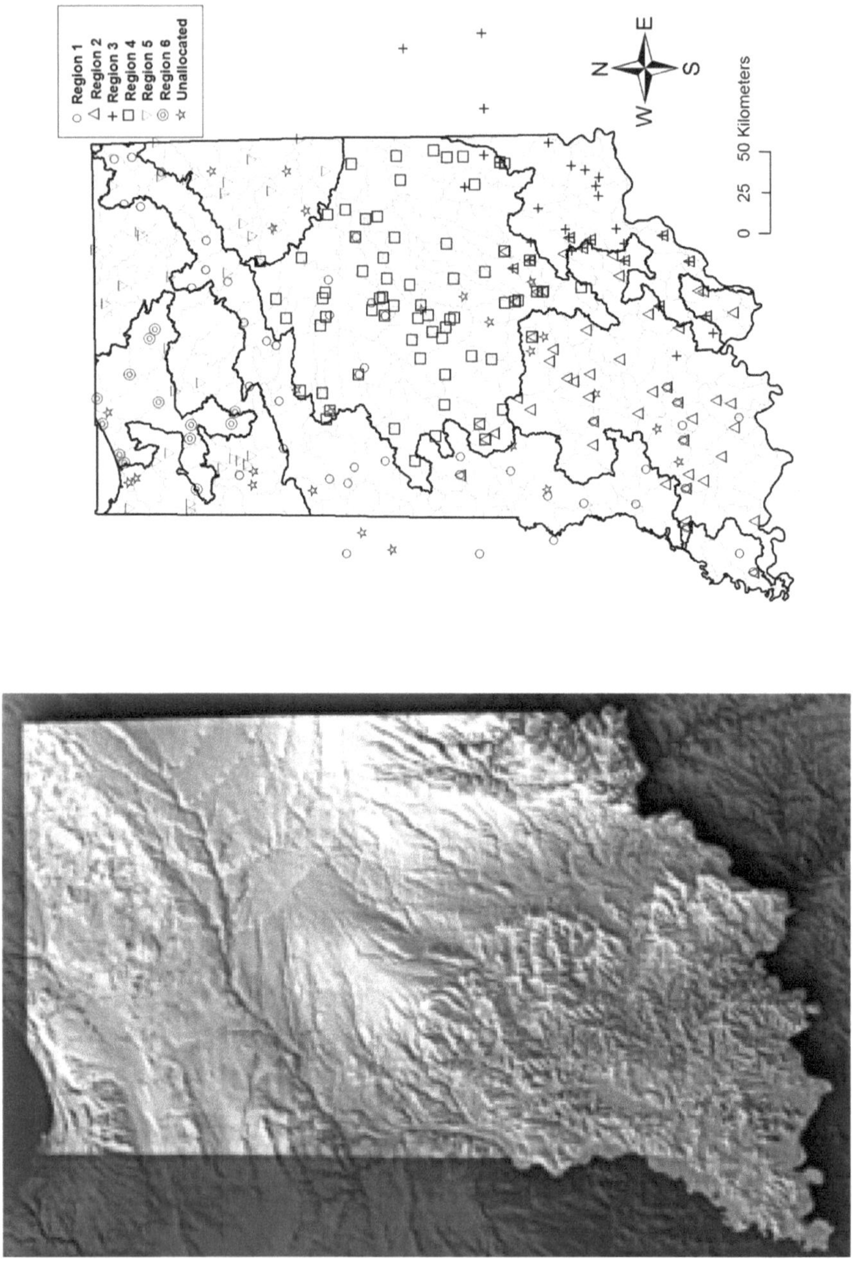

Fig. 4.3.18 Comparison of the hydrological regions identified in Indiana using SOFM with tapestry produced by union of geology and topography of Indiana

4.4 Regionalization by Two-Stage Clustering of SOFM

4.4.1 Introduction

The SOFM is an unsupervised learning technique that is useful for extracting the topological structure hidden in the higher dimensional data of input vectors that contain watershed attributes perceived as being important for regionalization. However, as demonstrated in the previous section, it is not always possible to interpret patterns in the output of SOFM. In this context, the second option that has been considered to form regions is clustering of nodes in 2-D feature map. A novel algorithm for regionalization of watersheds using two-level SOFM clustering is presented in this section.

4.4.1.1 Two-Level Clustering – Historical Perspective

The SOFM may be viewed as a nonlinear generalization of principal component analysis (Ritter, 1995). In contrast, clustering algorithms attempt to partition data into clusters or natural groups such that the data within a cluster are as similar as possible, and data belonging to different clusters are as dissimilar as possible. Therefore, an SOFM pursues a goal that is conceptually different from that of clustering (Pal et al., 1993, Wu and Chow, 2004). However, an SOFM can be successfully utilized as a first step in clustering algorithms. In the past decade this idea has been explored. Lampinen and Oja (1992) proposed a two-level SOFM, where outputs of the first SOFM are fed into a second SOFM as inputs. This model was shown to perform better than SOFM and classical K-means algorithms in classifying artificial data and sensory information from low-level feature detectors in a computer vision system. Murtagh (1995) proposed an agglomerative contiguity-constrained clustering method to merge (or group) the neighbouring nodes in the output from SOFM, based on a minimal distance criterion. The efficiency of this model was demonstrated for classification of Infrared Astronomical Satellite Point Source Catalog (IRAS PSC) data. Kiang (2001) extended the idea of merging neighbouring nodes of SOFM by using a minimum variance criterion as an alternative to minimum distance criterion. The performance of the method was tested with machine learning databases. Vesanto and Alhoniemi (2000) applied both hierarchical agglomerative and partitional K-means clustering algorithms to group the output from SOFM. Through experiments on real and artificial data sets, the model was shown to perform well compared to direct clustering of data using hierarchical agglomerative and partitional K-means clustering algorithms. More recently, Wu and Chow (2004) used cluster validity index based on inter- and intra-cluster density for merging the neighbouring nodes in the output from SOFM to obtain final clusters.

4.4.2 Algorithm for Fuzzy Clustering of Kohonen SOFM

The algorithm presented in this section has two levels. In the first level, the algorithm of SOFM, which has been presented in Section 4.2.1, is used to form a two-dimensional feature map. In the second level, nodes in the two-dimensional feature map are clustered by using Fuzzy C-Means (FCM) algorithm to form regions for flood frequency analysis. A schematic for the two-level model is presented in Fig. 4.4.1.

The m' output nodes in Kohonen layer, which are winners for at least one input vector, are considered for clustering by FCM algorithm. These m' winning output nodes ($m' \leq m$) are referred to as prototypes. The final weight matrix after the SOFM step is the $m' \times n$ data matrix $\boldsymbol{W}'$.

$$\boldsymbol{W}' = \begin{bmatrix} w_{11} & \dots & w_{1m'} \\ \vdots & \ddots & \vdots \\ w_{n1} & \dots & w_{nm'} \end{bmatrix} \tag{4.4.1}$$

Let $\boldsymbol{w}'_j$ denote the 'j-th' prototype in n-dimensional space i.e., $\boldsymbol{w}'_j = \left[w_{1j}, \dots, w_{nj}\right] \in \Re^n$, and let $\boldsymbol{V} = (\boldsymbol{v}_1, \dots, \boldsymbol{v}_c)$ represent a c-tuple containing c fuzzy cluster centroids. The FCM algorithm partitions the matrix $\boldsymbol{W}'$ into c overlapping subsets (or clusters) by minimizing the objective function in Eq. (4.4.2).

$$J(\boldsymbol{U}, \boldsymbol{V} : \boldsymbol{W}') = \sum_{i=1}^{c} \sum_{j=1}^{m'} (u_{ij})^{\mu} d^2(\boldsymbol{w}'_j, \boldsymbol{v}_i) \tag{4.4.2}$$

subject to the following constraints,

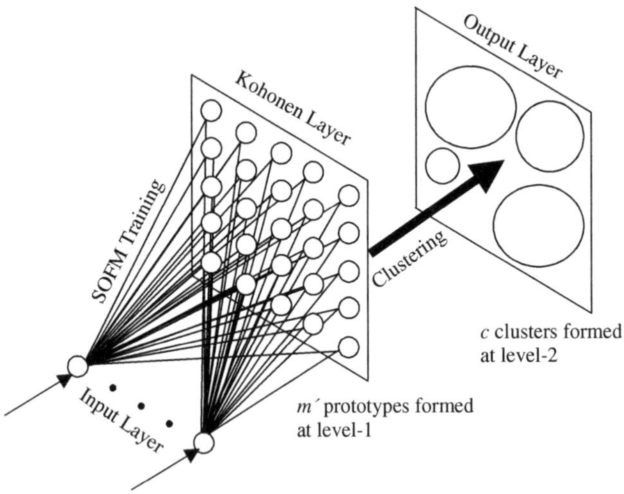

Fig. 4.4.1 A schematic of the two-level clustering process that is recommended for regionalization. The number of nodes in input layer is equal to the number of watershed attributes considered for regionalization

$$\sum_{i=1}^{c} u_{ij} = 1 \quad \forall j \in \{1, \ldots, m'\} \tag{4.4.3}$$

$$0 < \sum_{j=1}^{m'} u_{ij} < m' \quad \forall i \in \{1, \ldots, c\} \tag{4.4.4}$$

In Eqs. (4.4.2), (4.4.3) and (4.4.4), $u_{ij} \in [0, 1]$ and it denotes the degree of membership of the j-th prototype $\mathbf{w}'_j$ in the i-th fuzzy cluster depicted by its centroid $\mathbf{v}_i$; $\mathbf{U}$ is the fuzzy partition matrix which contains the membership of each prototype in each fuzzy cluster Eq. (4.4.5); the parameter $\mu \in [1, \infty]$ refers to the weight exponent for each fuzzy membership and is called fuzzifier; $d^2(\mathbf{w}'_j, \mathbf{v}_i)$ is the distance from j-th prototype $\mathbf{w}'_j$ to the centroid of i-th cluster $\mathbf{v}_i$. The general form of the distance measure is given by Eq. (4.4.6).

$$\mathbf{U} = \begin{bmatrix} u_{11} & \cdots & u_{1j} & \cdots & u_{1m'} \\ \vdots & & \vdots & & \vdots \\ u_{i1} & \cdots & u_{ij} & \cdots & u_{im'} \\ \vdots & & \vdots & & \vdots \\ u_{c1} & \cdots & u_{cj} & \cdots & u_{cm'} \end{bmatrix}_{c \times m'} \tag{4.4.5}$$

$$d^2(\mathbf{w}'_j, \mathbf{v}_i) = (\mathbf{w}'_j - \mathbf{v}_i)^{\mathrm{T}} \mathbf{A}_i (\mathbf{w}'_j - \mathbf{v}_i) \tag{4.4.6}$$

In Eq. (4.4.6), $\mathbf{A}_i$ is a positive definite, symmetric matrix associated with cluster i. The FCM algorithm used in this study consists of Euclidean distance measure, for which $\mathbf{A}_i = \mathbf{I}\ \forall i$, where $\mathbf{I}$ is a unit matrix. The weight exponent, μ, determines the fuzziness of the clusters. It controls the extent of membership shared among fuzzy clusters.

The iterative procedure of FCM algorithm (Bezdek, 1981) is summarized below:

(i) Initialize fuzzy partition matrix $\mathbf{U}$ (or fuzzy cluster centroid matrix $\mathbf{V}$) using a random number generator.

(ii) If the FCM algorithm is initialised with fuzzy partition matrix $\mathbf{U}$, adjust the initial memberships u_{ij}^{init} of $\mathbf{w}'_j$ belonging to cluster i by using Eq. (4.4.7) so that Eq. (4.4.3) is satisfied.

$$u_{ij} = \frac{u_{ij}^{init}}{\sum_{i=1}^{c} u_{ij}^{init}} \quad \text{for} \quad 1 \le i \le c, 1 \le j \le m' \tag{4.4.7}$$

If the FCM algorithm is initialised with fuzzy cluster centroid matrix $\mathbf{V}$ (containing c fuzzy cluster centroids $\mathbf{v}_1^{init}, \ldots, \mathbf{v}_c^{init}$), determine memberships u_{ij} of $\mathbf{w}'_j$ belonging to cluster i using Eq. (4.4.9) below with $\mathbf{v}_i^{init}$ replacing $\mathbf{v}_i$.

(iii) Compute the fuzzy centroid $\boldsymbol{v}_i$ for $i = 1, 2, \ldots, c$

$$\boldsymbol{v}_i = \frac{\sum_{j=1}^{m'} \left(u_{ij}\right)^{\mu} \boldsymbol{w}'_j}{\sum_{j=1}^{m'} \left(u_{ij}\right)^{\mu}} \tag{4.4.8}$$

(iv) Update the fuzzy membership u_{ij}

$$u_{ij} = \frac{\left(\frac{1}{d^2(\boldsymbol{w}'_j, \boldsymbol{v}_i)}\right)^{1/(\mu-1)}}{\sum_{i=1}^{c} \left(\frac{1}{d^2(\boldsymbol{w}'_j, \boldsymbol{v}_i)}\right)^{1/(\mu-1)}} \quad \text{for} \quad 1 \le i \le c, 1 \le j \le m' \tag{4.4.9}$$

Repeat steps (iii) and (iv) until change in the value of the memberships between two successive iterations becomes sufficiently small. At this point, following traditional methods of fuzzy cluster analysis (Ross, 1995, p. 398), the fuzzy partition matrix, $\boldsymbol{U}$ (shown in Eq. (4.4.5)), can be defuzzified to ultimately assign the prototypes to clusters. For instance, in the maximum-membership method, the largest element in each column of $\boldsymbol{U}$ is assigned a membership value of unity and all the other elements in the column are assigned a membership value of zero.

$$u_{kj} = \max_{1 \le i \le c} \{u_{ij}\} = 1; \quad u_{ij} = 0 \quad \text{for all} \quad i \ne k \tag{4.4.10}$$

In other words, a prototype is assigned to the cluster to which it has maximum resemblance. Alternatively in the nearest-center classifier method, each of the prototypes, $\boldsymbol{w}'_j$, is assigned to the cluster whose centroid is closest in terms of Euclidean distance.

$$\begin{aligned} \text{If } d_{kj} = \min_{1 \le i \le c} \{d_{ij}\} = \min_{1 \le i \le c} \left\| \boldsymbol{v}_i - \boldsymbol{w}'_j \right\| \quad \text{then} \quad u_{kj} = 1; \\ u_{ij} = 0 \quad \text{for all} \quad i \ne k \end{aligned} \tag{4.4.11}$$

As mentioned in Section 3.3.1, Hall and Minns (1999) used the aforementioned two defuzzification methods to form hard clusters in fuzzy cluster analysis. As most catchments often show characteristics from several regions, one cannot justify assigning a catchment to a single group.

In the two-stage SOFM algorithm, the prototypes obtained by using SOFM depict the catchments that are tightly linked to each other. The proximity of prototypes is determined by a threshold fuzzy membership T_j computed for each prototype j. In other words, a fuzzy cluster is formed by assigning to it the prototypes whose memberships in the cluster equal or exceed T_j computed by using Eq. (4.4.12).

$$T_j = \max \left\{ \frac{1}{c}, \frac{1}{2} \left[\max_{1 \le i \le c} (u_{ij}) \right] \right\} \tag{4.4.12}$$

In general, the choice of a threshold value to form fuzzy clusters is subjective. In the fuzziest partition, the memberships of a prototype in all the clusters would be equal to $1/c$. It is logical to assign a prototype to the cluster in which it has maximum membership. However, the decision becomes less clear when a prototype has nearly equal memberships in several clusters.

The FCM algorithm may converge to a local minimum of the objective function. The value of the objective function depends on initial guesses of cluster number, cluster centers, and fuzzy memberships. These *a priori* assumptions are necessary, but do not guarantee optimal partition. Over the past two decades, researchers have been developing several heuristic cluster validity criteria to address the choice of optimal number of clusters.

4.4.3 Example of Using Two-Level Fuzzy SOFM

The data from watersheds in Indiana, USA, which is described in Section 4.3.1, is provided as an input to the two-level fuzzy SOFM. A square grid is considered for the 2-D Kohonen layer (KL) in the model. Since regionalization of watersheds is carried out in 2-D, it is expected that the 2-D KL would be more revealing than the 1-D KL which is in use for regional flood frequency analysis. Further, to determine the architecture of the KL in the two-level model, the number of wins by each node in the KL of trained SOFM is noted by varying the grid size of KL from 6×6 nodes to 11×11 nodes, with an increment of one node for each edge.

4.4.3.1 Results from the Two-Level Fuzzy SOFM

It is seen that irrespective of the architecture chosen for KL, same groups of feature vectors are always mapped onto nearby nodes on the KL. Further, increase in the size of grid resulted in increase in the count of nodes with zero (or no) mapping on the KL. The sensitivity of mapping to grid size of Kohonen layer is seen to be less in the neighborhood of 8×8 lattice, hence it is chosen as the architecture of the KL.

To examine the sensitivity of results of the two-level model to variation in the value of fuzzifier, μ was varied from 1.1 to 2.5 with an increment of 0.1, as suggested by Pal and Bezdek (1995). Results obtained for Indiana data show that for a specified Kohonen grid, the value of objective function of FCM algorithm decreases with: (i) increase in the number of clusters for a specified value of fuzzifier and (ii) increase in the value of fuzzifier for a specified number of clusters. Furthermore, for a given number of clusters and fixed μ, the value of objective function increases with increase in the size of Kohonen lattice.

4.4.3.2 Validation of the Results

The optimal number of clusters in the data set was identified by computing fuzzy cluster validity indices for the partitions obtained from the second level of the two-level clustering model. The extended Xie-Beni index Eq. (3.4.16) indicated $c = 8$

as the optimal number of clusters for the value of μ equal to 1.7 (Fig. 4.4.2). For the choice of μ in the range 2.1–2.5, the values of the cluster validity index are quite high for some values of c, and are not included in Fig. 4.4.2 for clarity. The heterogeneity measures, which are described in Section 1.4, showed that the regions corresponding to the chosen partition are close to being homogeneous.

The fuzzy partition coefficient V_{PC}, fuzzy partition entropy V_{PE}, fuzziness performance index V_{FPI}, and normalized classification entropy V_{NCE} that have been used in the hydrologic literature (Bargaoui et al., 1998; Hall and Minns, 1999; Güler and Thine, 2004) are known to exhibit monotonic tendency, which is an undesirable characteristic of a validity measure. The values of these measures computed for clusters obtained by using the two-level clustering conform to this monotonic behaviour (Fig. 4.4.3). This suggests that these validity indices used in hydrologic literature may not be suitable for identifying optimal number of clusters. The V_{PC} exhibits monotonic decreasing tendency with increase in the value of fuzzifier, whereas V_{PE}, V_{FPI} and V_{NCE} exhibit monotonic increasing tendency with increase in the value of fuzzifier. The disadvantage of these indices is the lack of direct connection to any property of the data (Xie and Beni, 1991).

In cluster analysis using direct FCM, which is described in Chapter 3, all the 245 feature vectors were clustered by using FCM and the memberships of the 245 feature vectors in all the fuzzy clusters were used to compute the values of validity

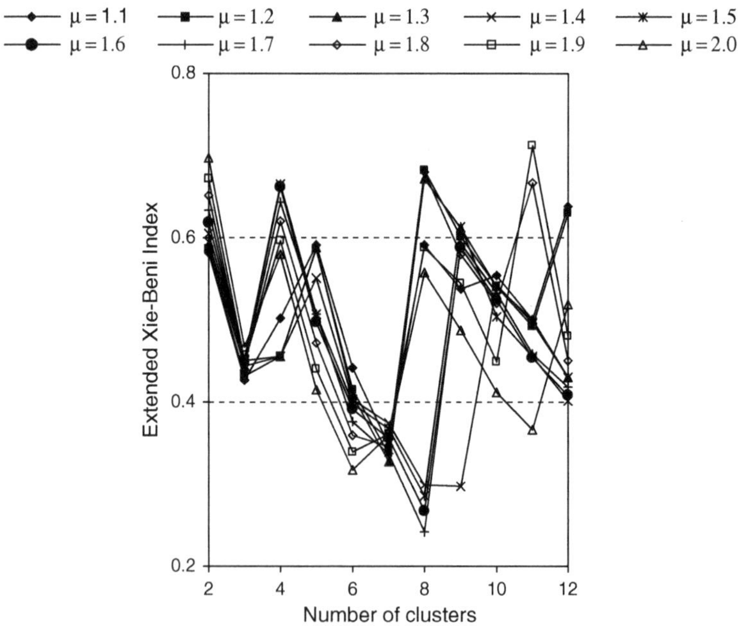

Fig. 4.4.2 The value of extended Xie-Beni index versus the number of clusters for the Indiana data. It is used to identify optimal partition provided by the two level fuzzy SOFM with map size of 8 × 8. The partition with the minimum value for the index is taken as the optimal partition. μ denotes weight exponent for fuzzy membership

indices for the clusters. However, in the proposed two-level clustering method, the m' prototypes, which were obtained from 245 feature vectors using SOFM, were clustered by FCM. Therefore, the memberships of the m' prototypes in the fuzzy clusters were used to estimate the values of validity indices.

To evaluate the relative performance of direct FCM and the two-level clustering algorithms in forming compact and well-separated clusters, the validity indices computed for clusters formed by using both the methods were compared. The validity measures are sensitive to the number of vectors used in their computation. Therefore, to enable comparison of FCM and the two-level clustering method, the cluster validity values for the clusters obtained by using the proposed two-level method were re-computed using 245 feature vectors (instead of m' prototypes) under the assumption that the memberships of each feature vector in all fuzzy clusters are the same as those of the prototype to which it belongs. For the clusters obtained from the two-level clustering method with $c = 8$ and $\mu = 1.7$, the recomputed value of extended Xie-Beni statistic was 0.358, which was less than the values of the statistic computed for clusters obtained with direct FCM for various combinations of c and μ (see Table 3.5.5).

The location of fuzzy clusters (plausible hydrologic regions) obtained from the two-level fuzzy SOFM is shown in Fig. 4.4.4. These hydrologic regions are adjusted following the procedure described in Section 2.4.3, and by using options listed in Section 1.4.1. In the two-level fuzzy SOFM, the knowledge of distribution of membership of a prototype among the fuzzy regions is useful in adjusting the regions to improve their homogeneity. In particular, there is no need to devote special effort for adjustments if the threshold fuzzy membership value is properly chosen to form the fuzzy clusters.

Sites are removed from clusters obtained by using the two-level model to form hydrologic regions which are statistically homogeneous. First, the discordancy values for all the sites in each cluster are estimated by using the discordancy measure described in Section 1.4.2. Table 1.4.1 provides critical values for the discordancy measure to declare a site unusual, and using these all the sites with high discordancy values were identified. Secondly, the heterogeneity measures (Eqs. 1.4.3–1.4.5) of the adjusted region are examined as they change with exclusion of each site from the region. A site is eliminated at a time with replacement. Thirdly, the discordant site, whose exclusion reduces the heterogeneity measures of the region by a significant amount, is identified and removed from the region after ensuring that the fuzzy membership of the prototype containing the site in the region is not high and the site discordancy is not due to sampling variability. The general finding of this step of adjustment is that prototypes with weak membership get eliminated.

A comparison of un-adjusted and adjusted fuzzy clusters is shown in Fig. 4.4.5 for two sample cases. While adjusting cluster-1 to form Region-4, twelve sites having memberships in the interval 0.2–0.4 are eliminated, whereas those eliminated from the intervals 0.4–0.6 and 0.6–0.8 are 2 and 1, respectively. Similarly, to form region 1 by adjusting cluster 3, the sites eliminated from the intervals 0.2–0.4 and 0.4–0.6 were 5 and 2 respectively. Only one site (out of 45 sites) is eliminated from the membership interval 0.6–1.0 to form region 4, whereas no site is eliminated

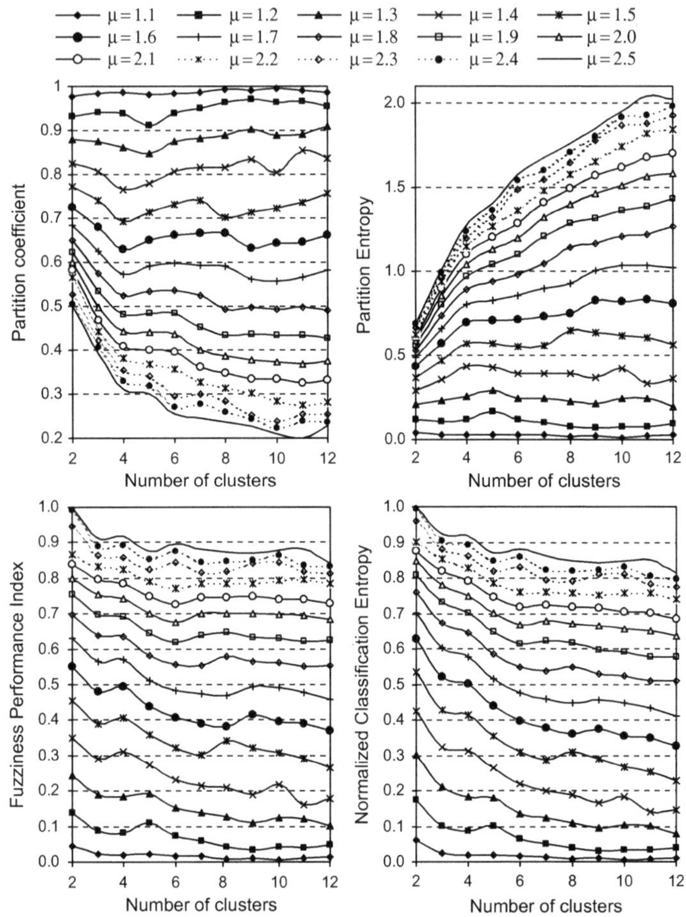

Fig. 4.4.3 Plots of fuzzy partition coefficient (V_{PC}), fuzzy partition entropy (V_{PE}), fuzziness performance index (V_{FPI}) and normalized classification entropy (V_{NCE}) values against the number of clusters for the Indiana watersheds obtained using the two-level fuzzy SOFM. The optimal partition corresponds to a maximum value of V_{PC} (or minimum value of V_{PE}, V_{FPI} and V_{NCE})

from the interval while forming region 1. Thus this technique only targets sites that have low membership in a region.

Hydrologic Regions 1, 2, 3, 4, 5 and 8 are formed by adjusting the clusters 3, 4, 5, 1, 7 and 8, respectively. To adjust the highly heterogeneous cluster 2, several sites were eliminated from the region. This amounts to splitting the highly heterogeneous cluster into two parts. The first part consists of a collection of highly heterogeneous sites forming hydrologic Region 6, whereas the second part consisting of the homogeneous sites constituted hydrologic Region 7. The cluster 6, which consists of only 5 sites, could not be considered as a potential region for flood frequency analysis. At the same time, it was not possible to dissolve the cluster by transferring them to other regions because four of those sites belong to a prototype, which does not

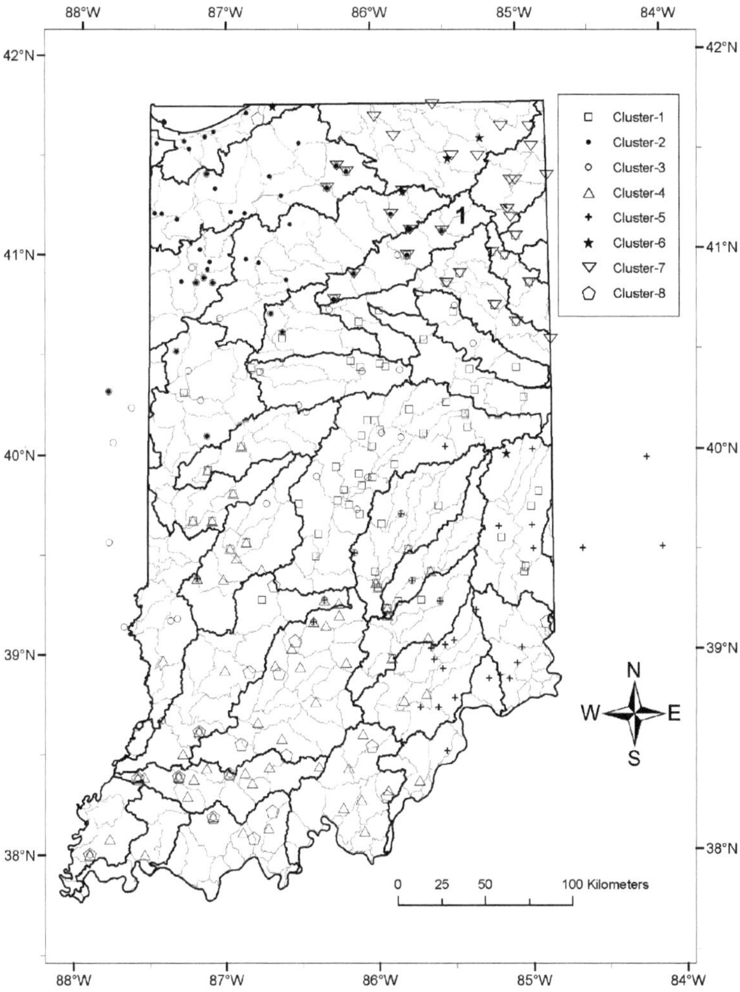

Fig. 4.4.4 Location of fuzzy clusters in Indiana obtained from the two-level fuzzy SOFM. The dark lines denote boundaries of eight digit watersheds, whereas the grey coloured lines are boundaries of 11 digit watersheds in Indiana

have reasonable membership in any other cluster. Hence the prototype remained unallocated.

4.4.3.3 Testing the Regions for Robustness

The hydrologic regions are further examined for their robustness. By specifying various threshold values, stations with record lengths significantly different from the rest of the group are removed and the region with the remaining stations is examined for homogeneity. The results of this analysis are presented in

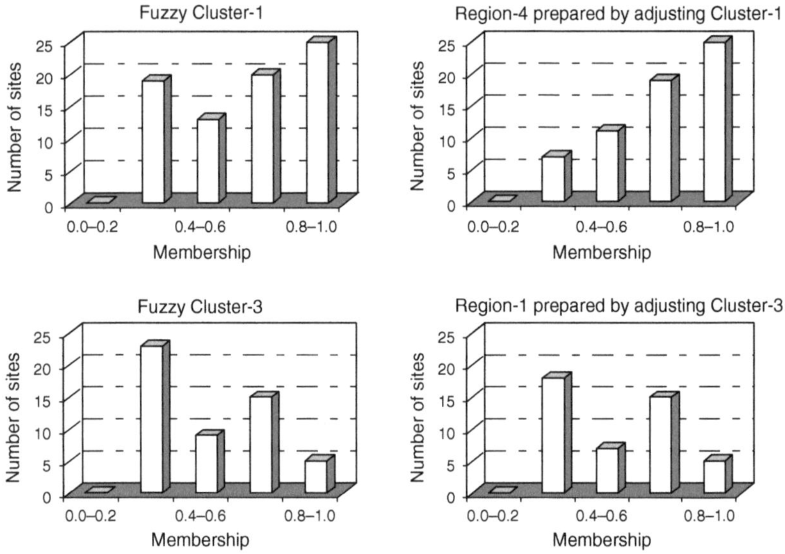

Fig. 4.4.5 Two typical fuzzy clusters obtained from the two-level fuzzy SOFM and the fuzzy regions formed by adjusting them using the guidelines of Hosking and Wallis (1997). The composition of a fuzzy cluster (or region) is shown as a histogram prepared with the memberships of sites in the cluster (or the region). It is assumed that the memberships of a site in fuzzy clusters (or regions) are same as those of the prototype to which the site belongs

Table 4.4.1 and indicate that all the homogeneous regions identified are indeed robust.

Sixteen sites, out of the 245 sites considered, could not be allocated to any region. Four of these 16 belong to the unallocated prototype, while another 5 of these 16 are eliminated to improve the homogeneity of clusters from being classified as ‘possibly homogeneous’ to ‘acceptably homogeneous’. The remaining unallocated sites are those that are highly discordant with sites in clusters where they have very strong membership.

The results presented in Table 4.4.2 indicate that except region 6, all the regions are acceptably homogeneous. Region 6 adjoining Lake Michigan is highly heterogeneous and consists of 11 watersheds in the Kankakee basin of Indiana all of which have high membership in region 6. The average record length per station in region 6 is 42-years, which is reasonably high.

The regions formed by the two-level clustering of SOFM are presented in Fig. 4.4.6. Except region 8, all the homogeneous regions identified have enough pooled data (Table 4.4.2). The region 8 could be merged with region 2 which contains it geographically. However, it was decided not to merge these two regions because all the prototypes of region 8 have very low membership in region 2.

The regions obtained by two-level fuzzy SOFM algorithm bear resemblance to those formed using hard and fuzzy clustering algorithms discussed in previous chapters.

Table 4.4.1 Results from the test of the regions for robustness

R	Condition	NS	RS	Heterogeneity measure			Region type
				H_1	H_2	H_3	
1	Entire region	45	674	0.65	−0.30	−0.93	Homogeneous
	Sites with RL≤10 are eliminated	19	421	0.77	0.70	0.71	Homogeneous
	Sites with RL≥25 are eliminated	38	471	0.33	−0.83	−1.79	Homogeneous
	Sites with RL≤10 and RL≥25 are eliminated	12	218	0.06	0.17	−0.08	Homogeneous
2	Entire region	55	1869	0.76	0.41	−0.37	Homogeneous
	Sites with RL ≤20 are eliminated	39	1656	0.38	0.82	0.23	Homogeneous
	Sites with RL≥50 are eliminated	40	984	0.38	−0.15	−0.66	Homogeneous
	Sites with RL≤20 and RL≥50 are eliminated	24	771	−0.14	0.32	0.23	Homogeneous
3	Entire region	29	608	−0.32	0.93	0.31	Homogeneous
	Sites with RL≤10 are eliminated	17	491	0.34	0.84	0.42	Homogeneous
	Sites with RL≥40 are eliminated	25	404	0.32	1.42	0.44	Homogeneous
	Sites with RL≤10 and RL≥40 are eliminated	13	287	1.25	1.88	1.05	Possibly Homogeneous
4	Entire region	62	2820	0.86	0.69	−0.46	Homogeneous
	Sites with RL≤30 are eliminated	48	2498	0.67	0.48	−0.41	Homogeneous
	Sites with RL≥60 are eliminated	50	1947	0.95	1.09	−0.10	Homogeneous
	Sites with RL≤30 and RL≥60 are eliminated	36	1625	0.71	1.23	0.20	Homogeneous
5	Entire region	24	1121	0.93	−0.11	−0.82	Homogeneous
	Sites with RL≤30 are eliminated	19	1012	1.23	0.22	−0.45	Possibly Homogeneous
	Sites with RL≥60 are eliminated	18	694	0.69	0.30	−0.43	Homogeneous
	Sites with RL≤30 and RL≥60 are eliminated	13	585	1.00	0.47	−0.10	Homogeneous
7	Entire region	25	990	0.82	0.42	1.27	Homogeneous
	Sites with RL≤20 are eliminated	23	963	0.97	0.30	1.16	Homogeneous
	Sites with RL≥55 are eliminated	20	683	0.90	0.65	0.98	Homogeneous
	Sites with RL≤20 and RL≥55 are eliminated	18	656	0.97	0.46	0.78	Homogeneous

R: Region; RL: record length; NS: Number of stations; RS: Region size in station-years

Table 4.4.2 Characteristics of the regions

Region number	NS	RS	Heterogeneity measure		
			H_1	H_2	H_3
1	45	674	0.65	−0.30	−0.93
2	55	1869	0.76	0.41	−0.37
3	29	608	−0.32	0.93	0.31
4	62	2820	0.86	0.69	−0.46
5	24	1121	0.93	−0.11	−0.82
6	11	467	13.77	6.05	2.41
7	25	990	0.82	0.42	1.27
8	16	188	0.54	−0.69	−1.99

NS: Number of stations; RS: Region size in station-years

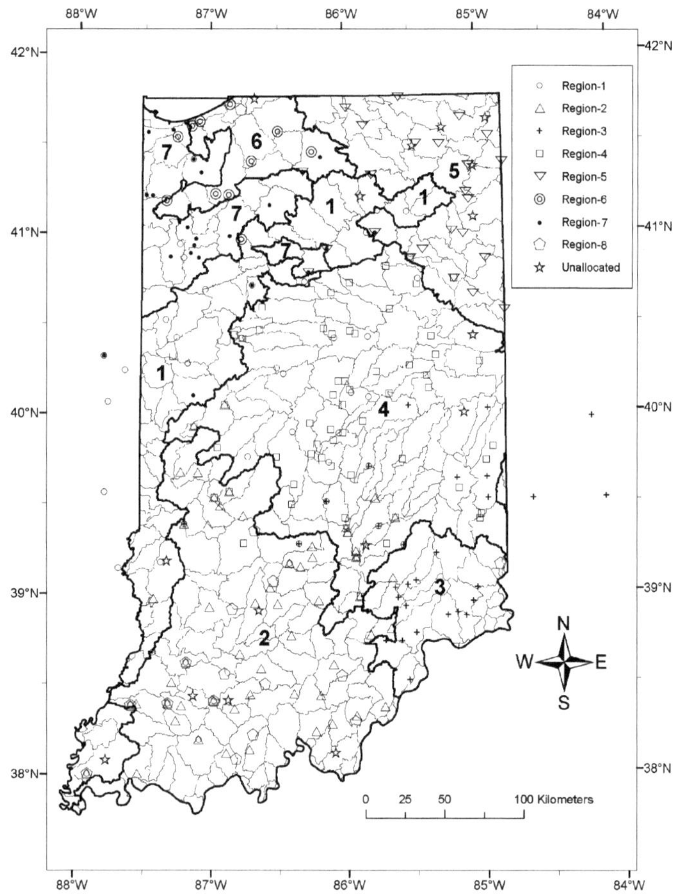

Fig. 4.4.6 Location of the hydrologic regions defined by using the two-level fuzzy SOFM. The dark lines denote fuzzy (soft) boundaries of regions, whereas the grey coloured lines are boundaries of 11 digit watersheds in Indiana. The region 8, which is thoroughly mixed with Region 2, is not marked by a soft boundary

4.5 Concluding Comments

A class of artificial neural networks called Self-Organizing Feature Maps (SOFMs) are described. Limitations associated with the use of one and two-dimensional classical SOFMs for regionalization are brought out by using data from watersheds in Indiana, USA. The feature maps generated did not reveal any information that would allow for selection of an appropriate number of clusters.

With a view to form clusters using SOFM, two options are considered. In the first option, 1-D feature map is used with the number of nodes in its output layer equal to the number of clusters to be formed. Alternatively, fuzzy clustering of the nodes in the 2-D feature map formed with SOFM is considered. Six clusters are identified as optimal partition using hard cluster validity measures and visual interpretation for the first option. On the other hand, eight clusters are identified as optimal partition using fuzzy cluster validity measures for the second option.

A promising technique for regionalization should require minimal effort for adjusting the plausible hydrologic regions (clusters) to form acceptably homogeneous regions. While the first option required considerable effort to adjust three of the six clusters that are formed, the second option yielded homogeneous regions with relatively minimal adjustments. Further, it is seen that the hydrologic regions identified with either of these options are consistent with those derived in Chapters 2 and 3. This is satisfying because there is ambiguity among practicing hydrologists regarding the appropriate procedure to be adopted for regionalization. The results suggest that Fuzzy clustering of 2-D SOFM can be a viable alternative to derive homogeneous regions for flood frequency analysis.

The extended Xie-Beni index performed reasonably well in identifying optimal number of clusters. In contrast, the fuzzy partition coefficient, fuzzy partition entropy, fuzziness performance index, and normalized classification entropy, exhibited monotonic increasing or decreasing tendencies with increase in the value of fuzzifier. Therefore they are not suitable for identifying optimal number of clusters.

Chapter 5
Effect of Regionalization on Flood Frequency Analysis

En-Ching Hsu, A. Ramachandra Rao, V.V. Srinivas

5.1 Introduction

Watersheds in a region can be classified into homogeneous groups by using the regionalization methods discussed in the previous chapters. Regionalization is, however, only a prelude to flood frequency analysis and several aspects related to flood estimation remain to be examined. Perhaps the most important of these is related to quantifying the improvement in flood quantile estimation brought about by regionalization of watersheds into hydrologically homogeneous groups.

In general, the error in estimation of flood quantiles depends on the method used for regional flood frequency analysis (RFFA). An inefficient method may have errors which could nullify any gains obtained by regionalization. Hence it is preferable to investigate more than one RFFA method in order to assess their relative advantages and to quantify the improvement in flood estimation brought about by regionalization. In this chapter performance of two commonly used methods of RFFA is investigated. The first method is based on the regional L-moment algorithm, whereas the second method is based on regional regression analysis.

Another important aspect is related to verifying whether a single distribution is acceptable for estimating flood quantiles in all the regions formed for flood frequency analysis. Bulletin 17 of the U.S. Water Resources Council (1976, 1977, 1981) recommends fitting log-Pearson type III (LP3) distribution to annual maximum streamflows. The stipulation that LP3 distribution must be used over a large area such as the continental United States presumes that it is the preferred distribution. In order to test the validity of such an assumption, the performance of LP3 can be compared with that of several other frequency distributions in fitting peak flow data of different homogeneous regions. If LP3 distribution is preferred to other distributions in such a comparative analysis then it has some justification behind using it. Also, the analysis would be useful to conclude whether any particular frequency distribution is preferred to fit peak flow data in all the regions.

Further, in the past decade, simple scaling methods have been developed in flood frequency analysis. It is found that within hydrologically homogeneous regions

---

E-C. Hsu
Purdue University, West Lafayette, IN 47906, U.S.A.

moments of peak flows scale with drainage area of watersheds according to log-log relations (Gupta and Waymire, 1990; Smith, 1992; Kumar et al., 1994; Gupta and Dawdy, 1995; Ribeiro and Rousselle, 1996; Blöschl and Sivapalan, 1997; Pandey, 1998; Cathcart, 2001; Eaton et al., 2002). If simple scaling results are valid in a hydrologically homogeneous region, then for any ungauged watershed in the region one can use the area of the watershed to estimate the moments of floods. These moments can then be used to estimate the parameters of any chosen distribution. These parameters may be used to estimate flood quantiles corresponding to different recurrence intervals. Consequently examining the behavior of scaling methods and how the scaling properties vary with homogeneity of watersheds is an interesting aspect, and is discussed in this chapter.

The material in this chapter is organized as follows. The regional index flood method for estimation of flood quantiles is discussed in Section 5.2. Following this, flood estimation by generalized least squares (GLS) regional regression analysis is discussed in Section 5.3. As both these methods are well known, the discussion is limited to data analysis and interpretation of results. A procedure for flood estimation based on both regional index flood method and GLS regional regression analysis is discussed in Section 5.4. As mentioned earlier, different flood estimation procedures are considered because conclusions based on only one method could be misleading. The performance of these procedures of flood estimation is compared in Section 5.5 by using split sample test. Among these methods, observed annual maximum flows are used in Method 1 for estimation of flood quantiles and hence it gives smallest errors. These are used as bench marks in the comparison of results.

The scaling behavior of annual maximum flows in different regions is examined in Section 5.6 by studying variations of the conventional statistical moments with the basin area. The importance of regionalization is brought out in a graphical form. Research in this area is progressing (Gupta and Waymire, 1990; Smith, 1992; Kumar et al., 1994; Gupta and Dawdy, 1995; Ribeiro and Rousselle, 1996; Blöschl and Sivapalan, 1997; Pandey, 1998; Cathcart, 2001; Eaton et al., 2002) and is of interest. The scaling concept may lead to better and simpler methods both for regionalization and for testing homogeneity of regions.

The plausibility of fitting a particular probability distribution to flood series in all the regions is tested in Section 5.7. It is of interest to note that a single probability distribution is not acceptable in all the regions of Indiana. Some concluding comments are presented in Section 5.8.

5.2 Regional Index Flood Method Based on L-Moments

5.2.1 Introduction

The basic idea behind the index flood method (Dalrymple, 1960), which has been in use for a long time, is that the frequency distributions of floods at the sites in a homogeneous region are identical except for a scaling factor known as the index-flood.

The index flood parameter reflects the important physiographic and meteorologic characteristics of a watershed. The L-moment based index flood method was proposed by Landwehr, Matalas and Wallis and popularized by Wallis and others (Hosking et al., 1985; Wallis, 1980; Wallis and Wood, 1985). An important requirement for the success of the index flood method is that data from hydrologically similar basins should be used (Lettenmaier et al., 1987).

Regional index flood methods based on probability weighted moments and L-moments have been studied, generally with Generalized Extreme Value (GEV) or Wakeby distributions (Hosking and Wallis, 1988; Jin and Stedinger, 1989; Landwehr et al., 1987; Potter and Lettenmaier, 1990; Wallis and Wood, 1985). These results, especially with GEV distribution have been demonstrated to be robust.

5.2.2 Regional L-Moment Method

Suppose that annual maximum flow data are available at N_R sites in a region, with site k having sample size n_k and observed data Q_{kj}, $j = 1, \ldots, n_k$. The first three L-moments $\hat{\lambda}_1(k)$, $\hat{\lambda}_2(k)$ and $\hat{\lambda}_3(k)$ at the site k are computed by using the unbiased probability weighted moment (PWM) estimators. The L-moments are subsequently scaled by using $\hat{\lambda}_1(k)$ as the index flood. The regional estimates of the normalized L-moments (or regional average L-moments) of orders 1, 2 and 3 are computed as

$$\hat{\lambda}_r^R = \frac{\sum_{k=1}^{N_R} w_k \left[\hat{\lambda}_r(k) / \hat{\lambda}_1(k)\right]}{\sum_{k=1}^{N_R} w_k}, \quad r = 1, 2, 3 \tag{5.2.1}$$

where r is the order of L-moment and w_k are the weights. A simple choice for w_k is n_k. In general the value of this parameter may depend on the heterogeneity of a region (Tasker and Stedinger, 1986, 1989) and some modification might be required.

The Eq. (5.2.1) indicates that the first order regional average normalized L-moment is 1.0. The normalized parameters for different 3-parameter probability distributions are computed by probability weighted moment method based on the first three normalized L-moments. Using these parameters, the quantiles $\hat{q}_T^R$ of the normalized regional distribution are estimated for various recurrence intervals T. Denote by F the nonexceedance probability ($F = 1 - 1/T$). The dimensionless quantile function $\hat{q}_T^R\ (F)$, known as *regional growth curve*, is common to every site in a region. The quantile estimates at site k, $\hat{Q}_T^k$, are obtained by multiplying $\hat{q}_T^R$ with the index flood value $\hat{\lambda}_1^k$ as

$$\hat{Q}_T^k = \hat{\lambda}_1^k\, \hat{q}_T^R \tag{5.2.2}$$

Since $\hat{\lambda}_1^k$ is the regressor, the confidence limit for the regional L-moment quantile estimate at site k is calculated by Eq. (5.2.3),

$$CL = \hat{Q}_T^k \pm t_{\alpha/2,N-2} \sqrt{MSE\left(\frac{1}{N} + \frac{\left(\hat{\lambda}_1^k - \bar{\lambda}\right)^2}{S_{\lambda\lambda}}\right)} \quad (5.2.3)$$

where N is the total number of observations of the annual peak flow, $\bar{\lambda}$ is the average of $\hat{\lambda}_1^k$ values, $S_{\lambda\lambda}$ is the sum of squares of errors $\sum_{k=1}^{N_R} (\hat{\lambda}_1^k - \bar{\lambda})^2$, MSE is the mean square of the errors, and $t_{\alpha/2,N-2}$ is the value of the student's t-distribution for a $100(1 - \alpha)$percent of confidence interval with N-2 degrees of freedom (Hines and Montgomery, 1980).

The advantage of estimating $\hat{q}_T^R$ by regional L-moment method is well documented (Hosking and Wallis, 1997). The importance of using data from a homogeneous region in index flood analysis is stressed by Lettenmaier et al. (1987). To adopt the index flood procedure for estimating flood frequency at ungauged sites one of the important variables which must be estimated for the sites is $\hat{\lambda}_1^k$. The usual practice is to estimate this variable by relating it to other watershed attributes that are easily available even for ungauged sites in the region.

5.2.3 At-Site and Regional Parameter Estimation

The regions that are formed by using regionalization methods discussed earlier are adjusted based on requirements of Indiana Department of Transportation, USA, to form contiguous regions 1 to 8 shown in Fig. 5.2.1. The heterogeneity measures computed for these regions show that region 6 adjoining the Lake Michigan is highly heterogeneous, whereas all other regions are either acceptably homogeneous or possibly homogeneous (Table 5.2.1).

For each site in a region, the conventional moments (mean, standard deviation, skewness and kurtosis), L-moments (l_1, l_2, l_3, l_4) and L-moment ratios ($t = l_2/l_1$, $t_3 = l_3/l_2, t_4 = l_4/l_2$) are estimated based on at-site records of annual maximum flows. Subsequently, these moments are used to compute parameters of six candidate distributions and flood quantiles for 2, 5, 10, 20, 25, 50, 100 and 200 year recurrence intervals based on method of moments and probability weighted moment procedures described in Rao and Hamed (2000).

Further, the regional average normalized L-moment ratios are derived from the at-site estimates of L-moments by Eq. (5.2.1). The normalized L-moments ratios are used to compute parameters and flood quantiles following the procedure discussed above. The normalized regional flood quantiles estimated for each region using the six distributions are presented in Table 5.2.2 for the eight recurrence intervals. The flood corresponding to normalized regional quantile is estimated by Eq. (5.2.2) considering the first at-site L-moment (l_1) as the index flood value.

The precision in estimation of regional flood quantiles is evaluated using variance ν^2 as an indicator. For a chosen distribution and exceedence probability $p = 1/T$, the variance is computed based on differences between the flood quantiles estimated

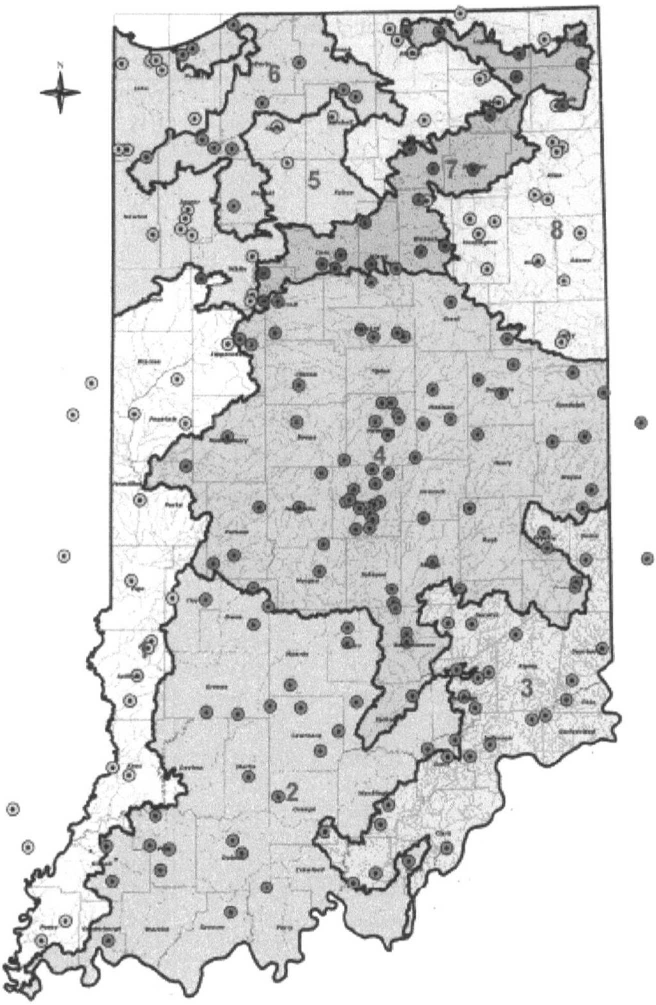

Fig. 5.2.1 Regions considered for flood frequency analysis

Table 5.2.1 Characteristics of the regions considered for flood frequency analysis. NS represents the number of stations

Region number	NS	Heterogeneity measure			Region type
		H_1	H_2	H_3	
1	21	0.66	−1.83	−2.40	Acceptably Homogeneous
2	30	1.17	−1.18	−2.00	Possibly homogeneous
3	24	0.26	0.53	0.12	Acceptably Homogeneous
4	72	0.79	−0.97	−1.45	Acceptably Homogeneous
5	18	1.18	−0.30	−0.09	Possibly homogeneous
6	12	14.68	5.42	2.47	Heterogeneous
7	22	1.56	0.04	−0.24	Possibly homogeneous
8	25	1.07	−0.59	−0.96	Possibly homogeneous

Table 5.2.2 Normalized regional quantile estimates obtained by using regional L-moment method

	T years	LN3	GM2	PT3	LP3	GEV	GLO
Region 1	2	0.9098	0.8478	0.9088	1.0036	0.6319	0.9209
	5	1.3367	1.7612	1.3469	1.0482	1.0522	1.3046
	10	1.6257	2.3350	1.6353	1.0696	1.3418	1.5809
	20	1.9066	2.8622	1.9069	1.0863	1.6286	1.8761
	25	1.9966	3.0252	1.9920	1.0910	1.7214	1.9774
	50	2.2767	3.5163	2.2513	1.1041	2.0131	2.3162
	100	2.5599	3.9897	2.5049	1.1153	2.3114	2.6984
	200	2.8481	4.4501	2.7545	1.1252	2.6177	3.1320
Region 2	2	0.8761	0.8298	0.8739	1.0023	0.5716	0.8916
	5	1.3556	1.7468	1.3737	1.0531	1.0362	1.3200
	10	1.7013	2.3349	1.7194	1.0785	1.3788	1.6438
	20	2.0512	2.8814	2.0538	1.0988	1.7363	2.0024
	25	2.1658	3.0514	2.1600	1.1046	1.8561	2.1280
	50	2.5305	3.5663	2.4871	1.1209	2.2452	2.5580
	100	2.9100	4.0658	2.8116	1.1353	2.6638	3.0587
	200	3.3066	4.5544	3.1350	1.1481	3.1156	3.6453
Region 3	2	0.8456	0.8257	0.8409	0.9989	0.4719	0.8656
	5	1.3352	1.7433	1.3635	1.0614	0.9369	1.2999
	10	1.7163	2.3347	1.7477	1.0947	1.3075	1.6479
	20	2.1217	2.8857	2.1306	1.1226	1.7187	2.0497
	25	2.2585	3.0573	2.2541	1.1307	1.8619	2.1942
	50	2.7046	3.5777	2.6393	1.1543	2.3458	2.7019
	100	3.1859	4.0835	3.0276	1.1756	2.8976	3.3163
	200	3.7055	4.5787	3.4195	1.1954	3.5287	4.0638
Region 4	2	0.8824	0.8250	0.8808	1.0040	0.6083	0.8964
	5	1.3748	1.7426	1.3903	1.0593	1.0885	1.3376
	10	1.7204	2.3346	1.7355	1.0859	1.4328	1.6648
	20	2.0642	2.8865	2.0657	1.1069	1.7843	2.0222
	25	2.1758	3.0584	2.1700	1.1128	1.9004	2.1464
	50	2.5276	3.5798	2.4898	1.1293	2.2724	2.5678
	100	2.8893	4.0867	2.8052	1.1435	2.6643	3.0525
	200	3.2630	4.5832	3.1179	1.1561	3.0785	3.6133
Region 5	2	0.9578	0.8657	0.9576	1.0061	0.7422	0.9623
	5	1.3213	1.7747	1.3235	1.0510	1.1097	1.2938
	10	1.5387	2.3336	1.5406	1.0713	1.3294	1.5106
	20	1.7340	2.8411	1.7337	1.0865	1.5242	1.7268
	25	1.7938	2.9971	1.7925	1.0907	1.5829	1.7979
	50	1.9725	3.4644	1.9668	1.1019	1.7550	2.0257
	100	2.1433	3.9118	2.1317	1.1113	1.9135	2.2670
	200	2.3084	4.3443	2.2894	1.1193	2.0599	2.5242
Region 6	2	0.9275	0.8831	0.9267	1.0017	0.5879	0.9538
	5	1.2558	1.7867	1.2645	1.0456	0.9097	1.2459
	10	1.4809	2.3309	1.4891	1.0677	1.1348	1.4458
	20	1.7014	2.8196	1.7018	1.0854	1.3600	1.6519
	25	1.7723	2.9688	1.7686	1.0904	1.4334	1.7210
	50	1.9940	3.4136	1.9727	1.1048	1.6657	1.9472
	100	2.2195	3.8366	2.1728	1.1174	1.9057	2.1940
	200	2.4502	4.2430	2.3703	1.1288	2.1545	2.4650

Table 5.2.2 (continued)

	T years	LN3	GM2	PT3	LP3	GEV	GLO
Region 7	2	0.8562	0.8262	0.8527	0.9966	0.5123	0.8713
	5	1.3460	1.7436	1.3704	1.0711	0.9811	1.3060
	10	1.7163	2.3347	1.7422	1.1121	1.3440	1.6496
	20	2.1027	2.8853	2.1087	1.1470	1.7376	2.0425
	25	2.2316	3.0568	2.2262	1.1573	1.8726	2.1830
	50	2.6482	3.5767	2.5911	1.1874	2.3224	2.6733
	100	3.0915	4.0818	2.9567	1.2150	2.8243	3.2614
	200	3.5640	4.5765	3.3240	1.2407	3.3859	3.9705
Region 8	2	0.9396	0.8722	0.9392	1.0031	0.6683	0.9477
	5	1.2937	1.7793	1.2989	1.0449	1.0212	1.2678
	10	1.5207	2.3328	1.5253	1.0651	1.2500	1.4880
	20	1.7339	2.8332	1.7336	1.0808	1.4662	1.7159
	25	1.8008	2.9866	1.7980	1.0853	1.5340	1.7925
	50	2.0055	3.4455	1.9923	1.0977	1.7410	2.0438
	100	2.2073	3.8837	2.1798	1.1084	1.9435	2.3189
	200	2.4082	4.3064	2.3622	1.1178	2.1422	2.6221

LN3: Three parameter log normal distribution. GM2: Two parameter gamma distribution. PT3: Pearson type 3 distribution. LP3: log Pearson type 3 distribution. GEV: Generalized extreme value distribution. GLO: Generalized logistic distribution.

for all the sites in the region by at-site frequency analysis and by regional frequency analysis.

$$\nu^2 = E\left[\hat{\lambda}_1 \hat{q}_T^R - \hat{\lambda}_1 \hat{q}_T^k\right]^2 \tag{5.2.4}$$

Higher ν^2 denotes high variability within the region and smaller variance indicates strong homogeneity within a region.

The at-site quantile estimates are plotted against the regional quantile estimates for each of the recurrence intervals considered and for all the candidate distributions to observe the goodness of fit. In general, the at-site and regional quantile estimates are found to be nearly equal for GEV, PT3 and LN3 distributions. Consequently, the points corresponding to each of these distributions are seen to lie close to 45 degree line for all the regions. In contrast, the points corresponding to LP3 and two-parameter gamma distributions are found to be widely scattered about the 45 degree line indicating that they are not suitable to fit the data. For brevity the results for 100-year recurrence interval are shown in Fig. 5.2.2.

The variance of estimation errors in the eight regions are shown in Fig. 5.2.3. The PT3 distribution has smaller variance than other distributions, especially for longer recurrence intervals. In general, the variance obtained for GEV and LN3 distributions is close to that obtained for PT3 distribution. Overall, GEV, PT3, and LN3 have good estimates for all the regions. One of the problems with LN3 distribution is that sometimes it does not yield convergent parameter estimates. The other issue is that although LP3 may not be a good candidate for regional index flood estimation,

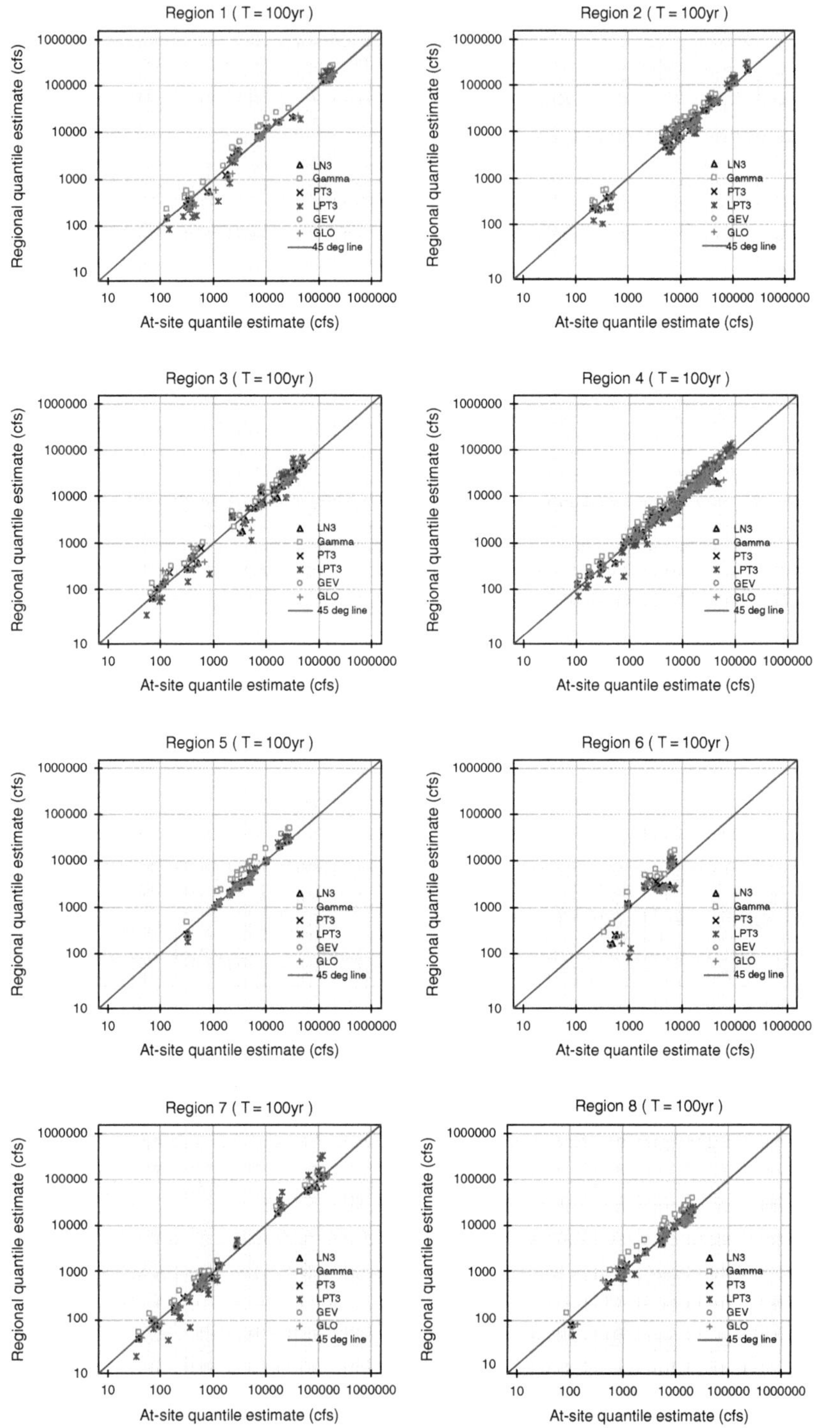

Fig. 5.2.2 At-site and regional flood quantile estimates for T = 100 year. LP3 is shown as LPT3, and GM2 is shown as Gamma

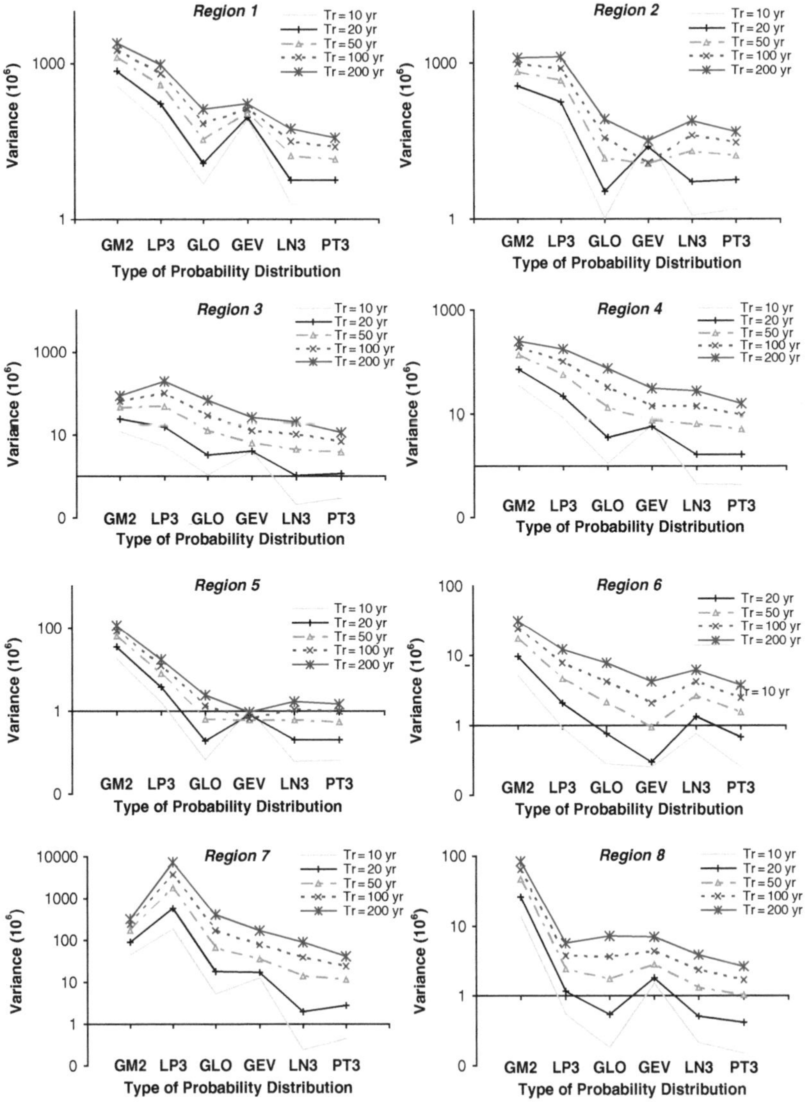

Fig. 5.2.3 Variance of the differences between at-site and regional estimates

it is widely used for engineering design in United States. Consequently, PT3, GEV and LP3 distributions are used in the following analysis.

Typical plots of the 95% confidence intervals for the regional L-moment flood quantile estimates are shown in Figs. 5.2.4–5.2.6. The confidence intervals are calculated by using Eq. (5.2.3) based on regression of the mean annual peak discharge, which is the first L-moment. In the figures the axes are logarithmic for better clarity. Further, both at-site and regional L-moment estimates of flood quantile are plotted

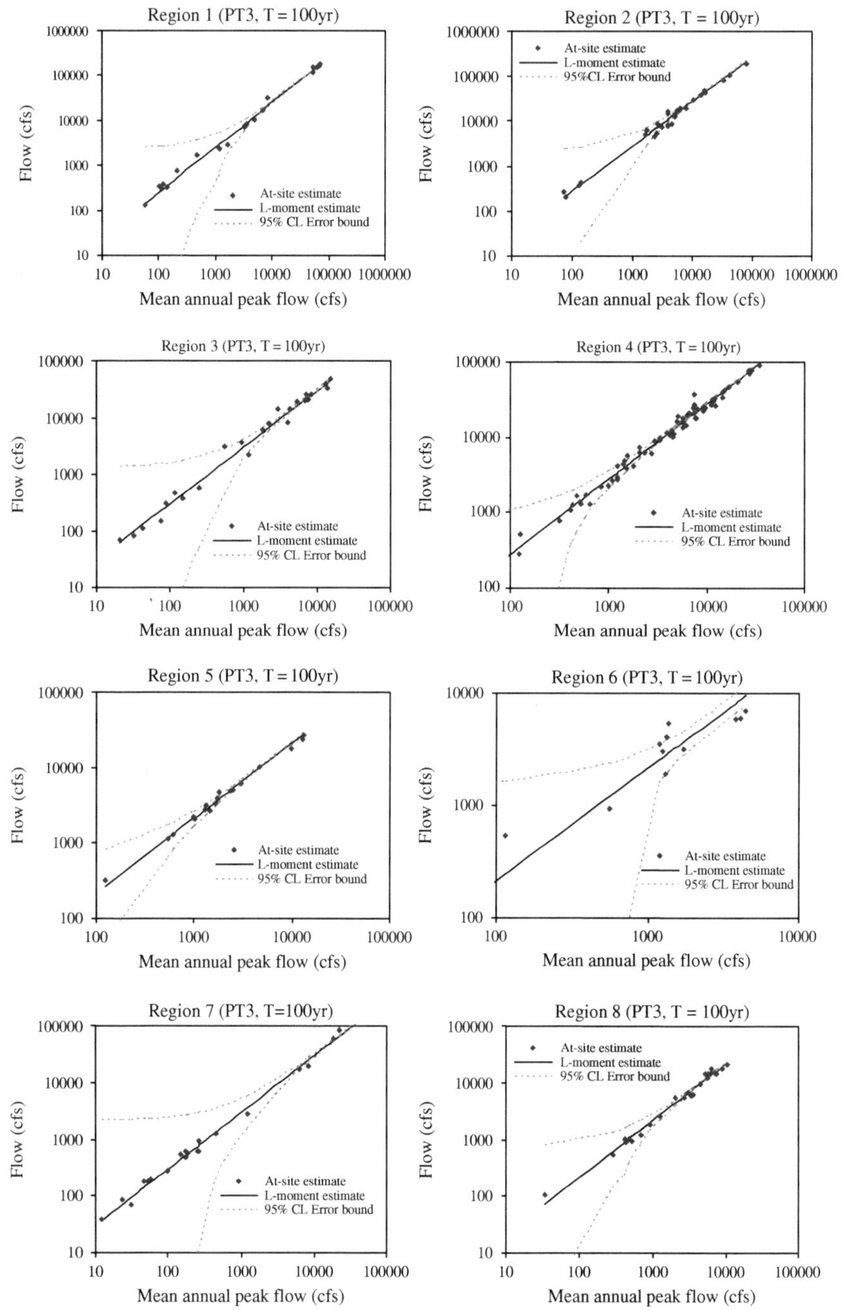

Fig. 5.2.4 Ninety-five percentage confidence interval error bounds for regional PT3 L-moment estimates

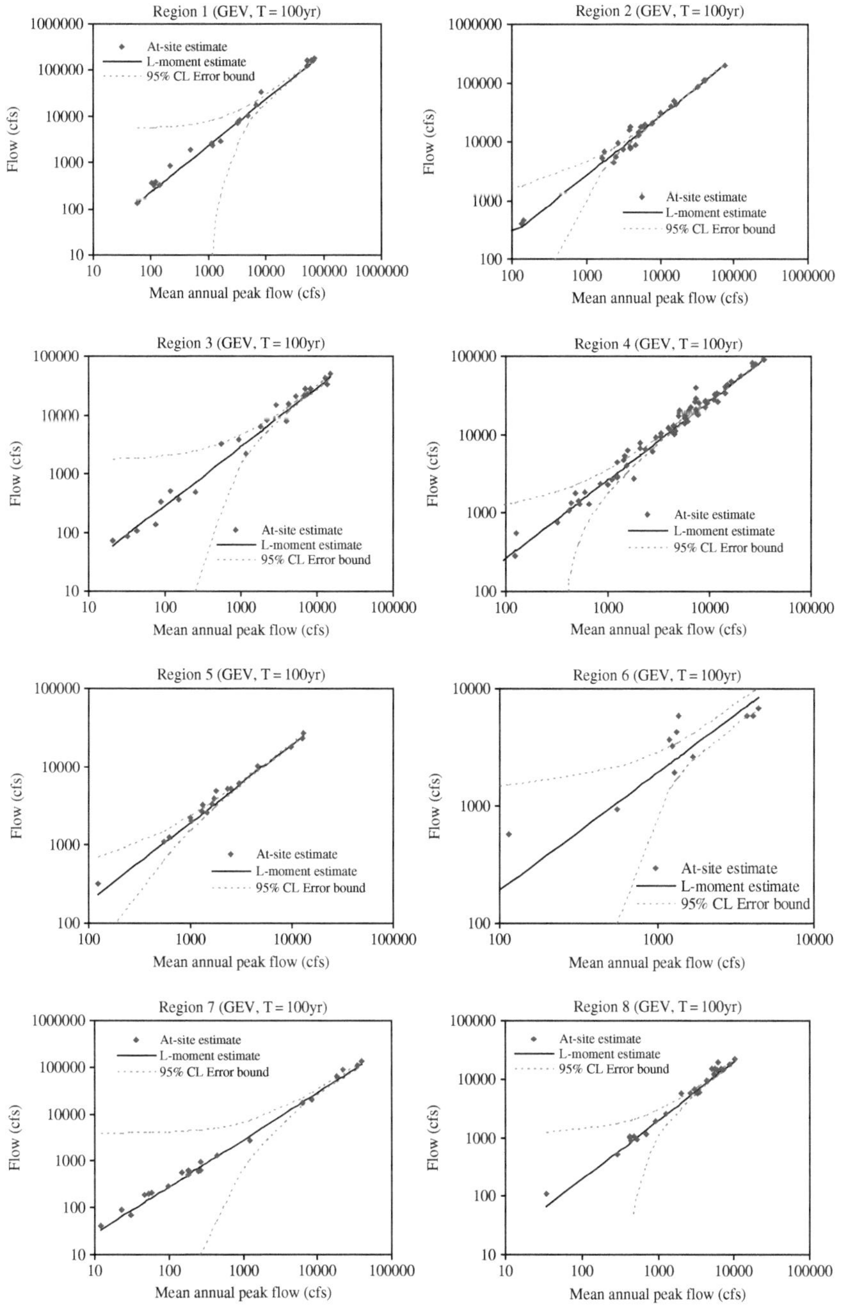

Fig. 5.2.5 Ninety-five percentage confidence interval error bounds for regional GEV L-moment estimates

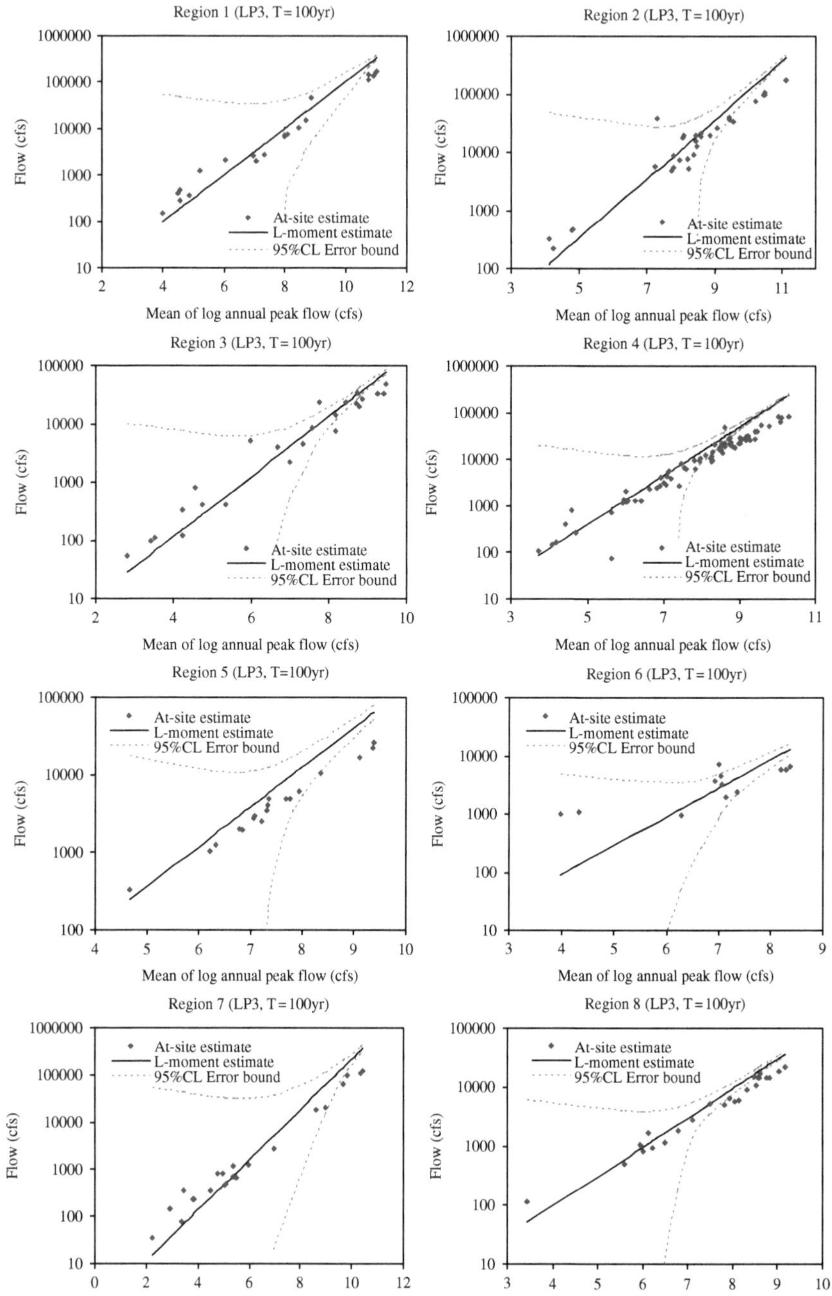

Fig. 5.2.6 Ninety-five percentage confidence interval error bounds for regional LP3 L-moment estimates

Table 5.2.3 Determination of optimal probability distributions for regional L-moment flood estimates of the entire series of data

Region No.	Candidate Probability Distributions	Optimal Distributions for Regional Estimates
1	PT3, GM2, LN3, GEV, LP3	PT3, LN3, GEV
2	GEV, LN3, PT3, GM2, GLO	GEV, PT3, LN3
3	LP3, GEV, LN3, GLO, PT3	PT3, LN3, GEV
4	GEV, LN3, LP3, PT3, GM2	PT3, LN3, GEV
5	GEV, LP3, LN3, PT3, GM2	GEV, PT3, LN3
6	LN3, PT3, GM2, GLO, GEV	PT3, GEV, LN3
7	PT3, GM2, LN3, LP3, GEV	PT3, LN3, GEV
8	LP3, GLO, GEV, LN3, PT3	PT3, LN3, GLO

as ordinate, whereas index flood value is abscissa. It can be inferred from the figures that the regional L-moment estimates obtained from LP3 are inferior to those from PT3 and GEV. Further, it is seen that the confidence intervals obtained for LP3 distribution are much wider than those obtained for PT3 and GEV distributions.

The candidate probability distributions are determined for each region based on the mean-square-error (MSE) of regional L-moment method. The order shown in Table 5.2.3 begins with the distribution having the minimum MSE. Optimal distributions for regional L-moment flood estimates are obtained from the variances of regional estimates. PT3, GEV and LN3 are good probability distributions for regional L-moment flood estimates.

An important conclusion from the results shown herein is that a single distribution is not applicable for all the regions even where the regions are adjacent to each other. Secondly, the LP3 distribution is clearly inferior to others for regional flood estimation in Indiana, USA.

5.3 Regional Regression Analysis

5.3.1 Introduction

In general, at-site information is used for frequency analysis of floods at locations of interest. However, in cases where data are inadequate or missing at the location of interest, regional regression relationships are used for flood quantile estimation. Regional regression is an idea in which the flood characteristics are related to the geographical or hydrological attributes which are measurable for any location in a watershed. Generalized Least Square (GLS) regression is introduced by Stedinger and Tasker (1985) to develop these relationships. The GLS method takes the data consistency (lengths of record and correlation) and geographical distance into account. Details of GLS regression are found in the reference cited above. The GLS regression relationships for different regions are discussed in this section.

5.3.2 GLS Regional Regression Results

To identify the governing hydrological attributes influencing peak flows from Indiana watersheds, first of all, the square of the correlation coefficient R^2 between each hydrological feature and the at-site quantile estimates are calculated. For brevity the results are shown for PT3 and GEV distributions in Tables 5.3.1 and 5.3.2 respectively. The drainage area (A) and slope (S) are found to be the primary factors affecting floods in Indiana. For the secondary factor, wet area (W) which is the percentage of area of lakes and ponds, and urbanization factor (U) which is the percent of urban area in the region are considered. These variables show smaller

Table 5.3.1 R^2 values for the relationship between the individual hydrological attributes and PT3 flood quantile estimates

Region number	Attribute	T = 10yr	T = 20yr	T = 50yr	T = 100yr	T = 200yr
1	Drainage area (mi^2)	0.978	0.975	0.970	0.967	0.964
	Slope (%)	0.856	0.850	0.841	0.836	0.831
	W(%)	0.380	0.306	0.260	0.232	0.211
	U (%)	0.280	0.223	0.191	0.172	0.158
2	Drainage area (mi^2)	0.935	0.935	0.933	0.931	0.929
	Slope (%)	0.729	0.731	0.732	0.733	0.733
	W(%)	0.183	0.166	0.160	0.156	0.154
	U (%)	0.288	0.230	0.196	0.175	0.160
3	Drainage area (mi^2)	0.978	0.976	0.973	0.969	0.966
	Slope (%)	0.753	0.749	0.744	0.739	0.736
	W(%)	0.096	0.067	0.054	0.046	0.041
	U (%)	0.042	0.029	0.023	0.020	0.018
4	Drainage area (mi^2)	0.935	0.934	0.932	0.931	0.929
	Slope (%)	0.397	0.394	0.390	0.388	0.386
	W(%)	0.234	0.213	0.203	0.196	0.192
	U (%)	0.008	0.001	0.000	0.001	0.002
5	Drainage area (mi^2)	0.939	0.938	0.937	0.936	0.935
	Slope (%)	0.453	0.456	0.458	0.460	0.461
	W(%)	0.028	0.037	0.043	0.048	0.053
	U (%)	0.150	0.165	0.170	0.173	0.175
6	Drainage area (mi^2)	0.797	0.750	0.689	0.644	0.601
	Slope (%)	0.466	0.405	0.333	0.285	0.243
	W(%)	0.074	0.091	0.112	0.127	0.141
	U (%)	0.228	0.193	0.188	0.186	0.187
7	Drainage area (mi^2)	0.907	0.903	0.899	0.896	0.894
	Slope (%)	0.534	0.529	0.523	0.520	0.517
	W(%)	0.001	0.007	0.011	0.014	0.017
	U (%)	0.377	0.379	0.380	0.380	0.381
8	Drainage area (mi^2)	0.795	0.792	0.788	0.785	0.781
	Slope (%)	0.674	0.677	0.680	0.682	0.683
	W(%)	0.000	0.003	0.007	0.009	0.012
	U (%)	0.005	0.000	0.002	0.006	0.010

Table 5.3.2 R^2 values for the relationship between the individual hydrological attributes and GEV flood quantile estimates

Region number	Attribute	T = 10yr	T = 20yr	T = 50yr	T = 100yr	T = 200yr
1	Drainage area (mi^2)	0.982	0.978	0.971	0.964	0.956
	Slope (%)	0.865	0.856	0.842	0.831	0.818
	W(%)	0.389	0.311	0.260	0.227	0.201
	U (%)	0.284	0.226	0.191	0.169	0.153
2	Drainage area (mi^2)	0.935	0.935	0.933	0.929	0.923
	Slope (%)	0.723	0.728	0.732	0.733	0.733
	W(%)	0.177	0.162	0.159	0.159	0.160
	U (%)	0.296	0.235	0.197	0.172	0.153
3	Drainage area (mi^2)	0.975	0.978	0.974	0.967	0.956
	Slope (%)	0.753	0.753	0.748	0.740	0.730
	W(%)	0.101	0.072	0.056	0.046	0.039
	U (%)	0.044	0.031	0.025	0.021	0.018
4	Drainage area (mi^2)	0.933	0.934	0.932	0.928	0.922
	Slope (%)	0.398	0.393	0.387	0.382	0.376
	W(%)	0.226	0.207	0.201	0.198	0.196
	U (%)	0.008	0.001	0.000	0.001	0.003
5	Drainage area (mi^2)	0.939	0.939	0.938	0.936	0.934
	Slope (%)	0.453	0.456	0.459	0.462	0.464
	W(%)	0.029	0.037	0.042	0.047	0.050
	U (%)	0.152	0.166	0.168	0.169	0.168
6	Drainage area (mi^2)	0.816	0.770	0.689	0.609	0.513
	Slope (%)	0.511	0.441	0.336	0.250	0.165
	W(%)	0.069	0.093	0.130	0.161	0.193
	U (%)	0.228	0.199	0.207	0.216	0.227
7	Drainage area (mi^2)	0.911	0.906	0.899	0.894	0.889
	Slope (%)	0.541	0.534	0.524	0.517	0.509
	W(%)	0.001	0.006	0.011	0.014	0.016
	U (%)	0.375	0.377	0.379	0.381	0.383
8	Drainage area (mi^2)	0.799	0.796	0.789	0.781	0.771
	Slope (%)	0.673	0.677	0.681	0.682	0.682
	W(%)	0.000	0.003	0.006	0.009	0.012
	U (%)	0.006	0.000	0.002	0.007	0.013

correlation with floods. The correlation coefficients are quite low for region 6 which is heterogeneous and region 8 which is close to being acceptably homogeneous (Table 5.2.1).

Models described by Eqs. (5.3.1)–(5.3.3) are developed by using GLS regional regression to fit the relationships between floods quantiles and hydrological features of watersheds.

$$\text{Model I: } Q_T = aA^b \tag{5.3.1}$$

$$\text{Model II: } Q_T = aA^bS^c \tag{5.3.2}$$

$$\text{Model III: } Q_T = aA^b S^c (1 + W)^d \tag{5.3.3}$$

The probability distributions used for regional regression are generalized extreme value (GEV), Pearson type III (PT3) and log-Pearson type III (LP3). Of these distributions GEV and PT3 fit the observed data well and also provide stable results in regional flood evaluation. The LP3 distribution is considered because it is widely used in engineering design in United States. However, from the previous results, it is not a good distribution to estimate regional flood values for Indiana watersheds and this aspect should be kept in mind.

The regression coefficients computed for the three models by the GLS method are summarized in Tables 5.3.3–5.3.5 for PT3, GEV and LP3 distributions respectively. There are two sub-tables in each table and they refer to recurrence intervals of 100 and 200 years. In each sub-table, the coefficients a, b, c, d and R^2 are given for each model and region. In the developed regression relationships the unit for drainage area is square miles, slope and wet area are in percentage and the regressed quantile flow is in cubic feet per second (cfs).

Table 5.3.3 GLS Regression coefficients computed for PT3 flood quantile estimates

Region number	Model	Parameters for T = 100 years				R^2
		a	b	c	d	
1	I	489.317	0.613			0.984
	II	57.265	0.817	0.648		0.988
	III	106.649	0.806	0.549	−0.282	0.990
2	I	896.268	0.573			0.959
	II	131.618	0.774	0.460		0.978
	III	219.760	0.750	0.380	−0.187	0.975
3	I	698.344	0.746			0.849
	II	83.356	0.952	0.521		0.919
	III	99.717	0.955	0.519	−0.191	0.913
4	I	334.991	0.702			0.901
	II	31.407	0.886	0.789		0.943
	III	27.817	0.882	0.800	0.110	0.940
5	I	130.743	0.663			0.939
	II	53.972	0.761	0.351		0.954
	III	54.480	0.758	0.343	0.008	0.954
6	I	639.416	0.303			0.569
	II	26.717	0.729	1.036		0.732
	III	19.759	0.704	0.942	0.264	0.790
7	I	158.534	0.733			0.985
	II	21.273	0.901	0.676		0.981
	III	130.208	0.909	0.591	−1.146	0.991
8	I	111.992	0.692			0.679
	II	368.409	0.564	−0.397		0.778
	III	543.001	0.739	−0.152	−0.956	0.707

Table 5.3.3 (continued)

Region number	Model	Parameters for T=200 years				R^2
		a	b	c	d	
1	I	546.361	0.610			0.983
	II	58.599	0.823	0.675		0.986
	III	114.985	0.810	0.566	−0.304	0.989
2	I	1007.864	0.570			0.959
	II	155.814	0.766	0.448		0.977
	III	245.104	0.744	0.378	−0.165	0.974
3	I	785.830	0.747			0.838
	II	88.529	0.959	0.534		0.912
	III	109.137	0.962	0.531	−0.218	0.905
4	I	367.385	0.704			0.897
	II	33.848	0.889	0.794		0.942
	III	28.799	0.883	0.808	0.151	0.939
5	I	143.074	0.660			0.937
	II	59.790	0.756	0.344		0.952
	III	58.519	0.753	0.336	0.027	0.952
6	I	793.937	0.278			0.500
	II	27.519	0.730	1.090		0.702
	III	20.111	0.706	0.997	0.266	0.768
7	I	175.509	0.735			0.984
	II	22.918	0.905	0.686		0.980
	III	137.604	0.915	0.602	−1.137	0.989
8	I	121.778	0.691			0.672
	II	427.940	0.556	−0.419		0.779
	III	628.520	0.730	−0.174	−0.952	0.697

Table 5.3.4 GLS Regression coefficients computed for GEV flood quantile estimates

Region number	Model	Parameters for T=100 years				R^2
		a	b	c	d	
1	I	502.034	0.613			0.983
	II	52.261	0.829	0.682		0.987
	III	101.327	0.818	0.577	−0.302	0.989
2	I	935.570	0.572			0.959
	II	142.691	0.770	0.450		0.977
	III	224.135	0.748	0.380	−0.165	0.974
3	I	699.260	0.758			0.837
	II	91.561	0.955	0.500		0.911
	III	106.380	0.959	0.501	−0.174	0.905
4	I	338.494	0.706			0.899
	II	29.401	0.896	0.815		0.942
	III	25.616	0.891	0.827	0.127	0.939
5	I	131.386	0.662			0.938
	II	55.997	0.756	0.338		0.953
	III	54.942	0.754	0.332	0.023	0.953

Table 5.3.4 (continued)

Region number	Model	Parameters for T=100 years				R^2
		a	b	c	d	
6	I	679.210	0.292			0.508
	II	23.975	0.740	1.095		0.712
	III	15.343	0.708	0.971	0.367	0.803
7	I	165.136	0.735			0.983
	II	21.175	0.906	0.691		0.980
	III	127.332	0.914	0.609	−1.134	0.990
8	I	113.807	0.691			0.673
	II	400.359	0.556	−0.420		0.779
	III	587.756	0.732	−0.174	−0.954	0.689
		Parameters for T = 200 years				
1	I	583.377	0.608			0.980
	II	48.663	0.844	0.750		0.982
	III	104.543	0.831	0.626	−0.346	0.985
2	I	1101.084	0.567			0.957
	II	191.505	0.751	0.419		0.973
	III	263.778	0.736	0.370	−0.117	0.971
3	I	827.871	0.765			0.806
	II	98.503	0.971	0.522		0.886
	III	121.376	0.976	0.522	−0.233	0.876
4	I	376.795	0.711			0.883
	II	31.130	0.904	0.830		0.935
	III	24.698	0.894	0.848	0.223	0.932
5	I	143.723	0.658			0.936
	II	63.126	0.749	0.325		0.950
	III	59.066	0.746	0.318	0.051	0.949
6	I	975.037	0.245			0.362
	II	22.440	0.751	1.218		0.667
	III	13.982	0.723	1.097	0.370	0.769
7	I	188.332	0.741			0.979
	II	22.107	0.920	0.721		0.975
	III	128.715	0.930	0.642	−1.121	0.984
8	I	127.852	0.686			0.657
	II	513.403	0.537	−0.464		0.777
	III	747.139	0.712	−0.218	−0.948	0.662

Table 5.3.5 GLS Regression coefficients computed for LP3 flood quantile estimates

Region number	Model	Parameters for T = 100 years				R^2
		a	b	c	d	
1	I	545.593	0.594			0.965
	II	43.115	0.837	0.762		0.971
	III	100.881	0.825	0.627	−0.400	0.976
2	I	1140.736	0.537			0.942
	II	225.895	0.707	0.388		0.966
	III	462.830	0.674	0.276	−0.264	0.962

Table 5.3.5 (continued)

Region number	Model	Parameters for T = 100 years				R^2
		a	b	c	d	
3	I	703.266	0.760			0.771
	II	115.414	0.934	0.447		0.823
	III	115.061	0.937	0.455	−0.031	0.822
4	I	343.147	0.699			0.857
	II	25.572	0.900	0.866		0.923
	III	24.071	0.899	0.872	0.051	0.923
5	I	128.672	0.662			0.933
	II	53.337	0.759	0.352		0.948
	III	45.829	0.759	0.350	0.091	0.948
6	I	1154.568	0.202			0.358
	II	36.271	0.671	1.082		0.588
	III	22.666	0.639	0.955	0.381	0.703
7	I	201.377	0.700			0.983
	II	16.612	0.909	0.838		0.973
	III	129.619	0.918	0.741	−1.296	0.984
8	I	113.236	0.690			0.707
	II	378.588	0.560	−0.403		0.804
	III	536.122	0.723	−0.175	−0.883	0.739
		Parameters for T = 200 years				
1	I	638.605	0.585			0.954
	II	39.569	0.851	0.835		0.959
	III	104.145	0.837	0.681	−0.452	0.966
2	I	1338.893	0.528			0.937
	II	309.724	0.682	0.352		0.959
	III	637.815	0.648	0.239	−0.267	0.956
3	I	810.942	0.766			0.690
	II	126.844	0.945	0.458		0.737
	III	126.029	0.950	0.468	−0.038	0.736
4	I	376.568	0.701			0.818
	II	26.199	0.908	0.886		0.899
	III	23.286	0.903	0.896	0.110	0.899
5	I	138.683	0.659			0.930
	II	58.230	0.755	0.345		0.944
	III	46.684	0.755	0.344	0.131	0.944
6	I	1713.317	0.148			0.199
	II	37.354	0.668	1.175		0.498
	III	21.930	0.639	1.042	0.409	0.629
7	I	235.898	0.699			0.968
	II	16.371	0.922	0.896		0.957
	III	130.285	0.933	0.798	−1.312	0.966
8	I	124.222	0.686			0.701
	II	453.553	0.547	−0.432		0.808
	III	634.820	0.710	−0.205	−0.874	0.724

Among the three regression models, the best-fitting GLS regional regression model has maximum R-square value. To observe the goodness-of-fit for the eight hydrological regions, the GLS regression estimates obtained from the best-fitting model are plotted against at-site quantile estimates in Figs. 5.3.1–5.3.3 for PT3, GEV and LP3 distributions respectively.

The ordinary least square (OLS) regression is also used to fit these data and the results are shown as dashed lines in these figures. Graphically the dashed lines are very close to the solid lines which result from the GLS regression. However, the result from GLS is slightly better than that from OLS in goodness-of-fit.

From the results shown in Tables 5.3.3–5.3.5 it is seen that the performance of models II and III is better than that of model I. The model II shows marginally better results for regions 2, 3, 4 and 8, whereas model III shows marginally better results for regions 1, 6 and 7. Models II and III performed equally well for region 5.

Ideally, flood quantile should be proportional to slope of watershed (S). But, region 8 has the opposite behavior with negative exponent (c) for the slope term (Tables 5.3.3–5.3.5). To avoid this unreasonable result, model I having only drainage area in regression relationship is considered for region 8. As for the other regions, the factors contributing to the best fit in regions 1, 6 and 7 are area (A), slope (S) and percentage wet area (%W); while the factors contributing to best fit in regions 2, 3 and 4 are area and slope. Region 5 could have area and slope, or area, slope and percent wet area.

Of the three models, it is better to chose model II which relates flood quantile (Q_T) to only drainage area and slope of watershed because often information on these attributes is available even for ungauged sites.

5.4 Combination of GLS Regional Regression and L-Moment Method

In order to use the regional L-moment method for estimating flood quantiles at target locations, information on the first moment (i.e., the mean annual peak flow) is needed. However, it is not possible to compute this statistic for an ungauged location. In such situations, GLS regression is a feasible approach to estimate flood quantiles for various recurrence intervals. The GLS regression relationships may be developed to relate the hydrological and geographical attributes of watersheds with mean (or logarithm of mean) annual peak flows at gauged sites in the region. The developed GLS regression equations find use in estimating the first moment of peak flows even at ungauged locations in the region. This information may be combined with the normalized regional quantiles determined using L-moment method to estimate desired flood quantiles. The mean of the logarithms of the annual maximum flows find use in estimation of flood quantiles by LP3 distribution.

Three GLS regression models are constructed for estimation of mean annual flood, MAF, based on the watershed attributes namely drainage area (A), slope

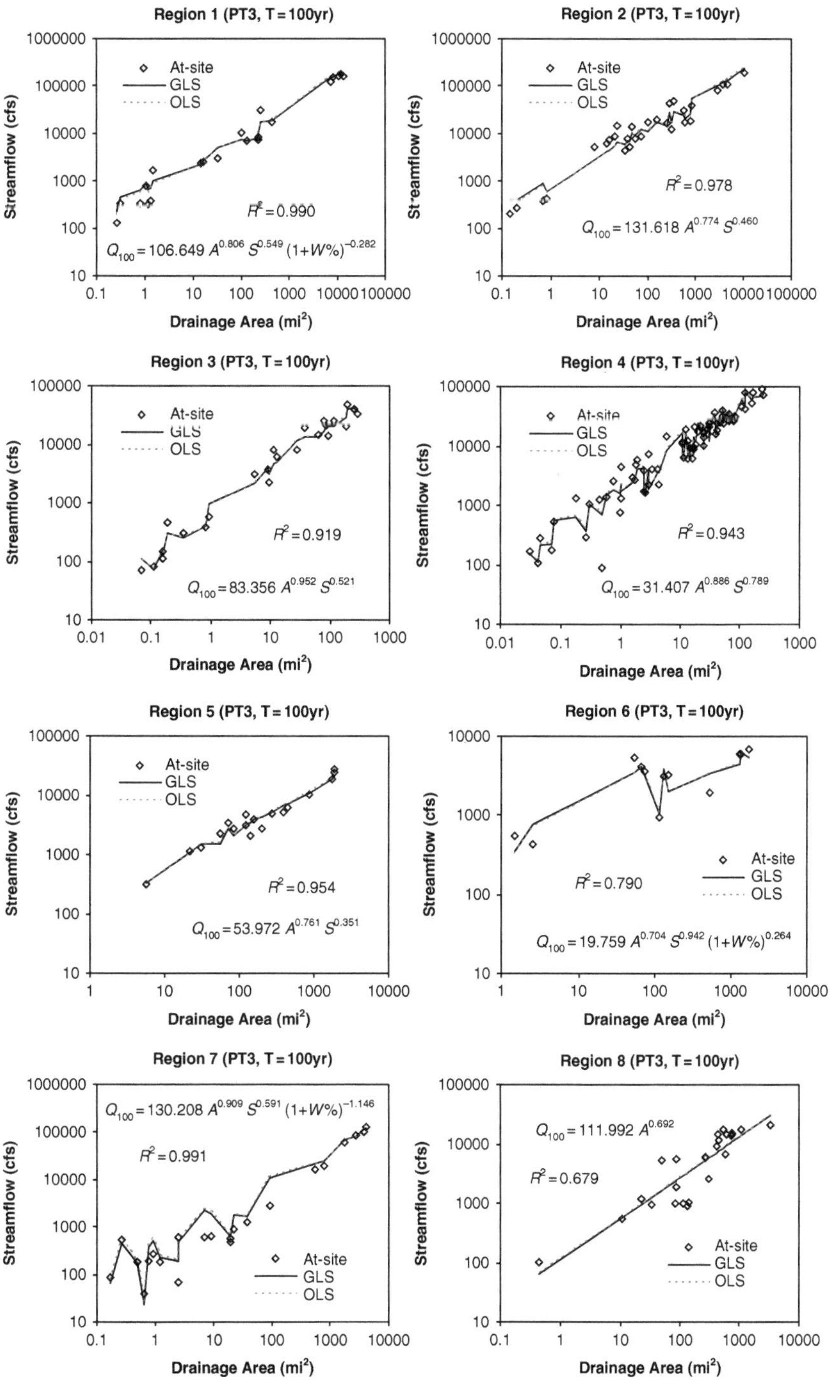

Fig. 5.3.1 GLS regional regression for PT3 (T = 100 years)

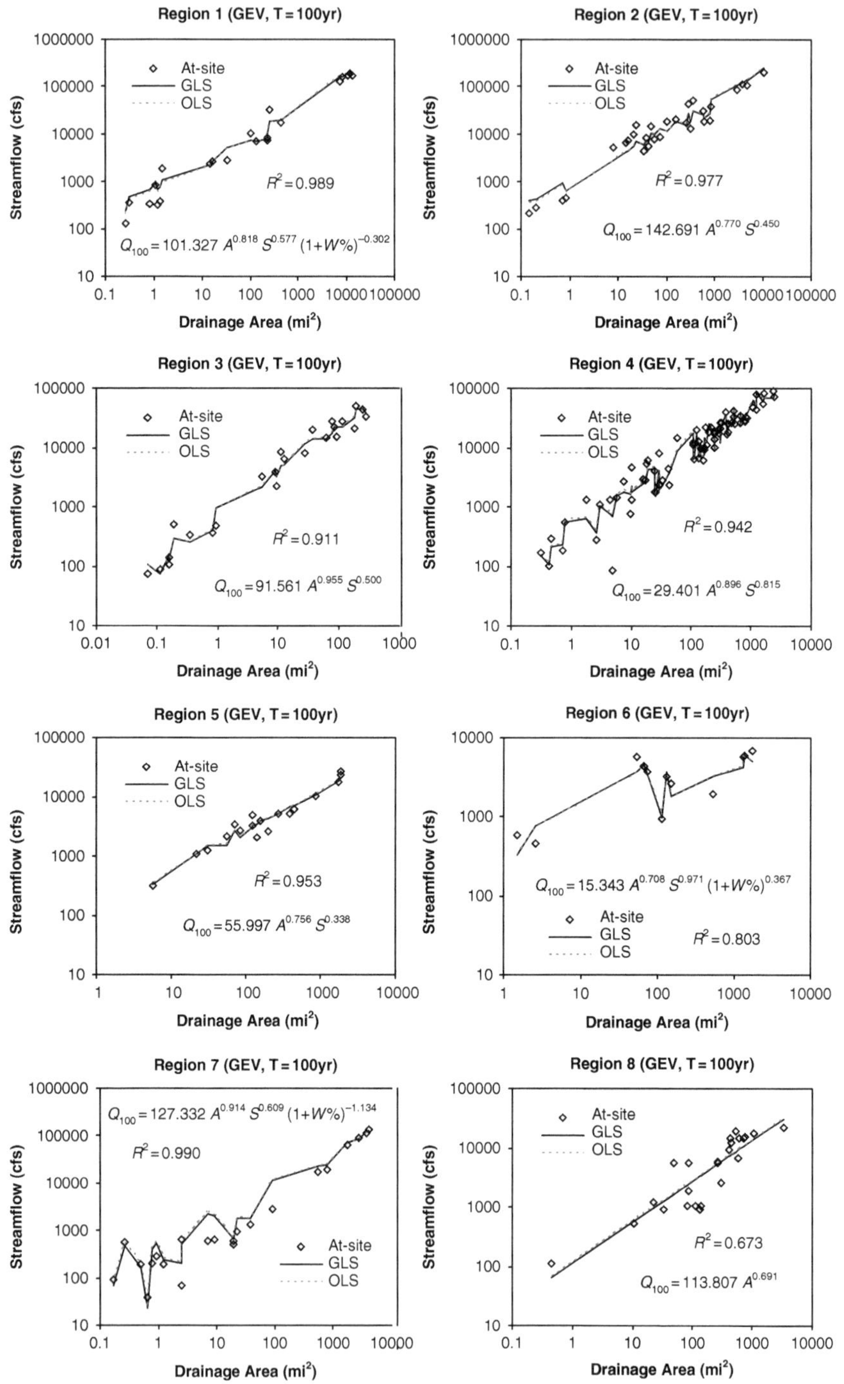

Fig. 5.3.2 GLS regional regression for GEV distribution (T = 100 years)

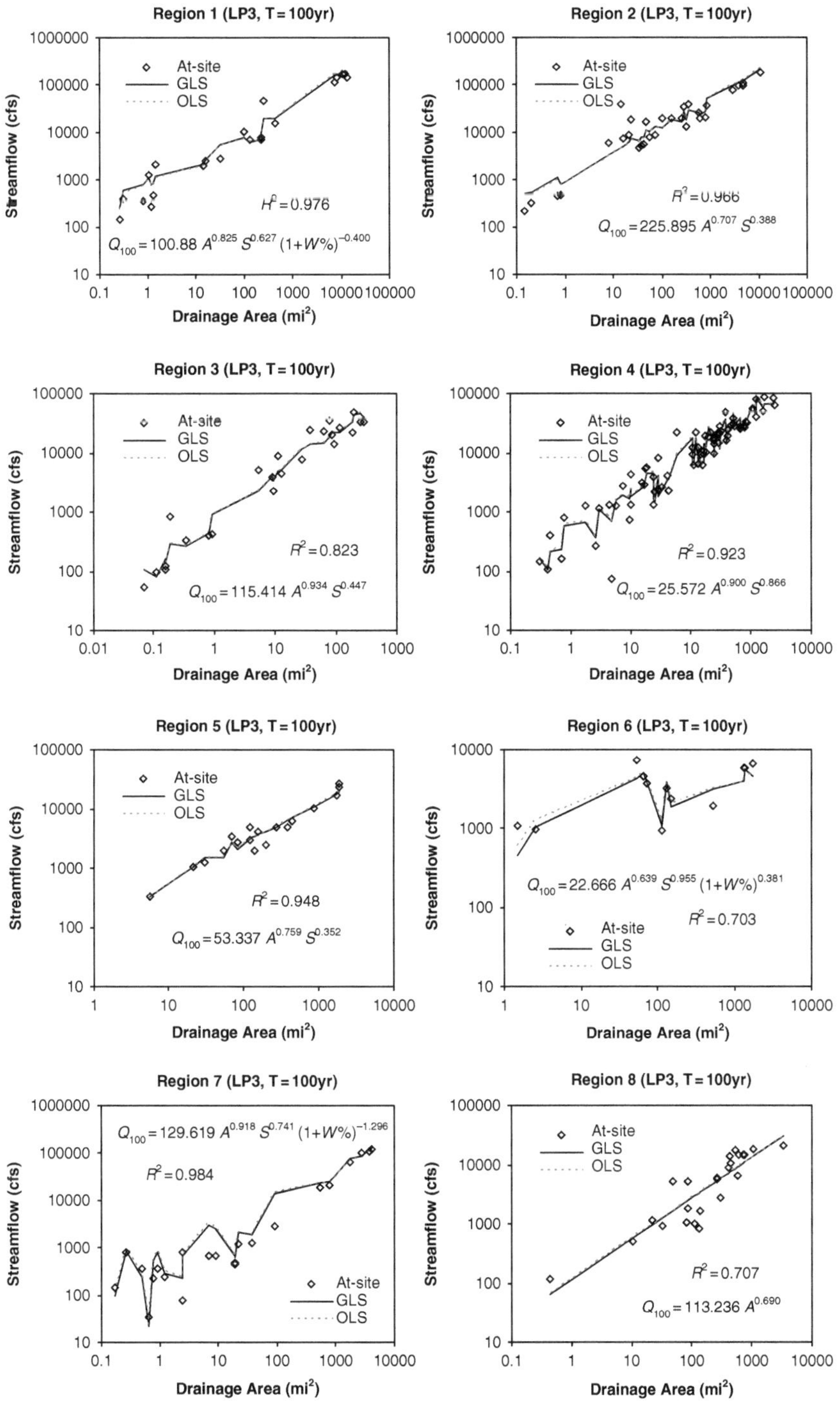

Fig. 5.3.3 GLS regional regression for LP3 (T = 100 years)

(S) and wet area percentage (W). Model I is based only on the drainage area, which is MAF $= aA^b$, Similarly, Model II considers area and slope, which is MAF $= aA^bS^c$, and Model III considers area, slope and wet area percentage, which is MAF $= aA^bS^c(1+W)^d$, where a, b, c and d are GLS regression coefficients. These coefficients for each region and each model are listed in Table 5.4.1. For the logarithms of peak flows, the results are presented in Table 5.4.2. In this case the unit of drainage area is square miles, slope and wet area are in percentage. If the regressed value is logarithm of mean annual peak flow, $Q_{\log}$, then it has to be transformed using exponential function to get mean annual peak flow as $exp(Q_{\log})$.

The error bounds for 95% confidence interval are calculated for all the flood estimates in each region. In Fig. 5.4.1 four series of data are plotted for each region. The observed mean annual peak flow, GLS-regressed mean annual peak flow, and 95% upper and lower confidence limits are shown. Simple log-linear fit is applied for each of them in order to show the trend of each data set. It is seen that the observed and fitted means are close to each other. Except for Regions 4 and 6, the two trend lines in the other six regions are almost overlapping.

Table 5.4.1 GLS regional regression for mean annual peak flows

Region number	Model	a	b	c	d	R^2
1	I	173.4825	0.6309			0.9881
	II	37.3649	0.7779	0.4602		0.9951
	III	42.5792	0.7763	0.4406	−0.0627	0.9953
2	I	299.5047	0.5894			0.961
	II	32.1958	0.8256	0.5314		0.9846
	III	81.8451	0.7822	0.3841	−0.3468	0.9793
3	I	226.0649	0.7443			0.8972
	II	41.8274	0.9055	0.4197		0.9335
	III	40.0037	0.9065	0.4231	0.0339	0.9333
4	I	145.918	0.6712			0.9091
	II	21.6708	0.8161	0.6508		0.937
	III	34.0161	0.8328	0.6172	−0.4202	0.9524
5	I	50.2466	0.7005			0.9556
	II	19.3921	0.8038	0.3927		0.9709
	III	25.7894	0.7971	0.3779	−0.1358	0.971
6	I	94.6589	0.5154			0.9026
	II	12.9361	0.7811	0.6608		0.9352
	III	11.5316	0.7621	0.6054	0.1367	0.9378
7	I	53.8668	0.7342			0.9845
	II	8.2252	0.8914	0.6306		0.9765
	III	52.7069	0.896	0.539	−1.1561	0.9892
8	I	49.8341	0.6986			0.7204
	II	96.4878	0.6277	−0.2204		0.7637
	III	143.2239	0.8076	0.0277	−0.9783	0.7449

Table 5.4.2 GLS regional regression for mean of logarithms of annual peak flows

Region number	Model	a	b	c	d	R^2
1	I	5.3345	0.0779			0.9645
	II	5.0811	0.0824	0.0152		0.9639
	III	4.7038	0.0857	0.033	0.0223	0.9645
2	I	5.7749	0.0728			0.894
	II	4.6272	0.0944	0.0557		0.9063
	III	5.896	0.0811	0.0178	−0.0796	0.9101
3	I	5.2372	0.1085			0.9546
	II	3.6401	0.1458	0.087		0.9694
	III	3.5076	0.1467	0.0896	0.0287	0.9703
4	I	5.2189	0.0874			0.9264
	II	4.1978	0.1041	0.073		0.9517
	III	4.5072	0.1075	0.0679	−0.0693	0.9583
5	I	4.7561	0.0868			0.9472
	II	3.7632	0.1137	0.079		0.9603
	III	3.9458	0.1149	0.08	−0.0316	0.9677
6	I	4.8775	0.0708			0.881
	II	3.4816	0.1175	0.0924		0.9253
	III	3.4933	0.1174	0.0899	−0.0017	0.9242
7	I	3.6987	0.1276			0.9461
	II	2.9408	0.1448	0.0874		0.9532
	III	3.9225	0.1416	0.0891	−0.1828	0.979
8	I	4.0622	0.1119			0.7936
	II	4.2363	0.1072	−0.013		0.795
	III	4.6362	0.1228	0.0269	−0.1297	0.9148

Figure 5.4.2 shows the histograms of the distribution of the drainage areas in each region. In regions 1, 3 and 7 there are quite a few small watersheds. In Regions 4 and 6 most of the drainage areas are larger than 100 square miles and there are a few small drainage areas with high variability of flow than in the other regions. Typically the data lengths in small watersheds are quite small. The accuracy of measurements is also less. These lead to reduction in accuracy in the estimates of mean annual flows also. The reduction in the accuracy of estimation in region 8 is not easy to explain.

In all regression methods the effect of flow variations in small drainage areas is not prominent. Also, the 95% confidence intervals are added to the logarithms of mean annual peak flow and examples of these are shown in Fig. 5.4.3. Regions 3, 5, 7 and 8 have perfect match between the trend lines, and for Regions 1, 2, 4 and 6 the trend lines are not as close.

By using the GLS regression, we can obtain the first L-moment (i.e., mean or log mean) for the location of interest. The T-year flood quantile $\hat{Q}_T^k$ at site k is computed by multiplying the first moment with the regional normalized flood quantile $\hat{q}_T^R$ estimated based on regional L-moment method. For LP3 distribution, the first L-moment is mean of the logarithms of the flows λ_k', and the estimated T-year quantile flood is $\hat{Q}_T^k$ (=$\exp(\lambda_k' \hat{q}_T^R)$ in unit cfs).

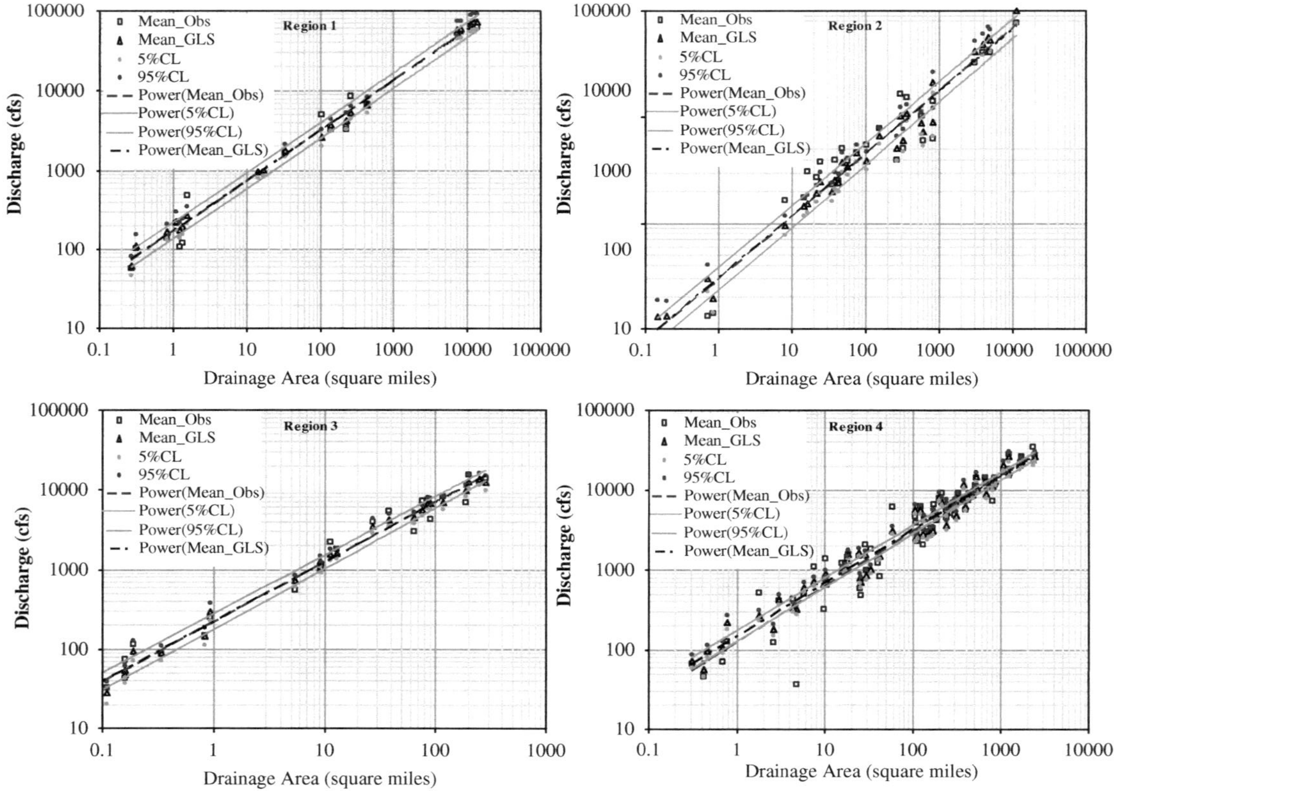

Fig. 5.4.1(a) At-site observed mean annual peak flows compared with GLS regression results and the 95% upper and lower confidence limits for Regions 1 to 4

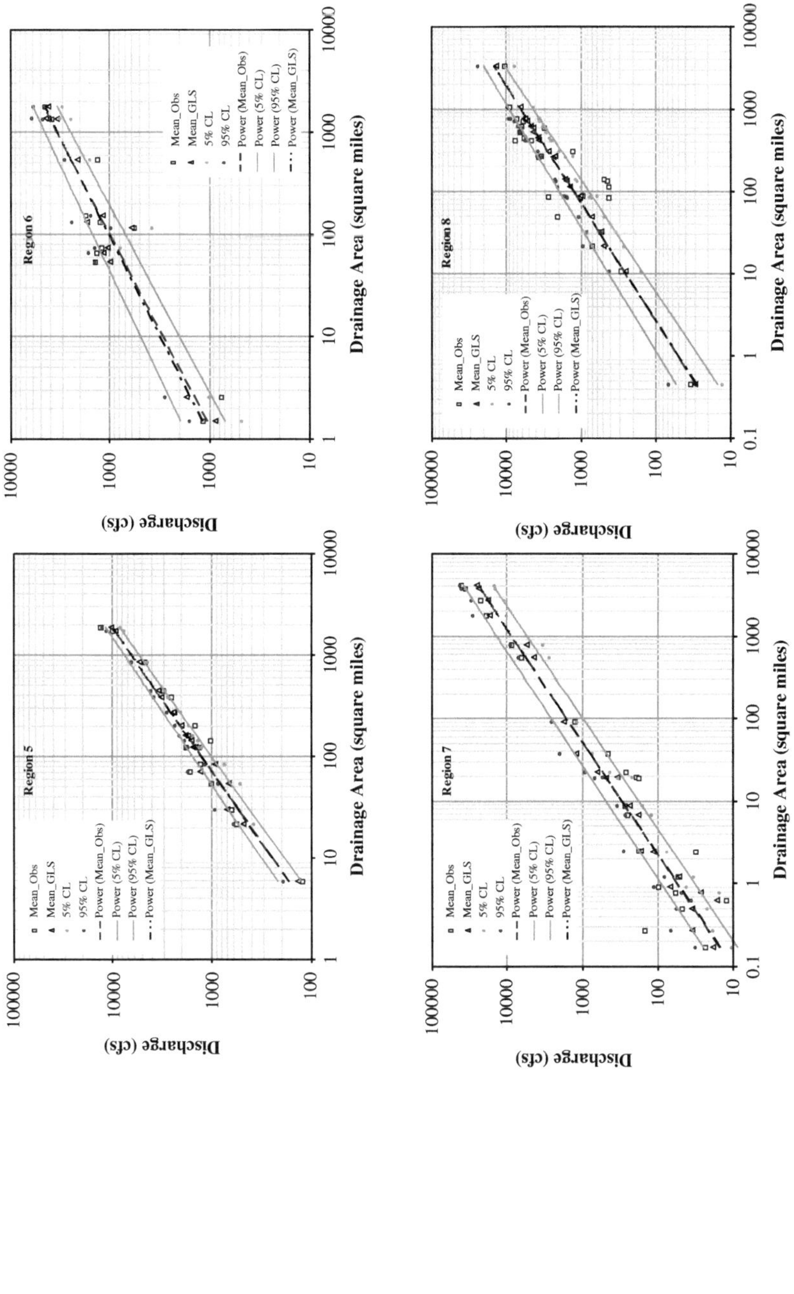

Fig. 5.4.1(b) At-site observed mean annual peak flows compared with GLS regression results and the 95% upper and lower confidence limits for Regions 5 to 8

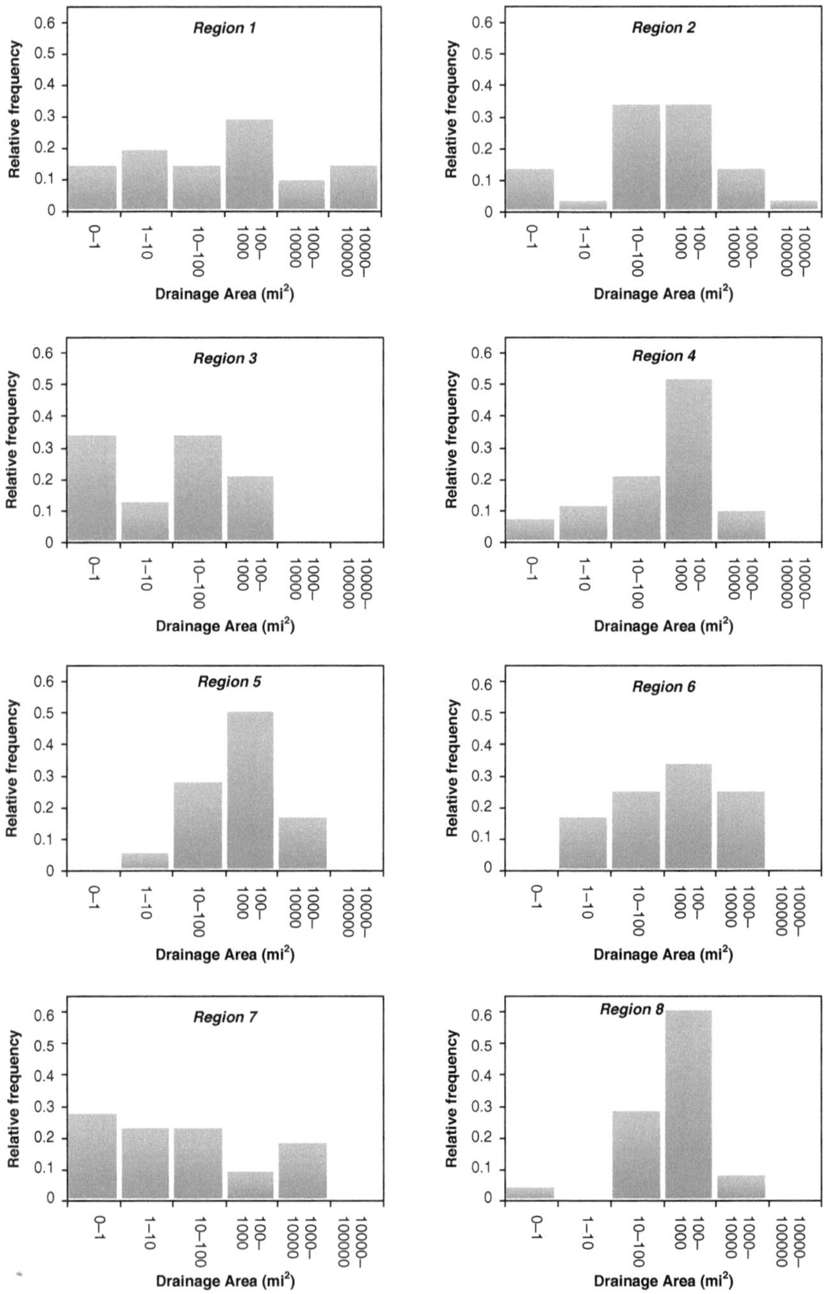

Fig. 5.4.2 Histographs of drainage areas for each region

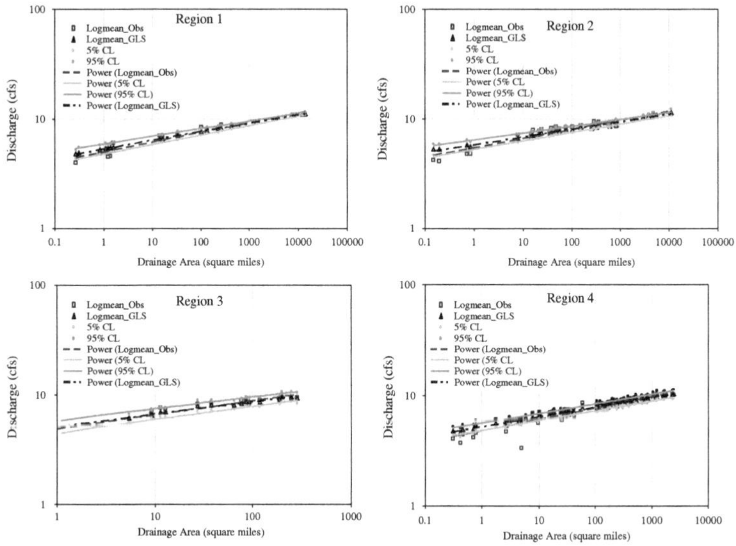

Fig. 5.4.3(a) At-site logarithms of mean annual peak flows compared with GLS regression results and the 95% upper and lower confidence limits for Regions 1 to 4

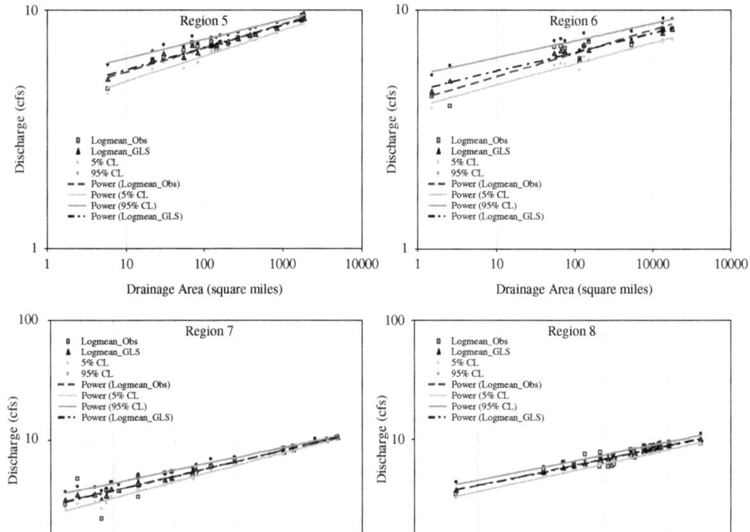

Fig. 5.4.3(b) At-site logarithm of mean annual peak flows compared with GLS regression results and the 95% upper and lower confidence limits for Regions 5 to 8

5.5 Comparative Analysis

There are three methods by which the procedures discussed in previous sections can be used to estimate flood quantiles. In the first method, the normalized regional quantiles can be used with the observed mean value of annual maximum flows at a site to compute the flood quantiles for specified recurrence intervals. In the second method, the mean annual maximum flows are estimated by GLS regression relationships based on hydrological characteristics at a site such as the watershed area and stream slope. These mean values are used with the normalized regional quantiles to estimate the flood quantiles. In the third method, the equations for quantiles derived by the GLS method are used directly to obtain flood magnitudes.

Split sample test is used to estimate the errors associated with each of these methods. The data from each region is divided into two parts. The first part known as calibration set has data from 75% of the watersheds in the region, whereas the second part known as validation set is chosen to have data from 25% of the watersheds that are selected to reflect the distribution of areas of watersheds in the region. For each method of flood quantile estimation, the calibration set is used to estimate the parameters of equations used for that method. The data from watersheds in the validation set are given as input to the equations developed using calibration set to estimate flood quantiles. The flood quantiles estimated for each of the sites in the validation set are compared to their respective at-site estimates to determine the error associated with the method. The procedures used in the three methods are schematically shown in Fig. 5.5.1.

5.5.1 Split Sample Test for the First Method

For each region the normalized regional quantiles are computed by regional L-moment method based on peak flow data of all the watersheds in the calibration set. Flood quantiles corresponding to various recurrence intervals are computed at each site in the validation set by multiplying the normalized regional quantiles with the index flood (i.e., first moment of the observed annual peak flows) for the site. Following this, the regional L-moment based flood estimates are compared to the at-site estimates of flood quantiles for all sites in the validation set. The measure to evaluate the error is the variance calculated by Eq. (5.2.4). The numbers of stations in regions 1–8 are 21, 30, 24, 72, 18, 12, 22 and 25 respectively and the number of stations in the validation sets prepared for these eight regions are 5, 8, 6, 18, 5, 3, 6 and 6 respectively. The split sample analysis is valid only in homogeneous or possibly homogeneous regions. The data from region 6, which is heterogeneous, is included only for the sake of completeness.

Typical results of the split sample validation are presented in Figs. 5.5.2 and 5.5.3. Figure 5.5.2 shows plots with the at-site quantile estimates as abscissa and the quantile estimates from the regional L-moment method as ordinate. The variance

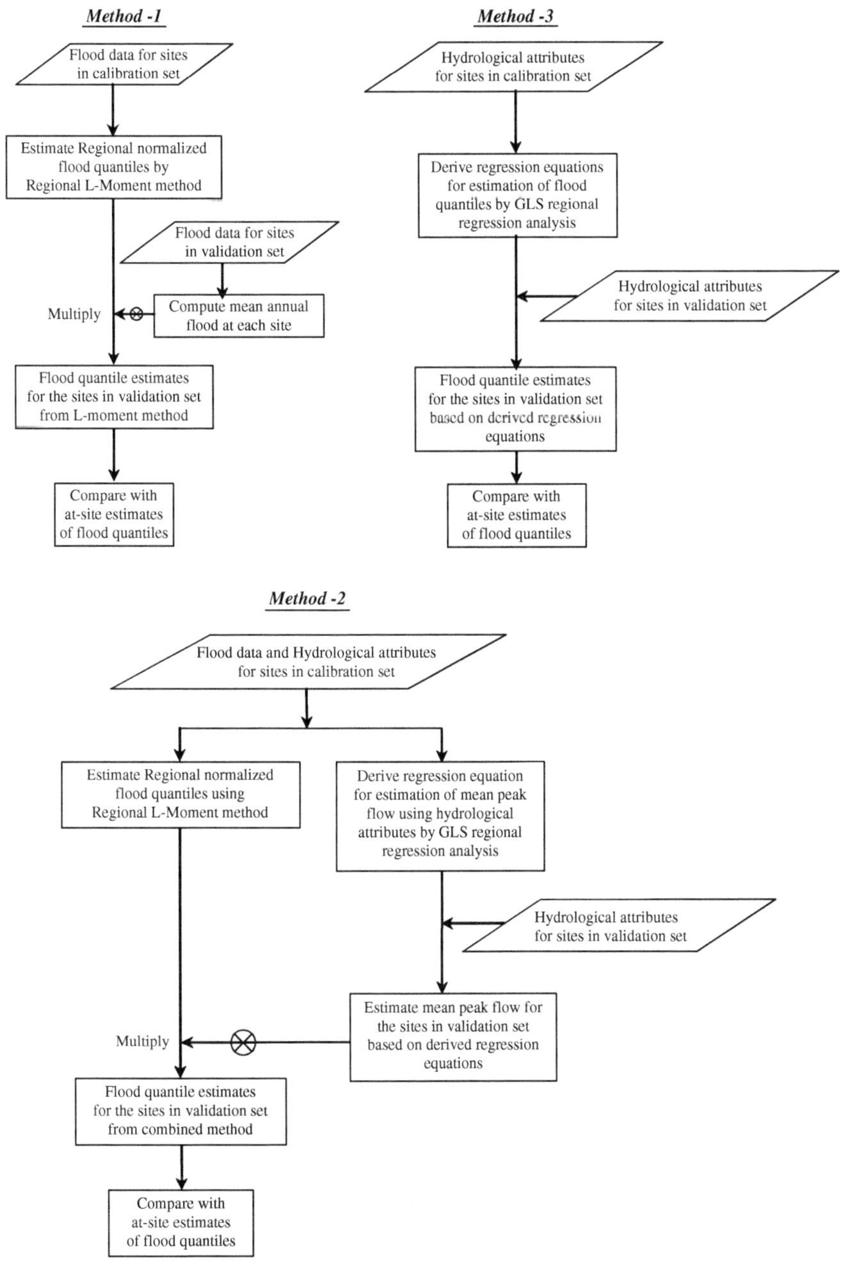

Fig. 5.5.1 Flowcharts of three methods considered for split sample test

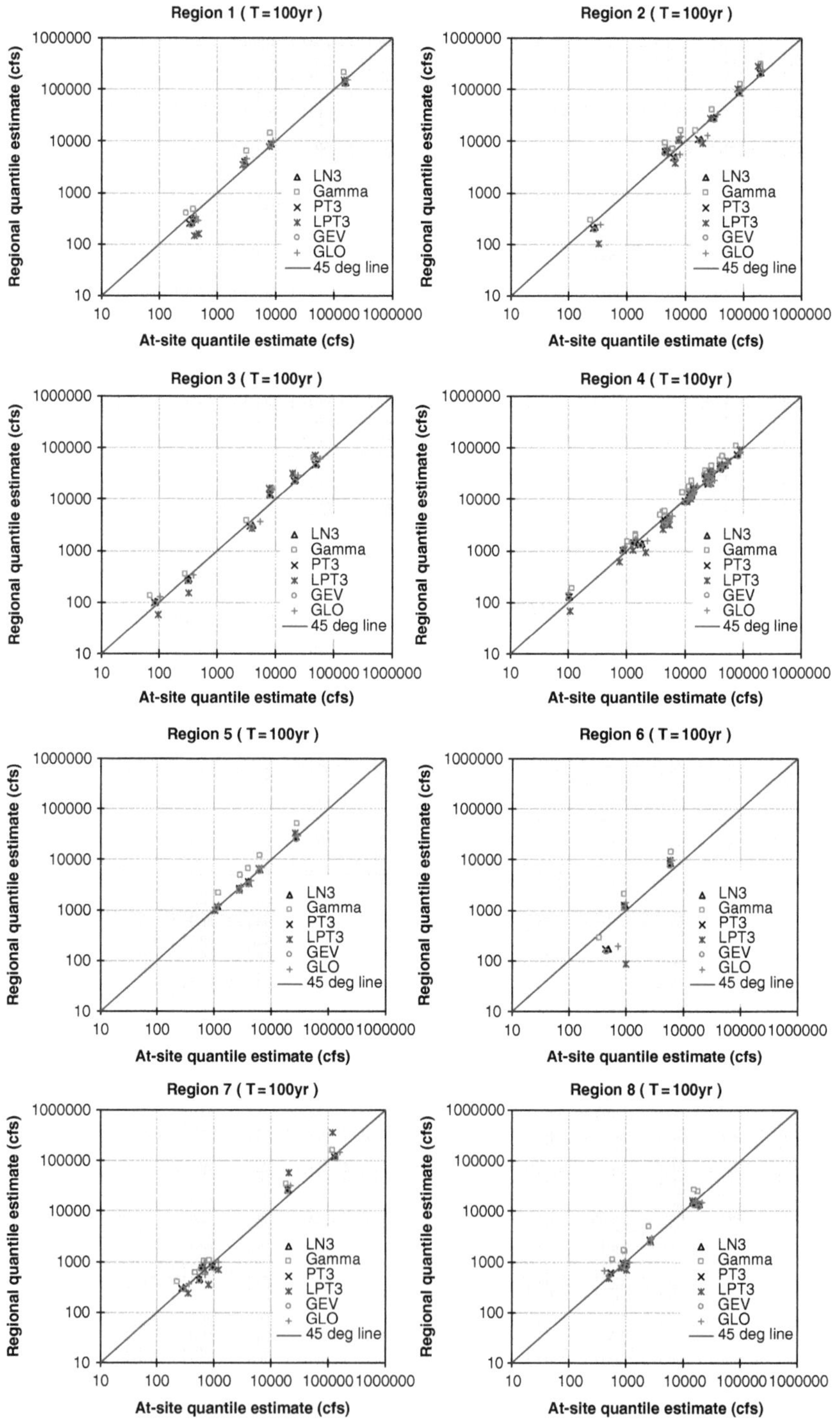

Fig. 5.5.2 Results of at-site and regional quantile floods from method 1 (T = 100 year). LP3 is shown as LPT3, and GM2 is shown as Gamma

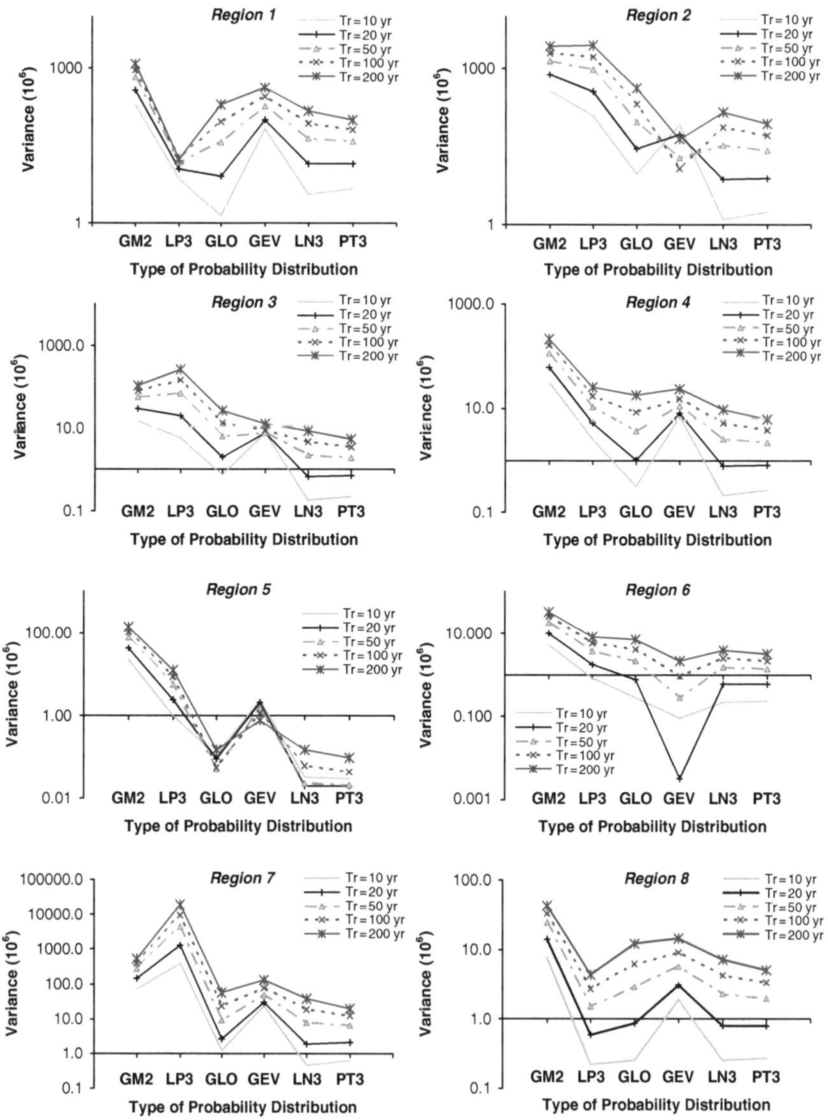

Fig. 5.5.3 Variance of the differences between at-site and regional estimates from method 1

of differences between at-site and regional estimates are shown in Fig. 5.5.3. These results show that by regional L-moment method Gamma distribution (GM2) overestimates the flood quantiles, whereas LP3 distribution underestimates them. The PT3, LN3 and GEV distributions produce consistent and better estimates. Also, results from Fig. 5.5.3 indicate that Gamma (GM2) and LP3 distributions are not good candidates for regional flood estimation. Results from PT3 distribution are

Table 5.5.1 Optimal probability distributions for regional flood estimates

Region No.	Candidate Probability Distributions	Optimal Distributions for Regional Estimates
1	PT3, GM2, LN3, GEV, LP3	LP3, PT3, LN3
2	GEV, LN3, PT3, GM2, GLO	GEV, PT3, LN3
3	GEV, LN3, LP3, GLO, PT3	PT3, LN3, GEV
4	GEV, LN3, LP3, PT3, GM2	PT3, LN3, GEV
5	GEV, LN3, LP3, PT3, GM2	PT3, LN3, GEV
6	LN3, PT3, GM2, GLO, GEV	GEV, PT3, LN3
7	PT3, GM2, LN3, GEV, LP3	PT3, LN3, GEV
8	LP3, GLO, GEV, LN3, PT3	LP3, PT3, LN3

Note: 1) Candidate probability distributions are determined from the mean-square-error for the 75% of data, and the order is beginning with the one having the minimum MSE. 2) Optimal distributions for regional estimates are obtained from the variances of L-moment regional estimates from the 25% of data.

better than those from other distributions. The optimal probability distributions for regional flood estimates from this method are summarized in Table 5.5.1. The results show that PT3 is the favored distribution for Regions 3, 4, 5 and 7, LP3 is preferred for Regions 1 and 8, and GEV is good for Regions 2 and 6. PT3 is acceptable for regions 1, 2, 6 and 8 as the second best distribution.

It is seen from Fig. 5.5.2 that both homogeneous and possibly homogeneous regions (1, 2, 3, 4, 5, 7, and 8) have flood quantile estimates close to 45 degree lines. On the other hand, the quantile estimates in region 6, which is heterogeneous, are farther away from the 45 degree line. It indicates that flood estimates for heterogeneous region are not accurate. Hence, once a region fails homogeneity tests, accurate estimation of flood quantile may not be possible.

The flood estimates are seen to deviate more from the 45-degree line with increase in recurrence interval to 200 years. This behavior is caused by extrapolation errors and because of short data lengths. Also, it is found that the prediction result is less stable for smaller flows which are mostly from small drainage areas. The hydrological responses from small watersheds are strongly affected by local events. Higher values of streamflow corresponds to larger drainage areas which bear more resemblance to the regional properties and are less influenced by local events. Similar conclusion is drawn from the results of methods 2 and 3.

5.5.2 Split Sample Test for the Second Method

In the second method GLS regional regression equations are developed between the first moments of annual peak flows and the characteristics of watersheds in the calibration set. Further, regional L-moment method is used for computing normalized regional flood quantiles for LN3, Gamma, PT3, LP3, GEV and GLO distributions based on observed peak flow data at the sites in the calibration set.

The mean annual peak flow (index flood) for each site in the validation set is computed by substituting its watershed characteristics in the developed regional regression equation. Subsequently, flood quantiles corresponding to various recurrence intervals are computed by multiplying the normalized regional quantiles (estimated by using data in calibration set) with the index flood computed for the site.

The flood quantiles computed for each of the sites in the validation set based on the second method are plotted against their at-site estimates for each of the six candidate distributions. Typical results presented in Fig. 5.5.4 show that PT3 distribution provides best estimates for Regions 3 and 5. The best estimates for Region 1 were obtained from GEV distribution and LP3 distribution gave the best estimates for Region 3.

For Regions 6, 7 and 8 the bias between at-site estimates and those based on the second method are found to be significant. The bias is considerable for small drainage areas in case of Regions 6 and 7. The bias in quantile estimates could be partly attributed to bias in estimate of index flood value for stations in these regions by GLS based regional regression equation. The correlation coefficients between hydrological attributes and the index flood value for sites in these regions are shown in Table 5.4.1.

5.5.3 Split Sample Test for the Third Method

To examine the accuracy of floods estimated by GLS regression method, the regional regression equations are developed between flood quantiles and the attributes of watersheds in the calibration set. This involves computation of the coefficients and exponents (for example, a', b', c', d' from equation $Q_T = a' A^{b'} S^{c'} (1 + W\%)^{d'}$).

For each site in the validation set the flood quantiles are estimated by inserting its watershed characteristics in the developed regional regression equations. Points are plotted in Fig. 5.5.5 using flood quantile estimate based on GLS regression equation as ordinate and the at-site quantile estimate as abscissa. If the points approach 45-degree line, it indicates a better capability to predict. In most cases GLS regression and PT3 quantile floods are in good agreement, except for some outlier points in Regions 6, 7 and 8. Similar results were obtained in the case of fitting flood quantiles based on GEV distribution. The errors noted for Region 2 (besides outliers in regions 6, 7 and 8) were more with LP3 distribution, than with PT3 and GEV distributions.

In summary, the third method does not indicate too many differences among PT3, GEV and LP3 probability distributions because the result of GLS regression is dominated by the correlation between watershed attributes and flood quantiles. The second method shows that both PT3 and GEV distributions exhibit similar performance, whereas LP3 yields poor result. Except for Region 6, the results from either GLS regression or combination method are quite reliable and follow the trend well. Region 7 may have more estimation errors for drainage areas less than 1000 square miles and Region 8 has more error for the drainage areas in the range of 500–5000 square miles.

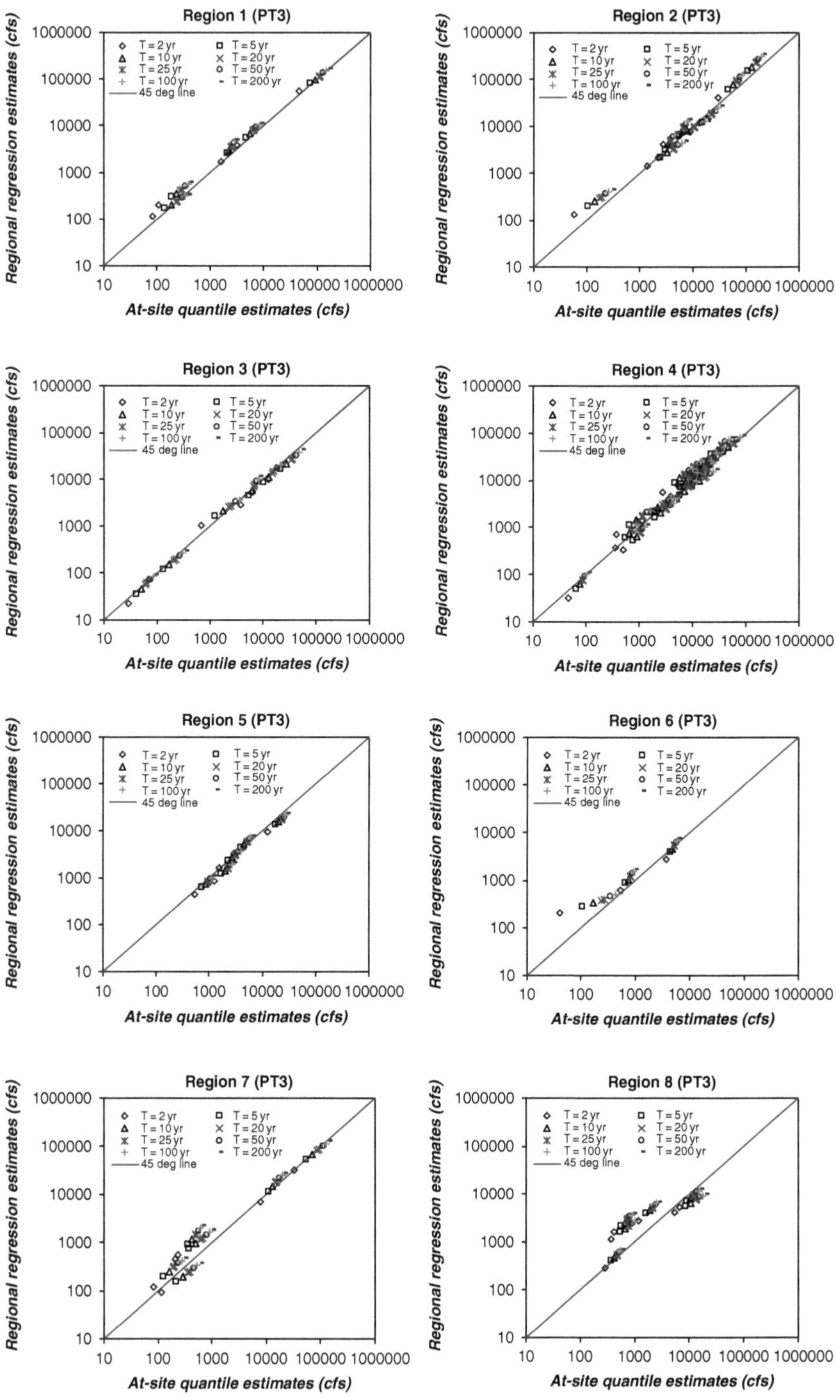

Fig. 5.5.4 At-site quantile floods and the quantile floods obtained by method 2 for the validation set with PT3 distribution

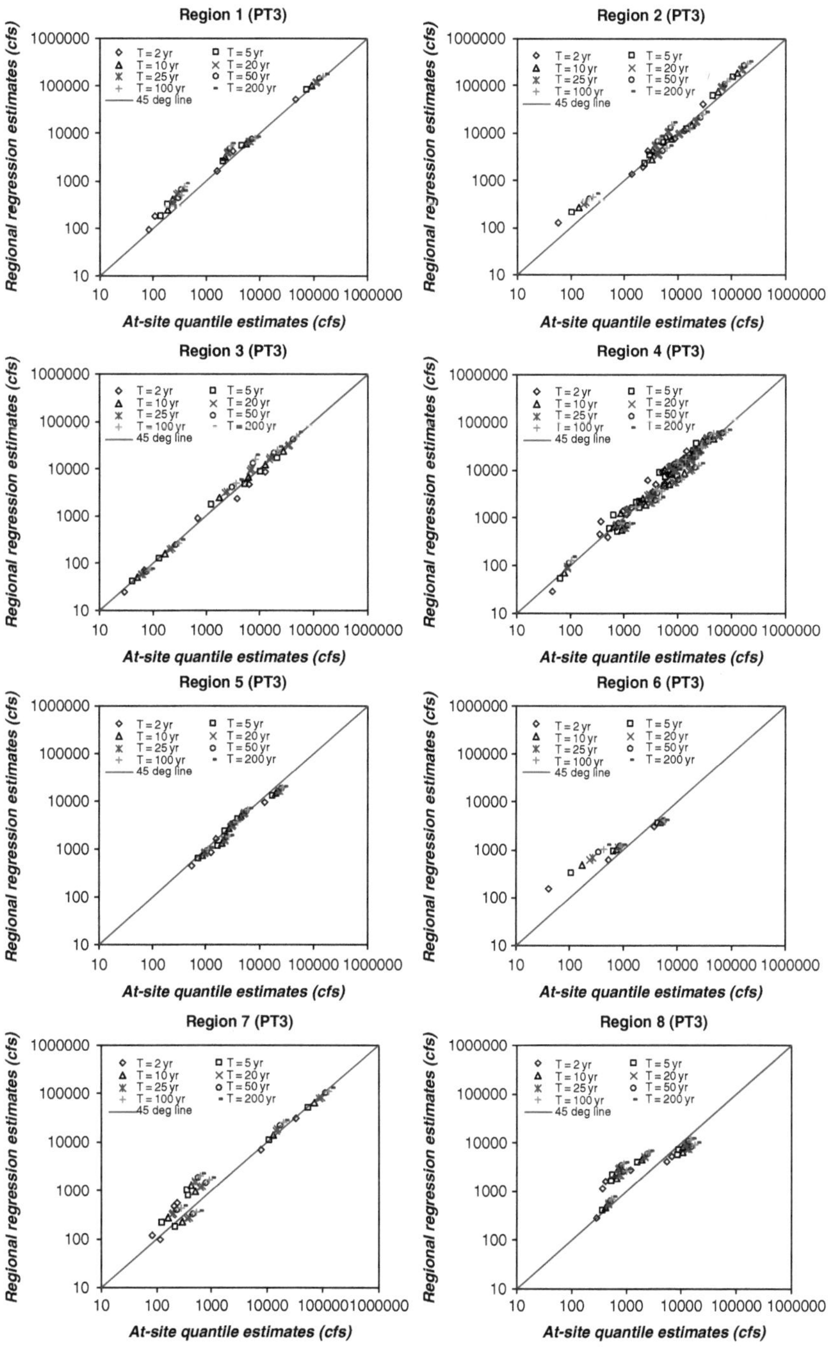

Fig. 5.5.5 At-site quantile floods and the quantile floods obtained by method 3 for the validation set with PT3 distribution

5.5.4 Comparison of the Three Methods

For each region, the average error between quantile estimates based on at-site and regional analysis is quantified as

$$e_j^{D,T,M}\,(\%) = \sum_{i=1}^{N_j} \frac{\left|Q_{AS}^{D,T}(i) - Q_M^{D,T}(i)\right|}{Q_{AS}^{D,T}(i)} \cdot DAR\,(i) \qquad j = 1, \ldots, K \tag{5.5.1}$$

where $e_j^{D,T,M}$ is the average error (in percentage) for region j with method M, probability distribution D (PT3, GEV or LP3), and recurrence interval T (2, 5, 10, 20, 25, 50, 100 or 200 years). N_j is the number of stations in Region j. $Q_{AS}^{D,T}(i)$ is the T-year flood quantile estimated at site i by using at-site frequency analysis with distribution D. $Q_M^{D,T}(i)$ is the T-year flood quantile estimated at site i by method M and distribution D. $DAR(i)$ is the drainage area ratio, which is ratio of the drainage area at site i divided by the sum of drainage areas at all the sites in the region.

The average error is computed by weighting error estimate at each site by its drainage area because it is not reasonable to give same weightage for errors from small and large drainage areas due to differences in their flood magnitudes. The percentage errors from small drainage areas are always larger and lead to misinterpretation if weights are not applied.

The average errors calculated for the calibration and validation data sets by each of the three methods of flood quantile estimation are presented in Tables 5.5.2 and 5.5.3. In general, the average errors estimated for calibration data set are smaller than those computed for validation data set over all the return periods. Further, for most of the regions flood quantiles are accurately predicted by the regional L-moment method (Method 1) than by GLS regression method (Method 3) and combination method (Method 2). Results with PT3 distribution are better than those from GEV and LP3 distributions. The performance of PT3 distribution is closely followed by that of GEV distribution. Except region 6, all the regions have less than 10% average errors with best fitting distribution in calibration. The poor result of region 6 is expected, because it is a heterogeneous region.

For Method 2, in which GLS regression is used to determine mean annual peak flow (index flood) from geographical attributes of watersheds, the average errors in quantile estimates are found to be more than those computed for Method 1. The error percentages are mostly between those estimated for methods 1 and 3.

As for the GLS regional regression method (Method 3), calibration and validation results show that PT3 distribution is the preferred distribution. In terms of performance, the PT3 distribution is closely followed by the GEV distribution. For regions 1, 4, 5, and 7 the average error noted in calibration for Method 3 is in the range 10–21%. Validation results suggest that GEV performs well for region 1, and PT3 is preferred distribution for regions 2, 3, 4, 5 and 6, whereas LP3 is preferred distribution for regions 7 and 8. For regions 1 and 7 error with PT3 is less than 11%, the same for regions 3, 4, 5 and 6 is in the range 16–26%.

Table 5.5.2 Errors obtained from calibration of the three methods of flood quantile estimation with PT3, GEV and LP3 distributions. Stdev is the standard deviation of errors

Method		Method 1			Method 2			Method 3		
Region	T (yrs)	PT3	LP3	GEV	PT3	LP3	GEV	PT3	LP3	GEV
1	2	1.23	1.8	31.52	11.4	20.31	23.94	8.09	8.33	7.01
	5	1.58	9.4	22.07	11.29	13.85	13.42	12.39	13.06	9.66
	10	1.74	15.1	18.25	11.4	26.33	9.24	14.54	17.19	12.46
	20	2	19.8	15.5	11.54	34.77	6.95	16.14	21.33	14.86
	25	2.22	21.1	14.76	11.59	36.96	6.82	16.46	22.66	15.68
	50	2.83	25.2	12.77	11.73	42.63	6.28	17.56	26.85	18
	100	3.37	28.9	11.15	11.87	47.07	5.6	18.39	30.97	20.19
	200	3.85	32.3	9.8	12.01	50.66	5.12	19.04	35.06	22.49
	Average	2.35	19.2	16.98	11.60	34.07	9.67	15.33	21.93	15.04
	Stdev	0.92	10.2	7.06	0.25	13.01	6.35	3.62	8.99	5.23
2	2	2.02	2.0	35.38	43.58	59.52	9.74	41.79	38.61	38.67
	5	1.24	9.5	23.14	45.81	15.37	15.31	45.59	41.77	42.29
	10	2.24	15.9	17.89	47.27	9.18	21.79	47.49	45.2	44.5
	20	3.14	21.8	13.98	48.5	19.68	27.23	48.76	49.01	46.64
	25	3.49	23.7	12.92	48.86	22.74	28.76	49.13	50.38	47.41
	50	4.45	29.4	9.97	49.88	30.62	33.09	50.09	54.61	49.55
	100	5.29	35.0	7.49	50.78	36.79	37.01	50.81	59.14	51.96
	200	6.03	40.6	5.37	51.59	41.75	40.66	51.47	63.92	54.47
	Average	3.49	22.2	15.77	48.28	29.46	26.70	48.14	50.33	46.94
	Stdev	1.67	12.9	9.75	2.66	16.30	10.61	3.18	8.61	5.14
3	2	6.82	5.4	45.18	27.64	30.94	49.06	27.73	27.16	41.39
	5	1.87	13.1	25.97	22.5	26.5	30.74	21.9	23.41	29.84
	10	5.08	23.1	18.24	19.99	33.22	26.63	20.71	20.62	24.46
	20	8.33	33.2	14.06	18.2	44.27	23.26	23.85	21.93	19.93
	25	9.29	36.5	13	17.7	47.12	22.34	25.13	23.81	19.5
	50	11.95	47.6	14.52	18.12	54.45	19.51	28.8	32.42	25.92
	100	14.23	60.0	18.08	19.92	60.07	19.08	31.96	41.84	33.49
	200	16.22	73.1	22.16	21.53	64.48	23.32	34.79	52.01	42.52
	Average	9.22	36.5	21.40	20.70	45.13	26.74	26.86	30.40	29.63
	Stdev	4.77	23.1	10.55	3.28	14.04	9.77	4.90	11.16	8.92
4	2	4.36	3.4	31.64	25.27	35.01	27.67	30.05	31.34	42.25
	5	1.77	8.9	19.21	23.71	16.21	17.09	22.29	25.86	28.56
	10	3.45	14.5	14.15	22.71	25.4	14.13	18.45	22.37	23.11
	20	5.48	20.0	11.21	22.32	34.84	13.77	15.66	19.22	18.59
	25	6.07	21.9	10.56	22.44	37.4	13.76	14.98	18.35	17.26
	50	7.68	27.5	9.76	22.78	43.94	14.27	13.42	16.58	13.86
	100	9.07	32.9	9.72	23.08	48.94	15.61	12.8	15.8	12.2
	200	10.31	38.1	11.19	23.75	52.91	17.16	12.48	16.48	12.51
	Average	6.02	20.9	14.68	23.26	36.83	16.68	17.52	20.75	21.04
	Stdev	2.88	11.8	7.54	0.97	12.07	4.65	6.04	5.47	10.24
5	2	3.25	2.9	24.81	17.86	17.31	28.54	17.26	17.49	17.4
	5	1.15	8.4	16.4	16.18	26.06	23.12	17.2	18.12	17.73
	10	3.56	14.1	12.22	14.59	32.28	20.5	16.38	16.95	16.8
	20	5.7	19.3	8.74	13.01	36.94	18.2	15.45	15.43	15.54
	25	6.35	20.9	7.71	13.04	38.2	17.5	15.25	15.02	15.15
	50	8.26	25.6	4.81	13.26	41.5	15.46	14.65	13.79	14.28

Table 5.5.2 (continued)

Method		Method 1			Method 2			Method 3		
Region	T (yrs)	PT3	LP3	GEV	PT3	LP3	GEV	PT3	LP3	GEV
	100	10.14	30.1	6.99	13.45	44	13.53	14.08	12.58	13.35
	200	11.98	34.5	9.88	13.76	45.96	12.66	13.5	12.97	13.17
	Average	6.30	19.5	11.45	14.39	35.28	18.69	15.47	15.29	15.43
	Stdev	3.68	10.7	6.45	1.76	9.69	5.28	1.39	2.09	1.77
6	2	7.54	1.7	40.14	23.13	20.36	41.69	17.00	15.25	15.80
	5	8.49	20.0	22.13	18.48	25.91	31.39	23.28	22.89	23.35
	10	18.1	33.1	10.57	20.78	28.19	25.58	24.97	26.26	25.76
	20	26.72	45.3	4.67	25.75	30.55	19.61	25.81	28.57	26.95
	25	29.38	49.0	8.21	28.22	31.18	19.37	25.98	29.16	27.19
	50	37.48	60.9	20.36	35.56	32.82	21.23	26.3	30.52	27.43
	100	45.1	72.7	33.04	42.48	34.09	29.11	26.43	31.36	27.09
	200	52.35	84.2	46.32	49.08	35.11	41.19	26.44	31.75	26.29
	Average	28.15	45.9	23.18	30.44	29.78	28.65	24.53	26.97	24.98
	Stdev	16.37	27.3	15.38	10.92	4.86	9.00	3.22	5.56	3.94
7	2	6.18	3.0	31.38	18.56	36.05	25.22	16.18	14.87	23.64
	5	2.67	38.3	17.66	15.46	11.09	17.61	11.53	11.91	17.55
	10	3.21	61.1	14.48	12.37	27.75	12.88	9.27	9.28	13.5
	20	5.93	81.9	13.68	9.88	41.12	9.76	7.76	7.27	9.74
	25	6.66	88.3	13.73	9.2	44.68	8.85	7.37	6.7	8.58
	50	8.61	108.1	14.73	8.37	53.97	10.54	8.01	10.79	5.99
	100	10.17	127.6	17.01	9.76	61.23	13.82	9.65	15.15	9.29
	200	11.45	147.3	19.79	10.91	67.02	17.47	11.03	19.55	13.72
	Average	6.86	81.9	17.81	11.81	42.86	14.52	10.10	11.94	12.75
	Stdev	3.10	47.3	5.89	3.52	18.26	5.43	2.88	4.38	5.70
8	2	2.06	2.2	27.77	39.91	40.12	28.95	40.84	39.02	41.39
	5	2.35	6.4	19.97	39.76	31.07	32.3	39.85	38.5	40.63
	10	4.09	8.8	17.09	39.89	28.7	33.82	39.41	37.38	39.56
	20	5.48	11.2	15.24	40.29	31.74	34.91	39.3	37.02	39.39
	25	5.86	12.0	14.79	40.39	33.66	35.2	39.24	36.85	39.31
	50	6.91	14.2	14.22	40.6	38.77	35.95	39.04	36.25	38.98
	100	7.84	16.2	14.22	40.77	42.91	36.52	38.86	35.71	38.69
	200	8.67	18.0	14.43	41.02	46.35	36.93	38.79	35.14	38.55
	Average	5.41	11.1	17.22	40.33	36.67	34.32	39.42	36.98	39.56
	Stdev	2.43	5.2	4.70	0.45	6.29	2.63	0.67	1.32	0.98

Table 5.5.3 Errors obtained from validation of the three methods of flood quantile estimation with PT3, GEV and LP3 distributions. Stdev is the standard deviation of errors

Method		Method 1			Method 2			Method 3		
Region	T (yrs)	PT3	LP3	GEV	PT3	LP3	GEV	PT3	LP3	GEV
1	2	4.4	2.1	27.2	15.07	25.33	20.61	11.84	12.48	10.63
	5	2.0	5.6	21.1	10.34	13.71	11.74	10.35	11.22	10.11
	10	5.2	6.5	20.0	7.89	28.04	9.69	9.11	11.46	9.32
	20	7.6	6.8	20.1	5.97	37.64	9.31	7.99	12.23	8.21
	25	8.3	6.7	20.3	5.44	40.12	9.36	7.55	12.55	7.88
	50	10.2	6.6	21.2	3.97	46.53	9.93	6.56	13.66	6.73

Table 5.5.3 (continued)

Method		Method 1			Method 2			Method 3		
Region	T (yrs)	PT3	LP3	GEV	PT3	LP3	GEV	PT3	LP3	GEV
	100	11.8	6.3	22.6	2.74	51.53	10.94	5.72	14.8	5.62
	200	13.2	5.8	24.1	1.68	55.56	12.26	4.9	15.94	4.51
	Average	7.8	5.8	22.1	6.64	37.31	11.73	8.00	13.04	7.88
	Stdev	3.8	1.6	2.5	4.39	14.22	3.75	2.35	1.64	2.15
2	2	4.3	3.9	36.8	33.42	52.36	14.99	37.03	36.06	36.12
	5	1.0	10.5	24.5	44.24	10.58	11.52	40.15	37.21	37.67
	10	2.7	20.4	18.3	50.89	7.64	23.19	42.49	40.02	40.02
	20	4.6	30.5	13.0	56.64	17.04	33.89	44.71	43.84	42.91
	25	5.4	33.8	11.3	58.38	19.38	37.31	45.37	45.28	44.1
	50	7.4	44.4	6.4	63.34	25.25	47.85	47.2	49.79	47.95
	100	9.3	55.1	2.0	67.74	29.62	58.49	48.74	54.62	52.31
	200	11.0	66.3	6.7	71.71	32.97	69.43	50.19	59.71	56.98
	Average	5.7	33.1	14.9	55.80	24.36	37.08	44.49	45.82	44.76
	Stdev	3.4	21.5	11.4	12.64	14.31	20.58	4.43	8.40	7.25
3	2	2.6	4.8	44.8	17.79	7.83	47.48	29.67	27.69	44.49
	5	1.7	9.2	29.2	17.78	35.6	37.11	19.03	23.28	33.37
	10	3.7	19.9	22.2	17.36	46.8	32.69	12.3	16.59	24.59
	20	5.4	30.9	16.5	17.58	54.53	29.33	11.14	12.48	16.52
	25	5.8	34.5	15.7	17.86	56.52	28.38	11.29	12.55	14.68
	50	7.2	46.3	14.0	18.67	61.65	26.4	11.88	12.81	12.36
	100	8.5	58.7	13.0	19.61	65.63	27.09	16.18	20.56	16.71
	200	9.6	71.7	13.7	20.55	68.81	28.19	22.36	35.5	33.3
	Average	5.6	34.5	21.1	18.40	49.67	32.08	16.73	20.18	24.50
	Stdev	2.8	23.5	11.0	1.13	19.97	7.14	6.64	8.33	11.49
4	2	2.8	2.1	29.2	26.93	37.76	20.45	38.4	38.47	50.96
	5	1.4	5.9	19.2	29.19	17.38	17.94	25.27	28.86	32.75
	10	3.0	9.0	15.6	30.21	25.24	19.3	20.99	22.87	25.21
	20	4.3	11.3	13.2	31	33.95	21.38	20.58	22.11	22.16
	25	4.7	12.0	12.6	31.23	36.51	22.07	20.5	22.05	21.86
	50	5.7	13.7	11.3	31.86	43.57	24.15	20.3	21.95	21.23
	100	6.7	15.0	12.4	32.41	49.02	26.15	20.2	21.96	21.05
	200	7.6	16.1	13.7	32.89	53.37	28.11	20.23	22.06	21.01
	Average	4.5	10.7	15.9	30.72	37.10	22.44	23.31	25.04	27.03
	Stdev	2.1	4.8	5.9	1.93	11.89	3.47	6.33	5.92	10.45
5	2	1.7	1.4	24.3	20.9	20.12	33.11	20.42	20.7	20.69
	5	2.2	5.5	18.6	21.02	34.58	27.92	22.55	23.48	22.98
	10	2.0	9.6	15.7	20.74	41.51	26.5	23.07	24.08	23.39
	20	1.6	13.3	13.4	20.57	45.76	25.62	23.39	24.41	23.59
	25	1.4	14.3	12.6	20.51	46.79	25.36	23.48	24.44	23.61
	50	1.1	17.7	10.6	20.29	49.29	24.59	23.62	24.4	23.54
	100	1.9	20.8	8.7	20.06	51.03	23.86	23.72	24.59	23.36
	200	2.6	23.7	6.9	19.82	52.25	23.17	23.76	25.02	23.09
	Average	1.8	13.3	13.8	20.49	42.67	26.27	23.00	23.89	23.03
	Stdev	0.5	7.6	5.6	0.41	10.74	3.14	1.12	1.36	0.97
6	2	6.8	0.2	41.2	22.30	19.13	47.86	15.49	13.56	13.06
	5	7.8	19.4	22.5	16.49	26.82	36.78	22.08	21.17	20.01
	10	16.9	31.3	10.9	12.93	30.53	30.43	25.02	26.14	23.91
	20	25.2	42.0	1.3	9.75	33.83	25.21	27.11	30.24	27.07

Table 5.5.3 (continued)

Method		Method 1			Method 2			Method 3		
Region	T (yrs)	PT3	LP3	GEV	PT3	LP3	GEV	PT3	LP3	GEV
	25	27.7	45.3	4.4	8.79	34.74	23.72	27.68	31.41	28
	50	35.1	55.3	15.6	5.95	37.13	19.14	29.18	34.54	30.49
	100	42.0	64.9	27.3	3.3	39.05	14.67	30.42	37.15	32.66
	200	48.5	74.2	39.5	5.81	40.63	10.2	31.47	39.28	34.62
	Average	26.2	41.6	20.3	10.67	32.73	26.00	26.06	29.19	26.23
	Stdev	15.3	24.3	15.1	6.31	7.09	12.22	5.22	8.60	7.08
7	2	3.7	2.8	38.0	3.18	11.54	33.35	5.96	7.46	8.9
	5	1.8	38.8	22.6	4.68	23.14	22.7	7.42	8.68	9.36
	10	3.0	70.5	17.0	8.47	36.12	20.06	9.33	8.48	10.02
	20	5.2	104.1	13.5	11.5	45.24	19.26	10.66	7.73	11.64
	25	5.8	115.4	13.6	12.34	47.69	19.25	11.03	7.36	12.14
	50	7.5	152.6	14.7	14.67	54.13	20.84	11.98	5.65	13.59
	100	9.0	193.3	16.5	16.63	59.25	23.7	12.81	5.11	15.04
	200	10.4	237.9	18.7	18.31	63.44	26.82	13.49	9.58	16.45
	Average	5.8	114.4	19.3	11.22	42.57	23.25	10.34	7.51	12.14
	Stdev	3.0	78.5	8.1	5.45	17.98	4.83	2.62	1.51	2.73
8	2	3.3	1.3	26.9	68.64	70.06	61.28	68.22	68.86	67.8
	5	2.4	3.9	22.0	73.1	65.75	66.4	72.13	71.66	70.71
	10	4.6	6.7	20.5	75.43	63.86	69.28	74.28	73.18	72.79
	20	6.4	9.2	19.8	77.29	62.45	71.87	75.97	74.48	74.85
	25	6.8	10.0	19.6	77.81	62.06	72.68	76.44	74.86	75.5
	50	8.2	12.3	19.3	79.29	61.02	75.13	77.72	76.02	77.54
	100	9.2	14.4	19.2	80.58	60.16	77.48	78.8	77.09	79.57
	200	10.2	16.4	19.3	81.73	59.44	79.75	79.72	78.13	81.62
	Average	6.4	9.3	20.8	76.73	63.10	71.73	75.41	74.29	75.05
	Stdev	2.8	5.1	2.6	4.28	3.46	6.02	3.79	3.01	4.58

For region 2 error with PT3 is 45% and for region 8 it is as high as 75%. The high error from region 8 could be attributed to poor correlation between flood quantiles and drainage area which makes the GLS regression equation not as reliable as for other regions. Nevertheless, the average errors computed for the method 3 are, in general, found to be more than those computed for the other two methods.

Overall, the average errors for LP3 distributions are quite high with all the three methods. Hence it can be concluded that LP3 is not preferable. Further, method 2 should be preferred less because it embeds the error from methods 1 and 3 and makes the result unreliable.

5.6 Simple Scaling in Regionalized Watersheds

In general, scaling implies that the properties associated with a process at different scales are related to each other by a transformation involving only the scale ratio between them.

Consider a regional flood process $\{Q(X)\}$ indexed on a parameter set X which characterizes the statistical spatial structure of peak flows. They are said to be simple scaling if the following equality holds:

$$\left\{\frac{Q(\lambda X)}{\lambda^{\theta}}\right\} \underline{\underline{d}} \{Q(X)\} \tag{5.6.1}$$

where $\lambda > 0$ is a scale parameter and θ is a scaling exponent; the equality is in the sense of probability distribution. Gupta et al. (1994) suggest taking X as representing the channel network in drainage basin. Since channel networks are proportional to drainage areas, X might be taken simply as the drainage area A. Parametric representation of peak flows by the drainage areas may seem insufficient, but is justified by the important preponderance of basin size in explaining variance of statistics related to flood peak discharges (Ribeiro and Rousselle, 1996). Assuming $X = 1$ and $\lambda = A$, Eq. (5.6.1) can be rewritten as

$$\left\{\frac{Q(A)}{A^{\theta}}\right\} \underline{\underline{d}} \{Q(1)\} \tag{5.6.2}$$

where $\{Q(1)\}$ represents the peak flows generated by an hypothetical basin with unit drainage area.

If the mean of peak flows is considered as a deterministic function $\mu = A^{\theta}$ and if $\{Q(1)\}$ corresponds to the dimensionless flood frequency curve, the index flood assumption is equivalent to simple scaling. Thus the above equation shows how the index flood assumption is closely related to simple scaling. The equality in distribution given by Eq. (5.6.2) is referred to as simple scaling.

Ribeiro and Rousselle (1996) obtained the following relationship for simple scaling using statistical moments,

$$\frac{E\left[Q\left(A\right)\right]^{h}}{A^{\theta\cdot h}} = E\left[Q\left(1\right)\right]^{h} \tag{5.6.3}$$

where h is the order of the statistical moments. This expression can be rewritten using log transform to show that h is proportional to the slope of the log–log trendline.

$$\log\left\{E\left[Q\left(A\right)\right]^{h}\right\} = \theta \cdot h \cdot \log\left(A\right) + \log\left\{E\left[Q\left(1\right)\right]^{h}\right\}. \tag{5.6.4}$$

Using the expression in Eq. (5.6.4), a relationship between any statistical moment $E[Q(A)]^{h}$ of order h, and A can be written as,

$$E\left[Q\left(A\right)\right]^{h} = Y \cdot A^{\theta\cdot h} \tag{5.6.5}$$

where Y is a coefficient that denotes the intercept of the power law. Y and θ can be determined through a simple regression analysis. For simple scaling to be valid for watersheds in a region, θ should be a constant. In the following discussion, only first three moments are used. The first moment, $E[Q(A)]$, is the mean. It is defined as,

$$E\left[Q\left(A\right)\right] = \mu_Q = \frac{1}{n}\sum_{i=1}^{n} Q_i. \tag{5.6.6}$$

where n is the sample size of peak flows at the site. The second central moment, $E[Q(A) - \mu_Q]^2$, is defined as,

$$E\left[Q\left(A\right) - \mu_Q\right]^2 = \sigma_Q^2 = \frac{1}{n-1}\sum_{i=1}^{n}\left(Q_i - \mu_Q\right)^2 \tag{5.6.7}$$

The third central moment, $E[Q(A) - \mu_Q]^3$, is defined as,

$$E\left[Q\left(A\right) - \mu_Q\right]^3 = \frac{n}{\left(n-1\right)\left(n-2\right)}\sum_{i=1}^{n}\left(Q_i - \mu_Q\right)^3 \tag{5.6.8}$$

The relationships between the first three conventional statistical moments of observed peak flows and the basin areas are developed for watersheds in each region of Indiana formed using SOFM in Chapter 4 (Fig. 4.3.15). Using these relationships, the first three moments can be estimated for ungauged watersheds.

The first, second and third moments estimated from observed peak flows at all the sites in a region are plotted against the basin areas of the respective sites (Fig. 5.6.1). It can be inferred from the results summarized in Table 5.6.1 that for regions 1 through 5, slopes of the regression lines plotted for second moment are approximately twice that of the first moment, and the slopes of the regression lines plotted for third order moment are nearly triple that of the first order moment. The log-linearity between the sample moments and basin area shows that the annual floods in a region scale with basin area. Nevertheless, the correlation for these relationships weakens as the moment order increases.

It is seen from Fig. 5.6.1 that for region 6 the magnitude of the regression coefficient (R^2) is very low for the regression line fitted between second and third order moments and basin areas. The estimated R^2 values for the second and third order moments for the region 6 are 0.6802 and 0.4518 respectively. This is due to the non homogeneity of Region 6. Therefore, in general, it can be concluded that simple scaling using statistical moments is valid for the homogeneous Regions 1–5, and not valid for Region 6 which is heterogeneous. Similar conclusions are drawn for regions obtained by other regionalization procedures discussed in previous chapters.

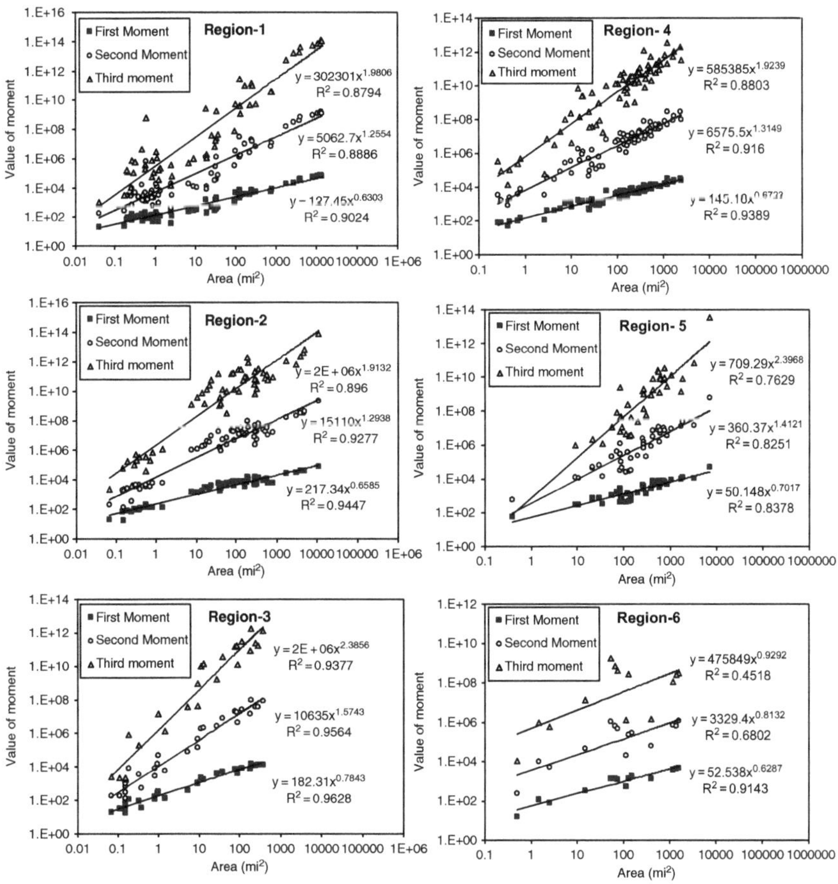

Fig. 5.6.1 Log–log plot prepared between conventional sample moments of annual peakflows and basin area for regions determined using SOFM clustering

5.7 Probability Distributions for Flood Frequency Analysis in Regionalized Watersheds

In flood frequency analysis, an assumed probability distribution is fitted to the available data to estimate the flood magnitude for a specified return period. The choice of an appropriate probability distribution is quite arbitrary, as no physical basis is available to rationalize the use of any particular distribution. The type of error which is associated with the wrong assumption of a particular distribution for the given data can be checked to a certain extent by using goodness-of-fit tests. These are statistical tests which may be used to evaluate the adequacy of distributions.

Even if an acceptable distribution is selected, proper estimation of parameters is important. Some of the parameter estimation methods may not yield good estimates, or may not even converge. Therefore, information about the parameter estimation method is also useful in practice.

Table 5.6.1 Characteristics of moments for regions 1 to 6. The 'ratio of slopes' for a region denotes ratio of slope of second (or third) moment to that of first moment for the region

Region number	Moment number	Intercept	Slope	R^2	Ratio of slopes
1	First	127.45	0.6303	0.902	–
	Second	5062.7	1.2554	0.889	1.99
	Third	302301	1.9806	0.879	3.14
2	First	217.34	0.6585	0.945	–
	Second	15110	1.2938	0.928	1.96
	Third	2.0E+06	1.9132	0.896	2.90
3	First	182.31	0.7843	0.963	–
	Second	10635	1.5743	0.956	2.01
	Third	2.0E+06	2.3856	0.938	3.04
4	First	145.16	0.6723	0.939	–
	Second	6575.5	1.3149	0.916	1.96
	Third	585385	1.9239	0.880	2.86
5	First	50.148	0.7017	0.838	–
	Second	360.37	1.4121	0.825	2.01
	Third	709.29	2.3968	0.763	3.42
6	First	52.54	0.6287	0.914	–
	Second	3329.4	0.8132	0.680	1.29
	Third	475849	0.9292	0.452	1.48

5.7.1 Parameter Estimation

Several methods can be used for parameter estimation. In the following discussion, the method of moments (MOM), the maximum likelihood method (MLM) and the probability weighted moment (PWM) are used for parameter estimation.

The maximum likelihood (ML) method is considered to be the most important method especially for large data sets since it leads to efficient parameter estimators with Gaussian asymptotic distributions. It provides the smallest variance of the estimated parameters, and hence of the estimated quantiles, compared to other methods. However with small samples the results may not converge.

The method of moments (MOM) is a relatively simple method and is more commonly used. It can also be used to obtain starting values for numerical procedures involved in ML estimation. However, MOM estimates are generally not as efficient as the ML estimates, especially for distributions with large number of parameters, because higher order moments are more likely to be highly biased for relatively small samples.

The PWM method gives parameter estimates comparable to the ML estimates. Yet in some cases the estimation procedures are much less complicated and the computations are simpler. Parameter estimates from samples using PWM are sometimes more accurate than the ML estimates. Further details on this topic are found in Rao and Hamed (2000).

5.7.2 Quantile Estimation

After the parameters of a distribution are estimated, quantile estimates (Q_T) which correspond to different return periods may be computed. The return period (T), is related to the probability of non-exceedence (F) by the relation,

$$F = 1 - \frac{1}{T} \tag{5.7.1}$$

where $F = F(Q_T)$ is the probability of having a flood of magnitude Q_T or smaller. The problem then reduces to evaluating Q_T for a given value of F. In practice, two types of distribution functions are encountered. The first type is that which can be expressed in the inverse form $Q_T = \phi(F)$. In this case, Q_T is evaluated by replacing $\phi(F)$ by its value. In the second type, the distribution cannot be expressed directly in the inverse form $Q_T = \phi(F)$. In this case numerical methods are used to evaluate Q_T corresponding to a given value of F.

5.7.3 Probability Distributions

There are many functions which fulfill the conditions to be satisfied by a probability density function. Four distributions which are commonly used in modeling flood data are used in this discussion. These are (1) Three parameter log normal distribution (2) Pearson type 3 distribution (3) log Pearson type 3 distribution and the (4) Generalized Extreme Value distribution. Details of these distributions are found in Rao and Hamed (2000).

To assess the reasonability of the selected distribution, several statistical tests like Chi-Square test and Kolmogrov-Smirnov test may be used. The Chi-square test and Kolmogrov-Smirnov tests are discussed below.

5.7.3.1 Chi-Square Test

In the chi-square test, data are first divided into k class intervals. The statistic χ^2 in Eq. (5.7.2) is distributed as chi-square with k - 1 degrees of freedom.

$$\chi^2 = \sum_{j=1}^{k} \frac{(O_j - E_j)^2}{E_j} \tag{5.7.2}$$

In Eq. (5.7.2), O_j is the observed number of events in the class interval j, E_j is the number of events that would be expected in the class interval from the theoretical distribution. If the class intervals are chosen such that each interval corresponds to an equal probability, then $E_j = n/k$ where n is the sample size, and Eq. (5.7.2) reduces to Eq. (5.7.3)

$$\chi^2 = \frac{k}{n}\sum_{j=1}^{k} O_j^2 - n \tag{5.7.3}$$

Class intervals corresponding to different values of probability F can be computed by taking the inverse of the distribution function and following the procedure similar to estimation of flood quantiles.

5.7.3.2 Kolmogorov-Smirnov Test

A statistic based on the deviations of the sample distribution function $F_n(x)$ from the completely specified continuous hypothetical distribution function $F_0(x)$ is used in this test. The test statistic D_n is defined in Eq. (5.7.4).

$$D_n = \max|F_n(x) - F_0(x)| \tag{5.7.4}$$

The values of $F_n(x)$ are estimated as n_j^c/n where n_j^c is the cumulative number of sample events in class j. $F_0(x)$ is then 1/k, 2/k, ... etc., similar to the chi-square test. The value of D_n must be less than a tabulated value of D_n at the required confidence level for the distribution to be accepted.

5.7.4 Data Analysis

The following nine distributions were selected as candidates for the best distribution suitable to each region in Indiana: Pearson Type III, Log Pearson Type III, Generalized Extreme Value, Log Normal III parameter, Gamma, Generalized Pareto, Logistic, Gumbel (Extreme value type I) and Weibull distribution. Pearson Type I, Extreme Value Type II, and Log Normal (II) distributions were not considered because the same distributions with three parameters were selected. Data sets from region 1 were selected to observe the results for all of the nine distributions. The plots of goodness of fit obtained for many data sets in the case of Gamma, Generalized Pareto, Logistic and Weibull distribution showed a very poor fit. Consequently, four distributions, Log Normal III (LN3), Log Pearson III (LP3), Pearson Type III (PT3) and Generalized Extreme Value (GEV), were chosen for further investigation.

The results from goodness-of-fit tests were ranked from 1 to 4. The distribution which showed best fit for the data is ranked 1, whereas the distribution which showed poorest fit for the data was ranked 4. The frequency distribution(s) that showed best fit for the data of each region are shown in Figs. 5.7.1–5.7.6. The results are also tabulated in Table 5.7.1.

As mentioned above, method of moments (MOM), maximum likelihood (ML) and probability weighted moments (PWM) were used to estimate parameters of the selected distributions. These parameters were used to calculate the flood quantiles corresponding to 10, 20, 50 and 100 year return periods. Standard errors corresponding to the observed values were also obtained. Results of goodness of fit at 95%

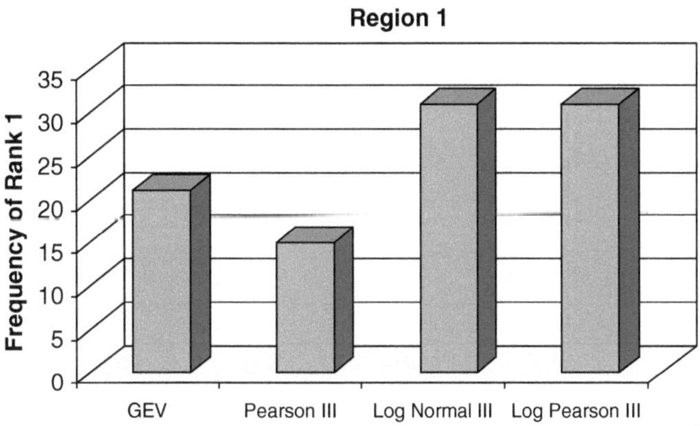

Fig. 5.7.1 Region 1 – Frequency of rank 1 for selecting the best distribution

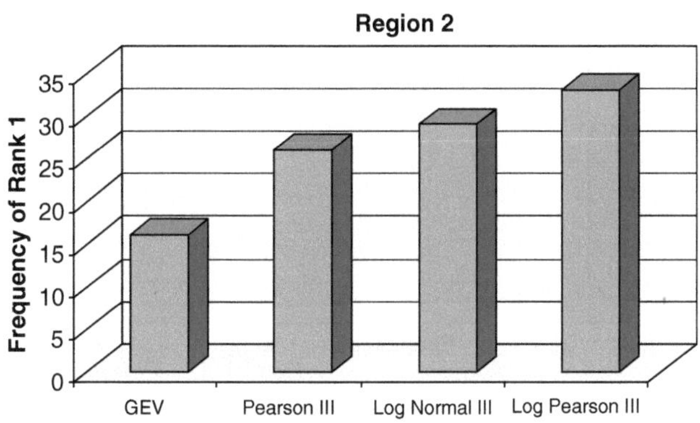

Fig. 5.7.2 Region 2 – Frequency of rank 1 for selecting the best distribution

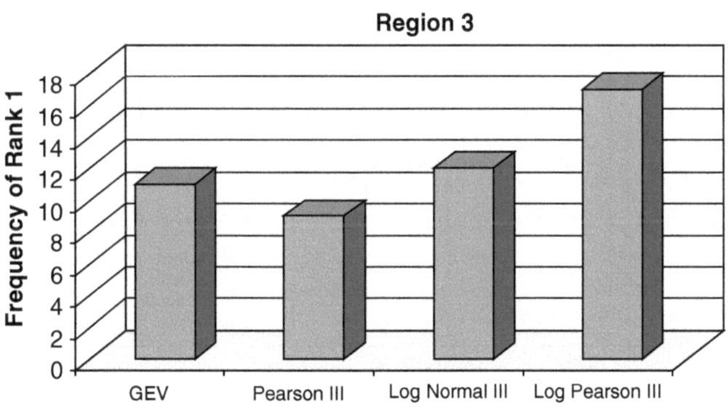

Fig. 5.7.3 Region 3 – Frequency of rank 1 for selecting the best distribution

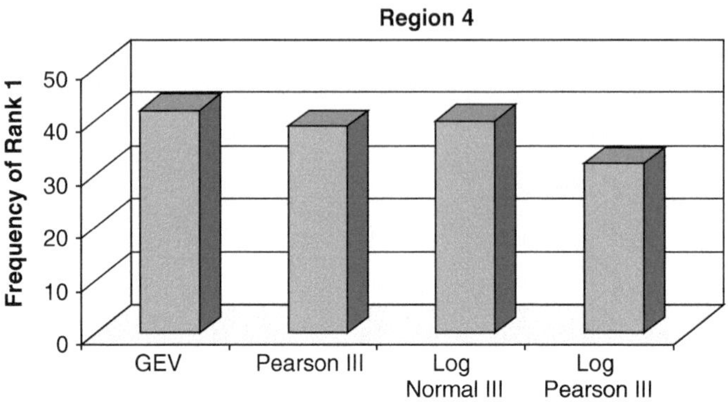

Fig. 5.7.4 Region 4 – Frequency of rank 1 for selecting the best distribution

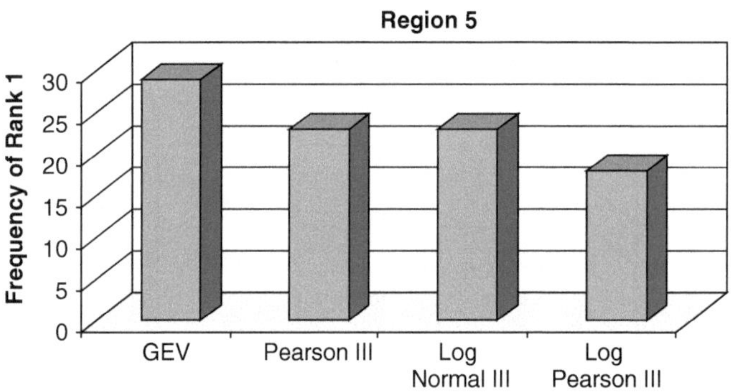

Fig. 5.7.5 Region 5 – Frequency of rank 1 for selecting the best distribution

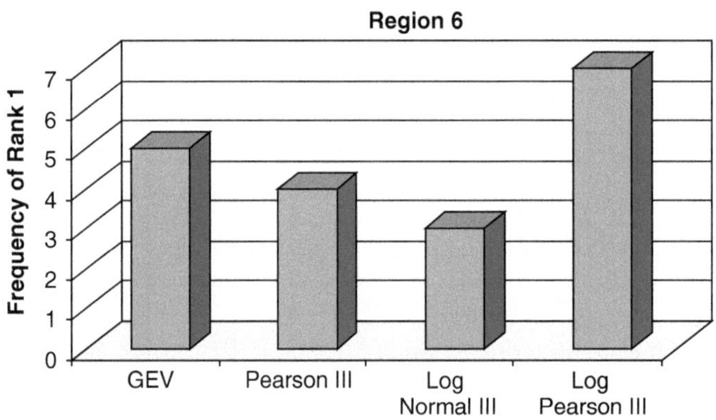

Fig. 5.7.6 Region 6 – Frequency of rank 1 for selecting the best distribution

Table 5.7.1 Selection of best distribution and method of parameter estimation for each Region

Region number	Number of stations	Rank 1	Rank 2	Rank 3	Rank 4	Best method of parameter estimation
1	62	LP3, LN3	–	GEV	PT3	ML
2	58	LP3	LN3	PT3	GEV	ML
3	30	LP3	LN3	GEV	PT3	MOM
4	73	GEV	LN3	PT3	LP3	ML
5	42	GEV	LN3	PT3	LP3	ML
6	14	LP3	GEV	PT3	LN3	ML

confidence limit were tabulated for each gage station in a region corresponding to each distribution and method of parameter estimation.

To select the best method of parameter estimation, the Chi-Square and the Kolmogorov-Smirnov test values for each distribution and gauging station, are compared for the three methods of parameter estimation. The method with the lowest value is given the highest rank, Rank 1. The method having highest frequency of Rank 1 in each region is selected as the best method of parameter estimation for that region. In most cases, maximum likelihood method turned out to be the best one. The final results are tabulated in Table 5.7.1.

The results in Figs. 5.7.1–5.7.6 and Table 5.7.1 were obtained based on observed data from all the watersheds in Indiana. In many of these watersheds the data were quite short. For example, in region 3, the number of observations is less than 30 in 17 out of 30 sites. The goodness-of-fit tests are not reliable for smaller samples. Therefore, sites having more than 30 station-years of peak flow data are screened to repeat the foregoing analysis. The procedure described above for ranking the distributions and the methods of parameter estimation are adopted and the results are shown in Figs. 5.7.7–5.7.12. The new rankings given to the distributions and the best method of parameter estimation are shown in Table 5.7.2 for each of the six regions.

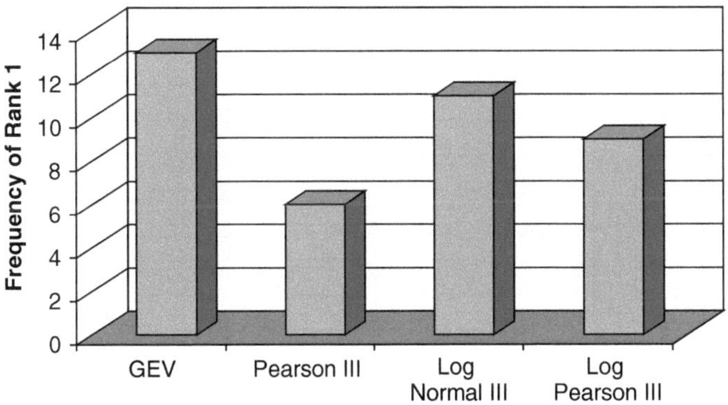

Fig. 5.7.7 Region 1 – Frequency of rank 1 for selecting the best distribution with more than 30 observations at each site

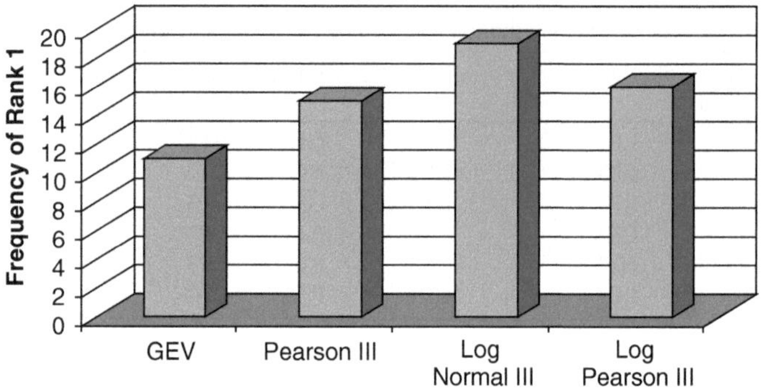

Fig. 5.7.8 Region 2 – Frequency of rank 1 for selecting the best distribution with more than 30 observations at each site

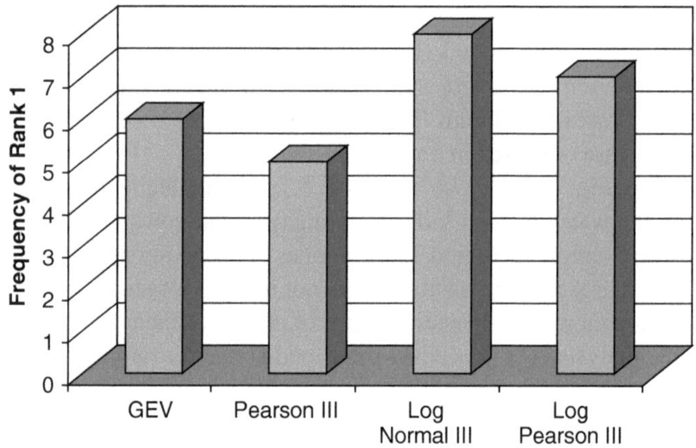

Fig. 5.7.9 Region 3 – Frequency of rank 1 for selecting the best distribution with more than 30 observations at each site

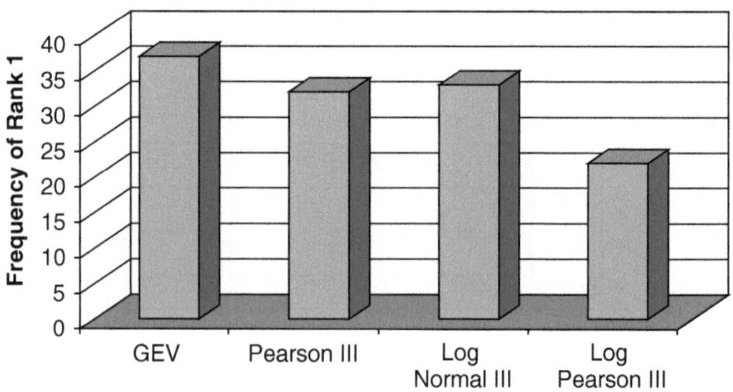

Fig. 5.7.10 Region 4 – Frequency of rank 1 for selecting the best distribution with more than 30 observations at each site

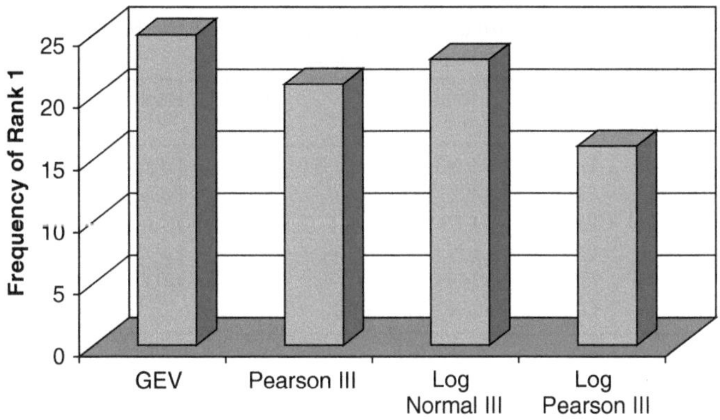

Fig. 5.7.11 Region 5 – Frequency of rank 1 for selecting the best distribution with more than 30 observations at each site

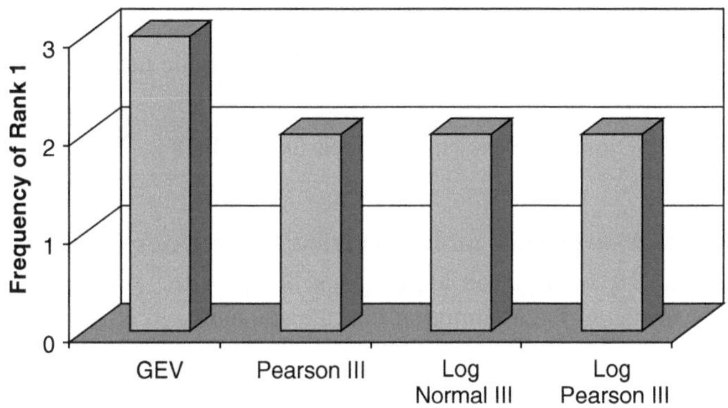

Fig. 5.7.12 Region 6 – Frequency of rank 1 for selecting the best distribution with more than 30 observations at each site

The significance of having longer data sequences in goodness-of-fit tests is clearly brought out by the results in Table 5.7.2. The GEV distribution is the best distribution with larger data sets, followed by Log Normal (III) distribution. Log Pearson (III) distribution which was selected as the best distribution in Table 5.7.1 is no longer in Rank 1 for any region.

The objective of the study was to select the probability distribution which best fits the data in each of the six regions in Indiana. Based on the results presented in Table 5.7.2 for regions 1, 4, 5 and 6, Generalized Extreme Value distribution comes out to be the best distribution. For regions 2 and 3, Log Normal (III) distribution is the best. The maximum likelihood method in found to be the best parameter estimation method.

Table 5.7.2 Selection of best distribution and method of parameter estimation for each Region, considering stations with more than 30 peak flow observations

Region number	Number of stations	Rank 1	Rank 2	Rank 3	Rank 4	Best method of parameter estimation
1	21	GEV	LN3	LP3	PT3	ML
2	30	LN3	LP3, PT3		GEV	ML
3	13	LN3	LP3	GEV	PT3	MOM
4	55	GEV	LN3, PT3		LP3	ML
5	36	GEV	LN3, PT3		LP3	ML
6	07	GEV	LN3, LP3, PT3			ML

5.7.5 Dimensionless and Standardized Quantile Measures

There are other ways to examine the behaviour of the quantile floods. Sveinsson (2002) introduces two types of measures. The first of these is a dimensionless quantile measure and the second one is standardized quantile measure. The assumption that the at-site population quantiles divided by their population mean are identical in a homogeneous region implies that the dimensionless quantile measure is constant in that region. The expression of the dimensionless quantile measure for site j is

$$\text{Dimensionless quantile measure (DQM)} = \frac{\hat{Q}_T^j}{\hat{\mu}_Q^j} \tag{5.7.5}$$

where $\hat{Q}_T^j$ is the flood quantile for return period T estimated by using any specified distribution and $\hat{\mu}_Q^j$ is the mean annual peak flow at site j. As for the standardized quantile measure, the assumption that the standardized at-site population are identical implies that the standardized quantile measures should not depend on the data from different stations. The expression of the standardized quantile measure for site j is

$$\text{Standardized quantile measure (SQM)} = \frac{\hat{Q}_T^j - \hat{\mu}_Q^j}{\hat{\sigma}_Q^j} \tag{5.7.6}$$

where $\hat{\sigma}_Q^j$ is the standard deviation of the annual peak flows at site j.

The dimensionless quantile measures calculated for all the sites in each of the eight regions using PT3 distribution are shown in Fig. 5.7.13. The standardized quantile measures calculated for the same by PT3 distribution are shown in Fig. 5.7.14. Each region has its own sub-plot for eight recurrence intervals of 2, 5, 10, 20, 25, 50, 100 and 200 years. The arrows indicate those stations which have high discordancy measures. The result shows that the measures are almost constant for all the stations for recurrence intervals less than 25 years. For recurrence intervals greater than 25 years, the statistics fluctuate considerably. These results bring out the fact that the homogeneity of dimensionless quantile measures depends on the recurrence intervals considered. For smaller recurrence intervals the homogeneity assumption may be acceptable, whereas for higher recurrence intervals the statistics for stations which are discordant to the other sites in a region show high variability.

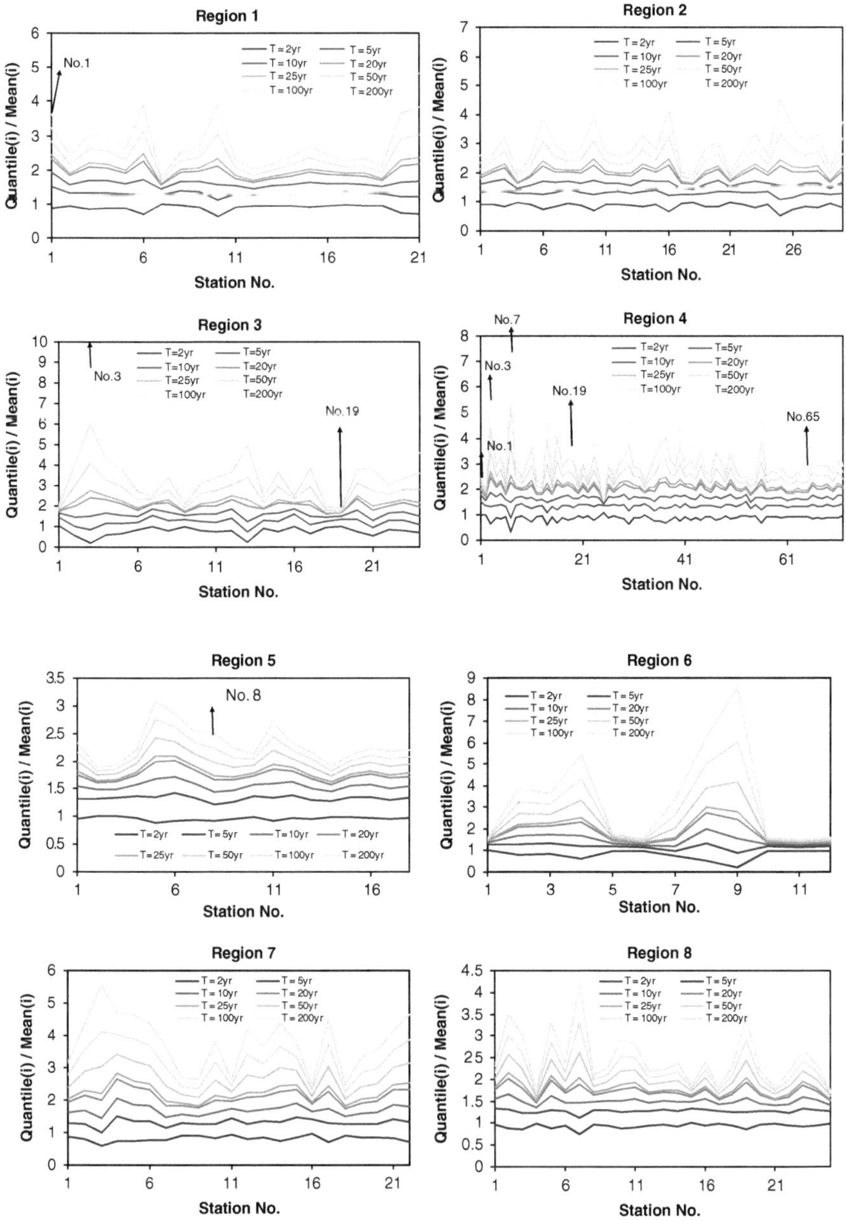

Fig. 5.7.13 Dimensionless quantile measures for each region

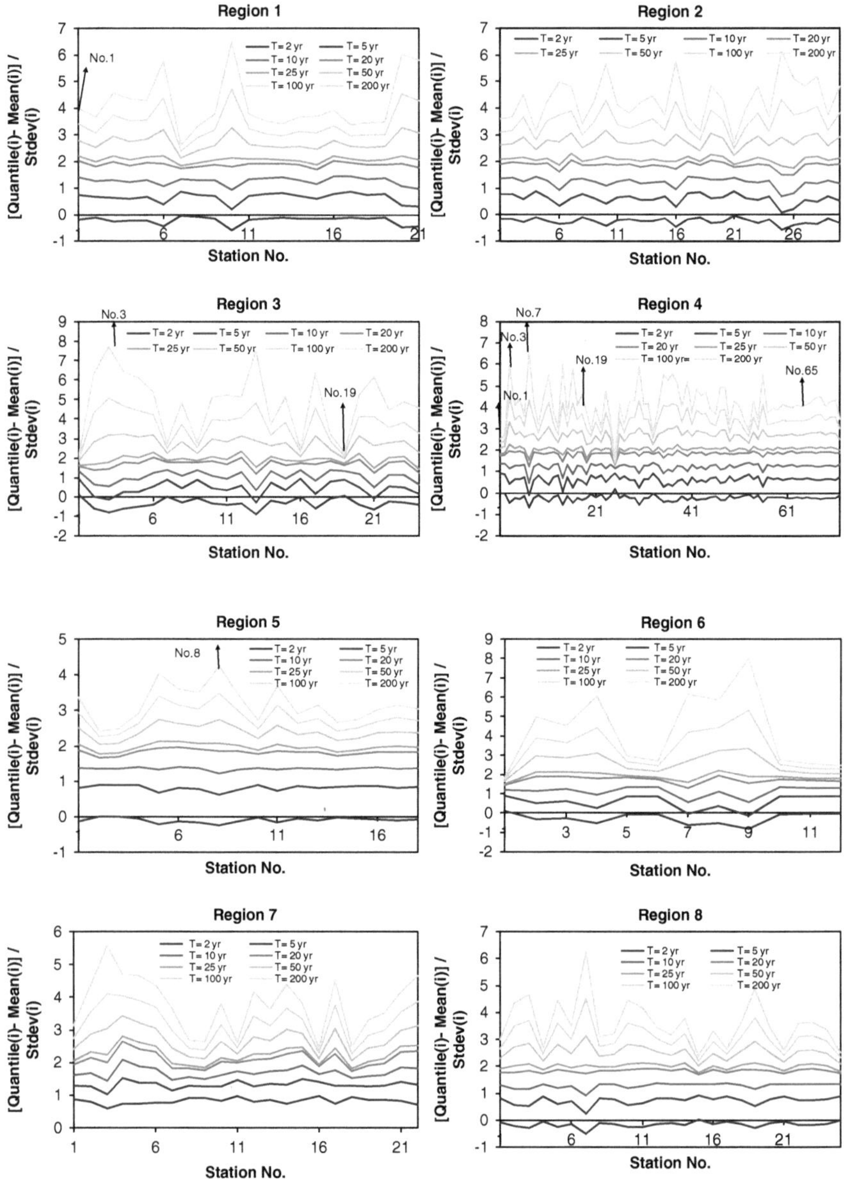

Fig. 5.7.14 Standardized quantile measures for each region

5.8 Concluding Comments

It is of interest to note that the flood estimates vary with the choice of distributions for the same region. These differences in flood estimates due to different distributions can be considerable. The differences depend on the method used to estimate the parameters of the distributions. These differences must be established and used to assess the accuracy of flood estimates. Procedures which yield the smallest errors of estimates must be used in flood estimation. Despite considerable advances in computational procedures these have not been put in to common practice. The results presented in this chapter highlight these issues.

The importance of regionalization in reducing the flood estimation errors is also brought out by these results. The smallest estimation errors are usually associated with the most homogeneous watersheds. The errors increase with heterogeneity. Consequently regionalization is an important first step in flood estimation. After regionalization the observed data must be used to develop flood estimation relationships. These relationships may be tested by using suitable techniques to assess estimation errors. The errors may be used in assessing flood estimation accuracy.

Similar considerations apply in selecting the probability distribution for flood estimation. Use of a single distribution may not be acceptable to an area such as a large state. We may have large unacceptable errors by using a single distribution. These issues must be carefully considered and used in practice.

Chapter 6
Concluding Remarks

6.1 General Remarks on Clustering Approach to Regional Flood Frequency Analysis

In flood frequency analysis, regionalization of watersheds is perceived mostly as a clustering problem. Identification of groups of watersheds (regions) having similar flood response was based on at-site flood statistics such as mean annual flood, coefficient of variation, skew, and kurtosis. The current practice is to form regions by seeking similarity in attributes that affect and depict floods in watersheds.

It is suggested not to form regions based on at-site flood statistics because: (1) the resulting clusters will not be useful for estimating floods at ungauged sites, when the requirement arises at a later time; (2) the regional average L-statistics of resulting clusters will not be much different from L-statistics of sites in those regions. Thus, little can be gained by using those regional statistics for estimating parameters of frequency distribution(s) to arrive at quantiles of floods for hydrologic design; (3) with at-site statistics there is a tendency to group together all the sites having similar outliers. An outlier for a site is a data point that is numerically far-off from the rest of the data at the site. The outlier could be due to random fluctuations caused by a localized meteorological event at the site that may not have affected its neighboring sites. If such an event is equally likely to affect any of its neighbors in the future, then it is incorrect to treat the site with outlier as different from its neighbors.

Further, the regions formed based on similarity in chosen watershed attributes need not be homogeneous in flood response. This is because it is impossible to collate data on exhaustive set of attributes for regionalization. Hence it is necessary to test homogeneity of identified regions. This aspect has not received the importance it deserves (e.g., Hall and Minns, 1999; Hall et al., 2002).

There are a few approaches to regionalization. Each approach has its strengths and limitations, and there are a number of issues in using them. One of the issues is that the use of different approaches leads to formation of different set of regions because the strategy used for grouping sites differs from one approach to another.

In practice, hydrologists have been trying different approaches as there are no established criteria by which the superiority of any particular approach can be clearly brought out. Further, there is a dearth of attempts to comprehend performance of various regionalization approaches ever since the contributions of Cunnane (1988), Bobée and Rasmussen (1995) and GREHYS (1996) in the past century.

Recently, increase in scientific knowledge and global concern about water resources assessment paved the way for creation of databanks of a variety of hydro-climatic and hydro-meteorological attributes that influence flood response of watersheds. In this scenario, clustering approach that is recognized for its effectiveness in classifying multivariate data has become a natural choice for grouping watersheds. Although cluster analysis is useful, there are several clustering techniques and their relative merits and de-merits in forming homogeneous regions for flood frequency analysis are not explored. This predicament makes it necessary to examine the performance of various clustering procedures in vogue.

Conventionally, hard clustering algorithms are used to form regions. To explore their merits, the performance of commonly used hierarchical clustering algorithms (single linkage, complete linkage and Ward's), a widely used partitional clustering algorithm (K-means), and hybrids of the hierarchical and partitional clustering algorithms is examined. The hybrid of Ward's and K-means algorithms is found to be better than that of the hierarchical and the partitional clustering algorithms considered. Plausible homogeneous hydrologic regions are identified by visual interpretation of partitions provided by the hard clustering algorithms, and by using six hard cluster validity indices, namely cophenetic correlation coefficient (CPCC), average silhouette width (ASW), Dunn's index, Davies-Bouldin index, Calinski Harabasz index, and Minimum Description Length. The CPCC is found to be inefficient, whereas the ASW performed reasonably well. The Dunn's index and Davies–Bouldin index are found to be effective in identifying optimal partition containing clusters that are close to being homogeneous. Optimal partition identified using the cluster validity indices is found to be very similar to the plausible hydrologic regions recognized by visual inspection of clusters.

One of the issues in the use of cluster analysis concerns partially or completely assigning watersheds to regions. In hard cluster analysis, a watershed is classified as belonging to one region or another. This type of analysis is acceptable if watersheds can be classified into regions in a stringent manner based on their attributes. However, application of hard clustering analysis to cases where the classification is rather vague is undesirable. Vagueness in region formation is unavoidable in grouping watersheds which partially resemble each other in terms of their attributes. The issue in this situation is whether a procedure such as fuzzy clustering can be used to achieve effective partition of watersheds. To explore this, fuzzy c-means (FCM) algorithm is used for regionalization of watersheds in Indiana and is found to be efficient in forming regions. Several fuzzy cluster validity indices are tested to examine their effectiveness in identification of optimal partition obtained from FCM algorithm. It is found that the fuzzy validity measures: partition coefficient, classification entropy, fuzziness performance index and normalized classification entropy that are in vogue in hydrologic literature are inefficient in deriving hydrologically homogeneous regions. This could be attributed to lack of direct connection of these indices to structure in multi-dimensional space of feature vectors prepared from watershed attributes. These measures are based only on membership values of feature vectors in various clusters. Xie-Beni, extended Xie-Beni, Fukuyama-Sugeno and Kwon's Indices have been tried as alternatives to the existing indices. Among these, extended Xie-Beni index is found to be quite useful for identification of optimal partition. Since the conclusions are drawn using only data sets of Indiana, USA, these could be premature. It is necessary to test these validity measures further with data sets from other parts of the world to support or reject the conclusions drawn based on the results presented in this book.

It is suggested that caution should be exercised in deciding optimal number of clusters based on the cluster validity indices because they are developed and validated on certain standard data sets (such as Iris data, and wine data found in the University of California, Irvine (UCI), machine learning repository) in applications other than regionalization of watersheds. Further, none of these indices is found to be suitable for identification of optimal partition in all types of standard data sets. Hence, further research is needed to explore the possibility of developing new validity indices exclusively for identification of optimal number of regions in regionalization studies.

As in the case of hard clustering, the regions obtained from the fuzzy cluster analysis are also adjusted to improve their homogeneity. The Fuzzy memberships of sites in clusters are found to be useful in adjusting the regions. Considerable effort is needed to adjust a region when clusters determined by hard cluster analysis or by defuzzification of partition provided by FCM algorithm are used to form hydrologic regions. In contrast, the effort needed to form homogeneous regions by adjusting fuzzy clusters derived from FCM algorithm is found to be smaller.

It is noted that the use of effective clustering procedures can provide regions that are close to being homogeneous. Consequently the effort involved in adjusting regions to make them homogeneous is reduced considerably. The subjectivity in region adjustment has been an issue of concern for several practising hydrologists, despite a large number of guidelines that have been framed for this purpose (see Section 1.4.1).

The linear Kohonen's self-organizing feature map (SOFM) has been applied as a clustering technique for regionalization in recent studies. However, specific patterns could not always be discerned in Kohonen lattice for grouping watersheds with SOFM, irrespective of its size and dimensionality (1-D or 2-D). It is demonstrated that SOFMs may, however, serve as a useful precursor to clustering algorithms. A novel two-level SOFM-based clustering approach is proposed for regionalization of watersheds. In the first level, the SOFM is used to form a two-dimensional feature map. In the second level, the output nodes of SOFM are clustered using FCM algorithm to form regions for flood frequency analysis. The optimal number of regions is determined by fuzzy cluster validity indices.

The two-level SOFM-based clustering algorithm is found to be efficient in determining homogeneous groups of watersheds. This is demonstrated through application to watersheds in Indiana. The knowledge of distribution of membership of winning output nodes of SOFM among the fuzzy regions is useful in adjusting the regions to improve their homogeneity. Thus the effort needed to adjust regions is smaller for the two-level fuzzy clustering than for the conventional hard, fuzzy and

SOFM clustering procedures. Several avenues should be explored to further refine these attempts to regionalize watersheds. Investigations in this direction can lead to identification of robust procedures for regionalization.

As regionalization is neither easy nor a simple exercise, an attempt is made to assess its importance in regional flood frequency analysis by examining if it improves accuracy in estimation of flood quantiles. Results show that the error in estimation of flood quantiles at target site by using regional information is smaller when target site belongs to a homogeneous group.

Practicing hydrologists in United States were using several distributions to estimate frequency of floods in different parts of the country. They resorted to log-Pearson type III (LP3) distribution following recommendations of the U.S. Water Resources Council (1976, 1977, 1981) for the continental United States published in Bulletin 17 (Griffis and Stedinger, 2007). The question of whether a single distribution can be used to fit peak flows in all the regions in a state is examined. The performance of LP3 distribution is compared with that of several other frequency distributions in modeling peak flow data of different regions in Indiana. Results show that a single distribution cannot be used for all the regions. Further, it is noted that the LP3 distribution yielded significant errors for several regions, thus indicating that it cannot be opted as a default choice for modeling floods.

Furthermore, it is noted that within hydrologically homogeneous regions in Indiana, moments of annual peak flows scale with drainage area according to log–log linear relations. Thus, for ungauged basins within a homogeneous region, it is possible to predict moments of annual peak flows fairly well. Subsequently, the moments can be used to estimate flood quantiles by using regional growth curve computed using index flood method. Alternatively, the moments can be used to compute parameters of any distribution and the corresponding flood quantiles. Future research should explore if the concept of scaling can be used to test homogeneity of regions.

6.2 Recent Developments

In the past decade there have been some interesting developments in the area of regional flood frequency analysis. New procedures that are useful to characterize regional frequency distribution of floods, and methodologies to perform frequency analysis of floods in the presence of non-stationarity are proposed. In this section some of those contributions are described briefly. Further, new avenues of research which are evolving are mentioned.

6.2.1 Tests of Regional Homogeneity

Viglione et al. (2007) compared heterogeneity measures based on L moment ratios (Hosking and Wallis, 1993) with the bootstrap Anderson-Darling test (Scholz and Stephens, 1987) and with the Durbin and Knott rank test (Durbin and Knott, 1972).

It is suggested that the Hosking and Wallis heterogeneity measure based only on coefficient of L-variation (p. 21, Hosking and Wallis, 1997) is preferable when regional skewness is low, while the bootstrap Anderson-Darling test is preferable for regions with higher skewness.

6.2.2 Methods for Characterizing Regional Frequency Distribution

LH moments were proposed by Wang (1997, 1998) as an alternative to the conventional L-moments for characterizing the upper parts of distributions and larger events in a sample. The idea underlying LH moments is that a distribution function which is inappropriate for describing complete data series may still be reasonable for describing the larger events in that data series. The LH moments are based on linear combinations of higher order statistics. Self-determined probability weighted moments (SD-PWM) were proposed by Haktanir (1997) as an extension of the probability weighted moments of Greenwood et al. (1979). Whalen et al. (2002) developed algorithms to simplify parameter estimation by SD-PWM. Moisello (2007) advocated the use of partial probability weighted moments (Wang, 1990) for regional analysis of hydrologic extremes.

6.2.3 Methods for Regional Frequency Analysis

Traditionally hydrologists have been using index flood procedure (Dalrymple, 1960) for combining information from different sites in a homogeneous region for regional frequency analyses. This procedure involves the use of scale factor (called index flood) to scale flood data at all the sites in a region, before proceeding to estimate at-site L-moment ratios and combining them to arrive at regional L-moment ratios. When the index flood method was proposed, the scale factor was taken to be the at-site population mean. However, since then the population statistic has been estimated by the at-site sample mean in several regionalization studies. Sveinsson et al. (2001) investigated the consequences of replacing a population characteristic with its sample counterpart, and proposed population index flood method as an analytical alternative to the traditional index flood procedure for regional frequency analyzes of extreme hydrologic events. Sveinsson et al. (2003) suggested methods for estimating the standard errors of at-site quantile estimators for two regional population index flood methods utilizing the generalized extreme value distribution with maximum likelihood estimation.

6.2.4 Goodness-of-fit Measures for Regional Frequency Analysis

Hosking and Wallis (1997) proposed a regional goodness-of-fit statistic based on L-moments for choosing a frequency distribution, from a number of candidate distributions, to fit the flood data in a homogeneous region. The quality of fit is judged by the difference between the theoretical value of L-kurtosis of the fitted candidate distribution and the sample estimate of the regional average L-kurtosis. To assess the significance of the difference, it is compared with sampling variability of the regional average L-kurtosis.

The distribution which may provide good fit to the regional data can also be determined by visual interpretation of L-moment ratio diagram, which is a plot of L-skewness against L-kurtosis of frequency distributions. A two-parameter distribution with a location and a scale parameter plots as a point on the diagram, whereas a three-parameter distribution having location, scale, and shape parameters plots as a line. Hosking and Wallis (1997) provided theoretical relationships between the L-moment ratios of various distributions that are useful to plot the theoretical L-moment ratio curves on the L-moment ratio diagram.

To select a distribution for regional frequency analysis, the sample L-moment ratios of sites in a region are plotted as points on the L-moment ratio diagram and the resulting scatter plot is compared with theoretical L-moment ratio curves of candidate distributions. The subjective methods that are in vogue for this task include those that are based on (i) sample average (i.e., comparison of point depicting regional average L-skewness and L-kurtosis with the theoretical L-moment ratio curves) and (ii) line of best-fit method (Vogel and Wilson, 1996).

In practice, all the regions delineated in a study area may not be homogeneous. For very heterogeneous regional data, exhibiting a large range in the distributions shape parameter, the curve of best-fit through sample L-moment ratios could be more useful for distribution selection than the goodness-of-fit statistic of Hosking and Wallis (Peel et al., 2001; Kroll and Vogel, 2003). Kroll and Vogel (2002) developed a performance measure named AWOD to alleviate the subjectivity and the effort required for interpretation of the L-moment ratio diagram. For each of the candidate distributions the AWOD statistic measures the average weighted orthogonal distance between the sample L-moment ratios of sites in a region and the theoretical L-moment ratio curve of the distribution on the L-moment ratio diagram. Among the candidate distributions, the distribution with smallest value of AWOD is chosen to fit the regional data. These ideas need to be tested with data from watersheds in different parts of the world before they are accepted.

6.2.5 Non-Stationary Flood Frequency Analysis

For estimation of flooding potential at the sites in a region, it is assumed that flood flows at the sites represent samples of independent and identically distributed realizations drawn from a stationary homogeneous stochastic process. These assumptions are not strictly valid (Klemeš, 2000). For example, natural and human-induced changes in global water and energy cycles could alter the magnitude and frequency of flood events. Also, the natural periodicity present in climate causes non-stationarity in the hydrologic time series (Rao and Hamed, 2003). However, the short historical records and the lack of mathematical framework for analyzing and modeling the dynamics of non-stationary processes have impaired studies in this direction (Sveinsson et al., 2003), especially on river basin scale.

A brief review of the current approaches to at-site and regional frequency analysis of dependent and/or non-stationary flood flows can be found in Khaliq et al. (2006). The development of models for non-stationary pooled frequency analysis is in formative state. Strupczewski and Kaczmarek (2001) and Strupczewski et al. (2001a,b) proposed non-stationary approach to at-site flood frequency analysis by assuming trend in the first two moments of probability distributions. Six probability distribution functions and four classes of time trends were selected for investigation. The probability distributions include the Normal, the two-parameter lognormal, the three-parameter lognormal, the Gamma, the Pearson type III, and extreme value type I distribution. The trends analyzed include (i) trend in the mean value, (ii) trend in the standard deviation, (iii) trend in the mean value and the standard deviation related by a constant value of the variation coefficient, and (iv) unrelated trend in the mean value and the standard deviation.

Cunderlik and Burn (2003) proposed a second-order non-stationary model to RFFA by assuming at-site non-stationarity in the first two moments of the peak flow time series. This model separates the regional flood quantile function into at-site time-dependent component comprising the location and scale parameters (i.e., mean and variance), and a regional component that was considered as time-invariant under the assumption of second order non-stationarity. Standardized annual maximum peak flow time series was decomposed into a 'trend component' and a 'residual time-dependent component', representing irregular fluctuations around the trend. The time varying location parameter was predicted from the 'trend component' based on regression, assuming trend to be a linear function of time. 'Transformed residual time series' was computed by taking absolute deviation of the time series of 'residual time-dependent component' about its mean value. The time varying scale parameter was predicted based on regression equation assuming trend in the 'transformed residual time series' to be a linear function of time. Furthermore, the trend in the 'transformed residual series' was removed from the 'residual time-dependent component' to obtain second order stationary time series having time invariant parameters.

6.2.6 Flood Frequency Analysis in Climate Change Scenarios

There is a need to update the existing methodologies for regional frequency analysis in parallel with developments in climate research. Recently, with growth in scientific consensus that current climate change is largely the result of human activities (Oreskes, 2004), scientists are devoting their efforts to explore implications of climate change on water which is one of the vulnerable resources of earth.

A group called Intergovernmental Panel on Climate Change (IPCC) has been established by the World Meteorological Organization (WMO) and the United Nations Environment Programme (UNEP), in 1988, with a view to assess scientific, technical and socio-economic information relevant for the understanding of climate change, its potential impacts and options for adaptation and mitigation. To date, the IPCC has published four comprehensive assessment reports. These reports essentially summarize the state of scientific knowledge on global climate change, its causes, impacts and possible response measures.

In its eighth session, the working Group II of the IPCC has summarized that the magnitude and frequency of floods are likely to increase in different parts of the world (IPCC, 2007). The increase in flood events is attributed to increase in snow melt, glacial lake outbursts, rise in sea-level, or increase in severity and frequency of storms. Warming in western mountains of North America is projected to cause increase in snow melt and more winter flooding. Increase in risk of inland flash floods and frequency of coastal flooding are projected for Europe. Increase in the severity and frequency of storms and coastal flooding are projected for Australia and New Zealand by 2050. Sea-level rise is projected to cause increased risk of flooding in low-lying areas of Latin America. Floods due to increase in Glacier melt in the Himalayas in the next two to three decades, and floods in mega-deltaic regions of South, East and Southeast Asia due to rise in sea level are projected to cause havoc.

In the past decade, following the developments in climate research (e.g., Waylen and Caviedes, 1986; Robson et al., 1998), a few researchers have attempted to study plausible impact of climate change on flood frequency (e.g., Olsen et al., 1999; Jain and Lall, 2000, 2001; Walker and Stedinger, 2000, among others). The findings reported in these studies are confined to a few selected stations with long flood record. Even with long records, the possibility of multiple causal factors for trends in a flood series makes it a challenging task to attribute the observed non-stationarities entirely to climate change. Moreover, to establish the nature and type of impact of a climate change signal on the flood response of watersheds in a region, research should address the trends in hydrologic time series at all the watersheds in the study region. In the absence of such an effort, the research findings would not be useful to propose general guidelines for estimating flood quantiles at ungauged sites and those with short records in a study region.

6.2.7 Simulation of Floods Using Output from GCMs

General Circulation Models (GCMs) are the most advanced tools currently available to simulate climatic conditions on earth hundreds of years into the future. The GCMs simulate climatic conditions for projected changes in large scale forcings (LSFs). Forcings in the climate sense are external boundary conditions or inputs to a GCM. In general, the LSFs could be natural or anthropogenic. Natural forcings include volcanic eruptions, variations in the solar radiation, the large-scale distribution of continents, oceans and ice, and large topographical systems. On the other hand, anthropogenic forcings are mostly decided based on IPCC climate scenarios which are developed to facilitate the scientific community to obtain projections for climate change for many decades into the future. A projection is a probabilistic statement that it is possible that something will happen to the response of global circulation in the future if certain boundary conditions develop. The set of boundary conditions that is used in conjunction with making a projection is often called a scenario, and each scenario is based on assumptions about how the future will develop (MacCracken, 2001). Examples of anthropogenic forcings include radiative forcing (which can result from changes in greenhouse gas concentrations and aerosol loading in the atmosphere), variations in stratospheric and/or tropospheric ozone and sulfate aerosols, future trends in population, economic growth, energy demand and land use change (IPCC, 1992).

In 1992, the IPCC worked out six alternative emissions scenarios termed IS92 a-f, which provided alternative emissions trajectories for the years 1990 through 2100 for various radiatively active greenhouse gases. In 2001, IPCC issued a Special Report on Emissions Scenarios (SRES) to replace IS92 scenarios (IPCC, 2001). Similarly the IPCC report released in the year 2007 has six scenario groups: A1B, A1FI, A1T, A2, B1 and B2. Detailed review of these scenarios can be found in IPCC (2007). Nevertheless, it is to be mentioned that these new scenarios represent a wider range of driving forces to reflect current understanding and knowledge about underlying uncertainties. Such scenarios are internally consistent patterns of plausible future climates, not predictions carrying assessed probabilities (Section 1.5, IPCC, 2001).

The GCMs provide climate variables as output at nodes of grid boxes covering the earth's surface. In general, the resolution of the present state-of-the-art GCMs is coarser than two degrees for both latitude and longitude, which is of the order of a few thousand square kilometers for grid box. However, the watershed scale, which is of interest to hydrologists, is of the order of a few hundred square kilometers. Furthermore, GCMs run on a sub-daily time step (hourly or daily). These high-frequency outputs are not reliable and therefore outputs are integrated in time to produce monthly or seasonal scale outputs that are considered to be more robust. Consequently, the temporal resolution of GCM outputs could be too coarse for hydrological studies at the basin scale. In the past decade, to deal with this problem of mismatch of spatial and temporal scales between the GCM output and the watershed-scale, a variety of regional climate models and statistical downscaling approaches have been developed.

The downscaling models can be used to obtain projections for hydro-meteorologic variables such as temperature, precipitation and wind speed which govern runoff and flood response of watersheds in the study region. The projected information on the hydro-meteorologic variables can then be routed through an appropriate rainfall-runoff model developed for each of the watersheds to yield projections for runoff, from which peak flows information can be extracted. One of the assumptions inherent in this analysis is that rainfall-runoff relationships remain unchanged with time for the watersheds.

Methodologies can also be developed to capture the relationship between global climate signals such as El Niño and La Niña, and flood events. The identified relationships would be useful to obtain flood forecast at target sites in the study region. The peak flows projected in each watershed can supplement the available at-site information. Flood quantile estimates based on an ensemble of projected floods corresponding to various scenarios and GCMs can reduce risk associated with hydrologic designs. However, this area of research still remains largely unexplored and provides an opportunity for future research.

The approaches discussed in this section are still evolving. But they deal with important questions. One must be aware of this research and extract the more reliable of the results in future practice.

References

Acreman MC, Sinclair CD (1986) Classification of drainage basins according to their physical characteristics: An application for flood frequency analysis in Scotland. Journal of Hydrology 84(3–4): 365–380.

Aldenderfer MS, Blashfield RK (1984) Cluster analysis. Sage Publications Inc., Beverly Hills, CA.

ASCE Task Committee on Application of Artificial Neural Networks in Hydrology (2000a) Artificial neural networks in hydrology, I: Preliminary concepts. Journal of Hydrologic Engineering, ASCE 5(2): 115–123.

ASCE Task Committee on Application of Artificial Neural Networks in Hydrology (2000b) Artificial neural networks in hydrology, II: Hydrologic applications. Journal of Hydrologic Engineering, ASCE 5(2): 124–137.

Backer E, Jain AK (1981) A clustering performance measure based on fuzzy set decomposition. IEEE Transactions on Pattern Analysis and Machine Intelligence 3(1): 66–75.

Ball G, Hall D (1967) A clustering technique for summarizing multivariate data. Behavioral Science 12: 153–155.

Bargaoui Z-K, Fortin V, Bobée B, Duckstein L (1998) A fuzzy approach to the delineation of region of influence for hydrometric stations. Revue des sciences de l'eau 11(2): 255–282 (In French).

Berkhin P (2002) Survey of clustering data mining techniques. Technical Report, Accrue Software, San Jose, CA.

Bezdek JC (1973) Fuzzy mathematics in pattern classification. Ph.D. dissertation, Cornell University, Ithaca, NY.

Bezdek JC (1974a) Numerical taxonomy with fuzzy sets. Journal of Mathematical Biology 1: 57–71.

Bezdek JC (1974b) Cluster validity with fuzzy sets. Journal of Cybernetics 3(3): 58–72.

Bezdek JC (1975) Mathematical models for systematics and taxonomy. In: Estabrook G (Ed.), Proceedings of 8th International Conference on Numerical Taxonomy. Freeman, San Francisco, CA, pp. 143–166.

Bezdek JC (1981) Pattern recognition with fuzzy objective function algorithms. Plenum Press, New York.

Bezdek JC (1987). Partition structures: A tutorial. In: Bezdek JC (Ed.), The Analysis of Fuzzy Information. CRC Press, Boca Raton, FL.

Bezdek JC, Ehrlich R, Full W (1984) FCM: The fuzzy c-means clustering algorithm. Computers and Geosciences 10(2–3): 191–203.

Bezdek JC, Pal SK (1992) Fuzzy models for pattern recognition. IEEE Press, New York.

Bhaskar NR, O'Connor CA (1989) Comparison of method of residuals and cluster analysis for flood regionalization. Journal of Water Resources Planning and Management 115(6): 793–808.

Bischof H, Leonardis A, Selb A (1999). MDL principle for robust vector quantization. Pattern Analysis and Applications 2(1): 59–72.

Black AR, Werritty A (1997) Seasonality of flooding: A case study of north Britain. Journal of Hydrology 195(1–4): 1–25.

Blöschl G, Sivapalan M (1997) Process controls on regional flood frequency: Coefficients of variation and basin scale. Water Resources Research 33: 2967–2980.

Bobée B, Rasmussen P (1995) Recent advances in flood frequency analysis. U.S. National Report to International Union of Geodesy and Geophysics (1991–1994). Reviews in Geophysics 33: 1111–1116.

Buishand TA (1989) Statistics of extremes in climatology. Statistica Neerlandica 43: 1–30.

Burn DH (1989) Cluster analysis as applied to regional flood frequency. Journal of Water Resources Planning and Management 115(5): 567–582.

Burn DH (1990a) An appraisal of the "region of influence" approach to flood frequency analysis. Hydrological Sciences Journal 35(2): 149–165.

Burn DH (1990b) Evaluation of regional flood frequency analysis with a region of influence approach. Water Resources Research 26(10): 2257–2265.

Burn DH (1997) Catchment similarity for regional flood frequency analysis using seasonality measures. Journal of Hydrology 202: 212–230.

Burn DH, Goel NK (2000) The formation of groups for regional flood frequency analysis. Hydrological Sciences Journal 45(1): 97–112.

Burn DH, Zrinji Z, Kowalchuk M (1997) Regionalization of catchments for regional flood frequency analysis. Journal of Hydrologic Engineering 2(2): 76–82.

Calinski T, Harabasz J (1974) A dendrite method for cluster analysis. Communications in Statistics 3: 1–27.

Castellarin A, Burn DH, Brath A (2001) Assessing the effectiveness of hydrological similarity measures for flood frequency analysis. Journal of Hydrology 241: 270–285.

Cathcart J (2001) The effects of scale and storm severity on the linearity of watershed response revealed through the regional L-moment analysis of annual peak flows. Ph.D. thesis, University of British Columbia, Canada.

Cavadias GS (1989) Regional flood estimation by canonical correlation. Paper Presented to the Annual Conference of the Canadian Society of Civil Engineering, St. John's, Newfoundland, Canada.

Cavadias GS (1990) The canonical correlation approach to regional flood estimation. In: Beran MA, Brilly M, Becker A, Bonacci O (Eds.), Proceedings of the Ljubljana Symposium on Regionalization in Hydrology. IAHS Publication No. 191, Wallingford, England, pp. 171–178.

Cavadias GS (1995) Regionalization and multivariate analysis: The canonical correlation approach. Proceedings of the UNESCO International Conference on Statistical and Bayesian Methods in Hydrological Sciences, Paris, p. 19.

Cavadias GS, Ouarda TBMJ, Bobée B, Girard C (2001) A canonical correlation approach to the determination of homogeneous regions for regional flood estimation of ungauged basins. Hydrological Sciences Journal 46(4): 499–512.

Chokmani K, Ouarda TBMJ (2004) Physiographical space based kriging for regional flood frequency estimation at ungauged sites. Water Resources Research 40: W12514, doi:10.1029/2003WR002983.

Choquette AF (1988) Regionalization of peak discharges for streams in Kentucky. Water Resources Investigation Report 87-4209. US Geological Survey, Louisville District, Louisville, KY.

Chow VT, Maidment DR, Mays LW (1988) Applied hydrology. McGraw-Hill Inc., New York.

Chowdhury JU, Stedinger JR, Lu L-H (1991) Goodness-of-fit tests for regional generalized extreme value flood distribution. Water Resources Research, 27(7): 1765–1776.

Cunderlik JM, Burn DH (2002a) The use of flood regime information in regional flood frequency analysis. Hydrological Sciences Journal 41(1): 77–92.

Cunderlik JM, Burn DH (2002b) Analysis of the linkage between rain and flood regime and its application to regional flood frequency estimation. Journal of Hydrology 261: 115–131.

Cunderlik JM, Burn DH (2003) Non-stationary pooled flood frequency analysis. Journal of Hydrology 276: 210–223. (Discussion: Markiewicz I, Strupczewski WG, Kochanek K, Singh VP (2006) Journal of Hydrology 330(1–2): 382–385).

Cunderlik JM, Burn DH (2006a) Switching the pooling similarity distances: Mahalanobis for Euclidean. Water Resources Research 42: W03409, doi:10.1029/2005WR004245.

Cunderlik JM, Burn DH (2006b) Site-focused nonparametric test of regional homogeneity based on flood regime. Journal of Hydrology 318: 301–315.

Cunderlik JM, Ouarda TBMJ, Bobée B (2004a) Determination of flood Seasonality from hydrologic records. Hydrological Sciences Journal 49(3): 511–526.

Cunderlik JM, Ouarda TBMJ, Bobée B (2004b) On the objective identification of flood seasons. Water Resources Research 40: W01520, doi:10.1029/2003WR002295.

Cunnane C (1988) Methods and merits of regional flood frequency analysis. Journal of Hydrology 100(1–3): 269–290.

Dalrymple T (1960) Flood frequency analysis. U.S. Geological Survey, Water Supply Paper 1543-A., U.S. Department of the Interior, Washington, DC.

Dave RN, Krishnapuram R (1997) Robust clustering methods: A unified view. IEEE Transactions on Fuzzy Systems 5(2): 270–293.

Davies DL, Bouldin DW (1979) Cluster separation measure. IEEE Transactions on Pattern Analysis and Machine Intelligence 1(2): 224–227.

Dixon WJ (Ed.) (1975) BMDP biomedical computer programs. University of California Press, Berkeley, CA.

Dunn JC (1973) A fuzzy relative of the ISODATA process and its use in detecting compact well-separated clusters. Journal of Cybernetics 3(3): 32–57.

Dunn JC (1974) Well separated clusters and optimal fuzzy partitions. Journal of Cybernetics 4: 95–104.

Durbin J, Knott M (1972) Components of Cramer–von Mises statistics. Journal of the Royal Statistical Society, Series B (Methodological), 34(2): 290–307.

Eaton B, Church M, Ham D (2002) Scaling and regionalization of flood flows in British Columbia. Canada. Hydrological Processes 16: 3245–3263.

Everitt B (1993) Cluster analysis, third edition. Halsted Press, New York.

Farris JS (1969) On the cophenetic correlation coefficient. Systematic Zoology 18: 279–285.

Fausett LV (1994) Fundamentals of neural networks: Architectures, algorithms, and applications. Englewood Cliffs, Prentice Hall, NJ.

Fill HD, Stedinger JR (1995) Homogeneity tests based upon Gumbel distribution and a critical appraisal of Dalrymple's test. Journal of Hydrology 166(1–2): 81–105.

Fisher NI (1993) Statistical analysis of circular data. Cambridge University Press, New York.

Frigui H, Krishnapuram R (1997) Clustering by competitive agglomeration. Pattern Recognition 30(7): 1109–1119.

Frigui H, Krishnapuram R (1999) A robust competitive clustering algorithm with applications in computer vision. IEEE Transactions on Pattern Analysis and Machine Intelligence 21(5): 450–465.

Fukuyama Y, Sugeno M (1989) A new method of choosing the number of clusters for the fuzzy c-means method. Proceedings of Fifth Fuzzy Systems Symposium, pp. 247–250 (In Japanese).

Gabriele S, Arnell N (1991) A hierarchical approach to regional flood frequency analysis. Water Resources Research 27(6): 1281–1289.

Gath I, Geva AB (1989) Unsupervised optimal fuzzy clustering. IEEE Transactions on Pattern Analysis and Machine Intelligence 11(7): 773–781.

Glatfelter DR (1984) Techniques for estimating magnitude and frequency of floods on streams in Indiana. US Geological Survey, Water Resources Investigations Report 84-4134.

Gordon AD (1981) Classification (monograph on applied probability and statistics). Chapman & Hall.

Gordon, AD (1999) Classification, second edition. Chapman and Hall/CRC, London.

Govindaraju RS, Rao AR (Eds.) (2000) Artificial neural networks in hydrology. Kluwer Academic Publishers, Holland, p. 329.

Greenwood JA, Landwehr JM, Matalas NC, Wallis JR (1979) Probability weighted moments: Definition and relation to parameters of distribution expressible in inverse form. Water Resources Research 15(5): 1049–1054.

GREHYS (Groupe de recherche en hydrologie statistique) (1996) Presentation and review of some methods for regional flood frequency analysis. Journal of Hydrology 186: 63–84.
Griffis VW, Stedinger JR (2007) Evolution of flood frequency analysis with bulletin 17. Journal of Hydrologic Engineering 12(3): 283–297.
Gu T, Dubuisson B (1990) Similarity of classes and fuzzy clustering. Fuzzy Sets And Systems 34: 213–221.
Guenoche A, Hansen P, Jaumard B (1991) Efficient algorithms for divisive hierarchical-clustering with the diameter criterion. Journal of Classification 8(1): 5–30.
Güler C, Thine GD (2004) Delineation of hydrochemical facies distribution in a regional groundwater system by means of fuzzy c-means clustering. Water Resources Research 40: W12503, doi:10.1029/2004WR003299.
Gupta VK, Dawdy DR (1995) Physical interpretations of regional variations in the scaling exponents of flood quantiles. Hydrological Processes 9: 347–361.
Gupta VK, Messa OJ, Dawdy DR (1994) Multiscaling theory of flood peaks: Regional quantile analysis. Water Resources Research 30(12): 3405–3421.
Gupta VK, Waymire E (1990) Multiscaling properties of spatial rainfall and river flow distributions. Journal of Geophysical Research 96(D3): 1999–2009.
Haktanir T (1997) Self-determined probability-weighted moments method and its application to various distributions. Journal of Hydrology 194: 180–200.
Halkidi M, Batistakis Y, Vazirgiannis M (2001) On clustering validation techniques. Journal of Intelligent Information systems 17(2/3): 107–145.
Hall MJ, Minns AW (1998) Regional flood frequency analysis using artificial neural network. In: Babovic V, Larsen LC (Eds.), Proceedings of the Third International Conference on Hydroinformatics (Copenhagen, Denmark), vol. 2. Balkema, Rotterdam, pp. 759–763.
Hall MJ, Minns AW (1999) The classification of hydrologically homogeneous regions. Hydrological Sciences Journal 44(5): 693–704.
Hall MJ, Minns AW, Ashrafuzzaman AKM (2002) The application of data mining techniques for the regionalization of hydrological variables. Hydrology and Earth System Sciences 6(4): 685–694.
Hartigan JA (1975) Clustering algorithms. John Wiley & Sons, New York.
Hartigan JA, Wong, MA (1979) Algorithm AS 136: A K-means clustering algorithm. Applied Statistics 28: 100–108.
Haykin S (2003) Neural networks: A comprehensive foundation. Fourth Indian Reprint, Pearson Education, Singapore, p. 842.
Hines W, Montgomery DC (1980) Probability and statistics in engineering and management science. John Wiley & Sons, New York.
Holgersson M (1978) The limited value of cophenetic correlation as a clustering criterion. Pattern Recognition 10(4): 287–295.
Hosking JRM (2005) Fortran routines for use with the method of L-moments. Version 3.04, IBM Research Division. T.J. Watson Research Center, Yorktown Heights, NY.
Hosking JRM, Wallis JR (1988) The effect of intersite dependence on regional flood frequency analysis. Water Resources Research 24(4): 588–600.
Hosking JRM, Wallis JR (1993) Some statistics useful in regional frequency analysis. Water Resources Research 29(2): 271–281 (Correction: 31(1), 1995, p. 251).
Hosking JRM, Wallis JR (1997) Regional frequency analysis: An approach based on L-moments. Cambridge University Press, New York.
Hosking JRM, Wallis JR, Wood EF (1985) An appraisal of the regional flood frequency procedure in the UK flood studies report. Hydrological Sciences Journal 30(1): 85–109.
Huang Z (1997) A fast clustering algorithm to cluster very large categorical data sets in data mining. Proceedings of SIGMOD Workshop on Research Issues on Data Mining and Knowledge Discovery, Tucson, Arizona.
Huang Z (1998) Extensions to the K-means algorithm for clustering large data sets with categorical values. Data Mining and Knowledge Discovery 2: 283–304.
Huang Z, Ng MK (2003) A note on K-modes clustering. Journal of Classification 20(2): 257–261.

Indiana Department of Natural Resources (IDNR), Division of Water (2001) Input and output files of flood frequency analysis. 402 W. Washington St, Room N264, Indianapolis, IN 46204.

IPCC (1992) Leggett J, Pepper WJ, Swart R (Eds.), Climate change 1992, the supplementary report to the IPCC scientific assessment. Intergovernmental Panel on Climate Change, Cambridge University Press, Cambridge, UK.

IPCC (2001) McCarthy JJ, Canziani OF, Leary NA, Dokken DJ, White KS (Eds.), Climate change 2001: Working group II: Impacts, adaptation and vulnerability. Contribution of Working Group II to the Third Assessment Report of the Intergovernmental Panel on Climate Change, p. 1000.

IPCC (2007) Climate change 2007: Impacts, Adaptation and Vulnerability, Working Group II Contribution to the Intergovernmental Panel on Climate Change, Fourth Assessment Report, Summary for Policymakers, April 2007.

Jain AK, Dubes RC (1988) Algorithms for clustering data. Prentice-Hall Inc., Englewood Cliff, NJ.

Jain S, Lall U (2000) The magnitude and timing of annual maximum floods: Trends and large-scale climatic associations for the Blacksmith Fork River, Utah. Water Resources Research 36(12): 3641–3652.

Jain S, Lall U (2001) Floods in a changing climate: Does the past represent the future? Water Resources Research 37(12): 3193–3205.

Jain AK, Murty MN, Flynn PJ (1999) Data clustering: A review. ACM Computing Surveys 31(3): 264–323.

Javelle P, Ouarda, TBMJ, Lang M, Bobée B, Galéa G, Grésillon J-M (2002) Development of regional flow-duration–frequency curves based on the index-flood method. Journal of Hydrology 258(1–4): 249–259.

Jin M, Stedinger, JR (1989) Flood frequency analysis with regional and historical information. Water Resources Research 25(5): 925–936.

Jingyi Z, Hall MJ (2004) Regional flood frequency analysis for the Gan-Ming river basin in China. Journal of Hydrology 296: 98–117.

Kalkstein LS, Corrigan P (1986) A synoptic climatological approach for geographical analysis: Assessment of sulfur dioxide concentrations. Annals of the Association of American Geographers 76(3): 381–395.

Kaufman L, Rousseeuw P (1990) Finding groups in data: An introduction to cluster analysis. Wiley, New York.

Khaliq MN, Ouarda TBMJ, Ondo J-C, Gachon P, BobéeB (2006) Frequency analysis of a sequence of dependent and/or non-stationary hydro-meteorological observations: A review. Journal of Hydrology 329: 534–552.

Kiang MY (2001) Extending the Kohonen self-organizing map networks for clustering analysis. Computational Statistics and Data Analysis 38: 161–180.

Klemeš V (2000) Tall tales about tails of hydrological distributions I. Journal of Hyrologic Engineering ASCE 5(3): 227–231.

Kohonen T (1982) Self-organized formation of topologically correct feature maps. Biological Cybernetics 43: 59–69.

Kohonen T (1997) Self-organizing maps, second edition. Springer Verlag, Berlin.

Kohonen T, Oja E, Simula O, Visa A, Kangas J (1996) Engineering applications of the self-organizing map. Proceedings of the IEEE 84(10): 1358–1384.

Krishnapuram R, Keller J (1996) The possibilistic c-means algorithm: Insights and recommendations. IEEE Transactions on Fuzzy Systems 4(3): 385–393.

Kroll CN, Vogel RM (2002) Probability distribution of low streamflow series in the United States. Journal of Hydrologic Engineering 7(2): 137–146.

Kroll CN, Vogel RM (2003) closure to "Probability distribution of low streamflow series in the United States" by Kroll CN, Vogel RM. Journal of Hydrologic Engineering 8(5): 297–298.

Kumar P, Guttorp P, Foufoula-Georgiou E (1994) A probability weighted moment test to assess simple scaling. Stochastic Hydrology and Hydraulics 8: 173–183.

Kwon SH (1998) Cluster validity index for fuzzy clustering. Electronics Letters 34(22): 2176–2177.

Lampinen J, Oja E (1992) Clustering properties of hierarchical self-organizing maps. Journal of Mathematical Imaging and Vision 2(2–3): 261–272.

Lance GN, Williams WT (1966) Computer programs for hierarchical polythetic classification ('similarity analysis'). Computer Journal 9: 60–64.

Landwehr JM, Tasker GD, Jarrett RD (1987) Discussion of relative accuracy of log Pearson III procedures, by J.R. Wallis and E.F. Wood. Journal of Hydraulic Engineering 111(7): 1206–1210.

Lecce SA (2000) Seasonality of flooding in North Carolina. Southeastern Geographer 41(2): 168–175.

Lettenmaier DP, Wallis JR, Wood, EF (1987) Effect of regional heterogeneity on flood frequency estimation. Water Resources Research 23(2): 313–324.

Lu L-H, Stedinger JR (1992) Sampling variance of normalized GEV/PWM quantile estimators and a regional homogeneity test. Journal of Hydrology 138(1–2): 223–245.

MacCracken M (2001) Prediction versus projection – Forecast versus possibility. Guest editorial, WeatherZine 26.

MacQueen J (1967) Some methods for classification and analysis of multivariate observations. In: Le Cam LM, Neyman J (Eds.), Proceedings of the Fifth Berkeley Symposium on Mathematical Statistics and Probability, vol. 1. University of California Press, Berkeley, CA, pp. 281–297.

Mardia KV (1972) Statistics of directional data. Academic Press, San Diego, California.

Moisello U (2007) On the use of partial probability weighted moments in the analysis of hydrological extremes. Hydrological Processes 21: 1265–1279.

Mosley MP (1981) Delimitation of New Zealand hydrological regions. Journal of Hydrology 49: 173–192.

Murtagh F (1983) A survey of recent advances in hierarchical-clustering algorithms. Computer Journal 26(4): 354–359.

Murtagh F (1995) Interpreting the Kohonen self-organizing feature map using contiguity-constrained clustering. Pattern Recognition Letters 16(4): 399–408.

Nathan RJ, McMahon TA (1990) Identification of homogeneous regions for the purposes of regionalisation. Journal of Hydrology 121: 217–238.

Natural Environment Research Council (NERC) (1975) Flood studies report, vol. I, Hydrological Studies, NERC, London.

Ng R, Han J (1994) Efficient and effective clustering methods for spatial data mining. Proceeding of the 20th VLDB conference, Santiago, Chile.

Obermayer K, Sejnowski TJ (Eds.) (2001) Self-organizing map formation foundations of neural computation. MIT Press, Cambridge, MA.

Oliveira JVD, Pedrycz W (Eds.) (2007) Advances in fuzzy clustering and its applications, Wiley, New York, p. 454.

Olsen JR, Stedinger JR, Matalas NC, Stakhiv EZ (1999) Climate variability and flood frequency estimation for the upper Mississippi and lower Missouri rivers. Journal of the American Water Resources Association 35(6): 1509–1523.

Oreskes N (2004) Beyond the ivory tower – The scientific consensus on climate change. Science 306(5702): 1686.

Ouarda TBMJ, Cunderlik JM, St-Hilaire A, Barbet M, Bruneau P, Bobée B (2006) Data-based comparison of seasonality-based regional flood frequency methods. Journal of Hydrology 330(1–2): 329–339.

Ouarda TBMJ, Girard C, Cavadias G, Bobée B (2001) Regional flood frequency estimation with canonical correlation analysis. Journal of Hydrology 254(1–4): 157–173.

Ouarda TBMJ, Hache M, Bruneau P, Bobée B (2000) Regional flood peak and volume estimation in Northern Canadian Basin. Journal of Cold Regions Engineering ASCE 14(4): 176–191.

Pakhira MK, Bandyopadhyay S, Maulik U (2004) Validity index for crisp and fuzzy clusters. Pattern Recognition 37: 487–501.

Pal NR, Bezdek JC (1995) On cluster validity for the fuzzy c-means model. IEEE Transactions on Fuzzy systems 3(3): 370–379.

Pal NR, Bezdek JC, Tsao EC-K (1993) Generalized clustering networks and Kohonen's self-organizing scheme. IEEE Transactions on Neural Networks 4(4): 549–557.
Pandey GR (1998) Assessment of scaling behaviour of regional floods. Journal of Hydrologic Engineering 3: 169–173.
Peel MC, Wang QJ, Vogel RM, McMahon TA (2001) The utility of L-moment ratio diagrams for selecting a regional probability distribution. Hydrological Sciences Journal 46(1): 147–155.
Potter KW, Faulkner EB (1987) Catchment response time as a predictor of flood quantiles. Water Resources Bulletin 23(5): 857–861.
Potter KW, Lettenmaier DP (1990) A comparison of regional flood frequency estimation methods using a resampling method. Water Resources Research 26(3): 415–424.
Qin AK, Suganthan PN (2004) Robust growing neural gas algorithm with application in cluster analysis. Neural Networks 17(8–9): 1135–1148.
Ralambondrainy H (1995) A conceptual version of the K-means algorithm. Pattern Recognition Letters 16: 1147–1157.
Rao AR, Ernst S, Jeong GD (2002) Regionalization of Indiana watersheds for flood flow predictions I. Results from L moment based method. Report No. 1, Joint Transportation Research Program, Project No. C-36-62K. School of Civil Engineering, Purdue University, West Lafayette, IN 47906, p. 191.
Rao AR, Hamed KH (2000) Flood frequency analysis. CRC Press, Boca Raton, FL, p. 350.
Rao AR, Hamed K (2003) Multi-taper method of analysis of periodicities in hydrologic data. Journal of Hydrology 279: 125–143.
Rao AR, Srinivas VV (2006a) Regionalization of watersheds by hybrid cluster analysis. Journal of Hydrology 318(1–4): 37–56.
Rao AR, Srinivas VV (2006b) Regionalization of watersheds by fuzzy cluster analysis. Journal of Hydrology 318(1–4): 57–79.
Reed DW, Jakob D, Robson AJ (1999) Selecting a pooling group. In: Robson AJ, Reed DW (Eds.), Statistical procedures for flood frequency estimation, Flood estimation handbook, vol. 3, Institute of Hydrology, Wallingford, UK (chapter 6, pp. 28–39).
Ribeiro J, Rousselle J (1996) Robust simple scaling analysis of flood peaks series. Canadian. Journal of Civil Engineering 23: 1139–1145.
Ribeiro-Corréa B, Cavadias GS, Clement B, Rousselle J (1995) Identification of hydrological neighbourhoods using canonical correlation analysis. Journal of Hydrology 173: 71–89.
Rissanen J (1989) Stochastic complexity in statistical inquiry. World Scientific: Series in computer science, NJ.
Ritter H (1995) Self-organizing feature maps: Kohonen maps. In: Arbib MA (Ed.), The Handbook of Brain Theory and Neural Networks, MIT Press, Cambridge, MA,, pp. 846–851.
Robson AJ, Jones TK, Reed DW, Bayliss AC (1998) A study of national trend and variation in UK floods. International Journal of Climatology 18(2): 165–182.
Romesburg HC (1984) Cluster analysis for researchers. Lifetime Learning Publications, Belmont, CA.
Ross TJ (1995) Fuzzy logic with engineering applications. McGraw-Hill, New York.
Roubens M (1982) Fuzzy clustering algorithms and their cluster validity. European Journal of Operational Research. 10(3): 294–301.
Rousseeuw PJ (1987) Silhouettes: A graphical aid to the interpretation and validation of cluster analysis. Journal of Computational and Applied Mathematics 20: 53–65.
Ruspini EH (1969) A new approach to clustering. Information and Control 15(1): 22–32.
Ruspini EH (1970) Numerical methods for fuzzy clustering. Information Sciences 2: 319–350.
Sato-Ilic M, Jain LC (2006) Innovations in fuzzy clustering: Theory and applications, Springer, Berlin, p. 152.
Savaresi SM, Boley DL, Bittanti S, Gazzaniga G (2002) Cluster selection in divisive clustering algorithms. Proceedings of the 2nd SIAM ICDM, Arlington, VA, pp. 299–314.
Scholz FW, Stephens MA (1987) K-sample Anderson-Darling tests. Journal of the American Statistical Association 82(399): 918–924.

Shu C, Burn DH (2004) Homogeneous pooling group delineation for flood frequency analysis using a fuzzy expert system with genetic enhancement. Journal of Hydrology 291(1–2): 132–149.

Smith JA (1992) Representation of basin scale in flood peak distributions. Water Resources Research 28: 2993–2999.

Sokal RR, Rohlf FJ (1962) The comparison of dendrograms by objective methods. Taxon 11: 33–40.

Sokal RR, Sneath PHA (1963) Principles of Numerical Taxonomy. W.H. Freeman and Co., San Francisco, CA.

SPSS (1988) Statistical package for the social sciences-X, user's guide. SPSS Inc., McGraw-Hill, New York.

Srinivas VV, Rao AR (2003) Regionalization of Indiana Watershed by Fuzzy Cluster Analysis. Interim report FHNA/JTRP-2002-2 Joint Transportation Research Program, School of Civil Engineering, Purdue University, W. Lafayette, IN 47907, pp. 123.

Srinivas VV, Rao AR (2002) Regionalization of Indiana Watersheds by hybrid cluster analysis. Interim Report No. 2 on Regionalization of Indiana Watersheds for Flood Flow Predictions (Phase I), submitted to Federal High Ways Authority, USA, under Joint Transport Research Project FHWA/IN/JTRP-2002/2 of Purdue University, West Lafayette, Indiana, USA.

Srinivas VV, Rao AR, Govindaraju RS (2002) A hybrid cluster analysis for regionalization. Proceedings of ASCE Environmental and Water Resources Institute (EWRI) Conference (CD ROM), Roanoke, Virginia, USA.

Srinivas VV, Tripathi S, Rao AR, Govindaraju RS (2008) Regional flood frequency analysis by combining self-organizing feature maps and fuzzy clustering. Journal of Hydrology 348(1–2): 148–166.

Stedinger JR, Tasker GD (1985) Regional hydrologic regression, 1. Ordinary, weighted and generalized least squares compared. Water Resources Research 21(9): 1421–1432.

Strupczewski WG, Kaczmarek Z (2001) Non-stationary approach to at-site flood frequency modelling II. Weighted least squares estimation. Journal of Hydrology 248(1–4): 143–151.

Strupczewski WG, Singh VP, Feluch W (2001a) Non-stationary approach to at-site flood frequency modeling I. Maximum likelihood estimation. Journal of Hydrology 248(1–4): 123–142.

Strupczewski WG, Singh VP, Mitosek HT (2001b) Non-stationary approach to at-site flood frequency modelling. III. Flood analysis of Polish rivers. Journal of Hydrology 248(1–4): 152–167.

Su MC, Chang HT (2000) Fast self-organization feature map algorithm. IEEE Transactions on Neural Networks 11(3): 721–733.

Sveinsson OGB (2002) Modeling of stationary and non-stationary hydrologic processes. Doctor of Philosophy Dissertation, Colorado State University, Colorado, USA.

Sveinsson OGB, Boes DC, Salas JD (2001) Population index flood method for regional frequency analysis. Water Resources Research 37(11): 2733–2748.

Sveinsson OGB, Salas JD, Boes DC (2003) Uncertainty of quantile estimators using the population index flood method. Water Resources Research 39(8): 1206, doi:10.1029/2002WR001594.

Sveinsson OGB, Salas JD, Boes DC, Pielke Sr. RA (2003) Modeling the dynamics of long-term variability of hydroclimatic processes. Journal of Hydrometeorology 4(3): 489–505.

Tamura S, Higuchi S, Tanaka K (1971) Pattern classification based on fuzzy relations. IEEE Transactions on Systems, Man, and Cybernetics 1(1): 61–66.

Tasker GD (1980) Hydrologic regression with weighted least squares. Water Resources Research 16(6): 1107–1113.

Tasker GD (1982) Comparing methods of hydrologic regionalization. Water Resources Bulletin 18(6): 965–970.

Tasker GD, Stedinger JR (1986) Estimating generalized skew with weighted least squares regression. Journal of Water Resources Planning and Management 112(2): 225–237.

Tasker GD, Stedinger JR (1989) An operational GLS model for hydrologic regression. Journal of Hydrology 111: 361–375.

Theodoridis S, Koutroubas K (1999) Pattern recognition. Academic Press, New York.

Thomas DM, Benson MA (1970) Generalization of streamflow characteristics from drainage basin characteristics. US Geological Survey Water-Supply Paper 1975. US Government Printing Office, Washington, DC.
Trauwaert E (1985) On the meaning of Dunn's partition coefficient for fuzzy clusters. International working paper, Vrije Universiteit Brussels, Brussels, Belgium.
Trauwaert E (1988) On the meaning of Dunn's partition coefficient for fuzzy clusters. Fuzzy Sets and Systems 25: 217–242.
U.S. Water Resources Council (1976) Guidelines for determining flood flow frequency. Bulletin 17, Hydrology Committee, Washington, DC.
U.S. Water Resources Council (1977) Guidelines for determining flood flow frequency. Bulletin 17A, Hydrology Committee, Washington, DC.
U.S. Water Resources Council (1981) Guidelines for determining flood flow frequency. Bulletin 17B, Hydrology Committee, Washington, DC.
Vesanto J, Alhoniemi E (2000) Clustering of the self-organizing map. IEEE Transactions on Neural Networks 11(3): 586–600.
Viglione A, Laio F, Claps P (2007) A comparison of homogeneity tests for regional frequency analysis. Water Resources Research 43: W03428, doi:10.1029/ 2006WR005095.
Vogel RM, Wilson I (1996) Probability distribution of annual maximum, mean, and minimum streamflow in the United States. Journal of Hydrologic Engineering 1(2): 69–76.
Walker FR, Stedinger JR (2000) Long-term variability in the arrival rate of flood events as evidenced by flood clustering. EOS transactions, American Geophysical Union, Spring Meeting, 81(19): S200.
Wallis JR (1980) Risk and uncertainties in the evaluation of flood events for the design of hydraulic structures. In: Siccita PE, Guggino E, Rossi G, Todini E (Eds.), Fondazione Politecnica del Mediterraneo, Catania, Italy, pp. 3–36.
Wallis JR, Wood EF (1985) Relative accuracy of log Pearson III procedures. Journal of Hydraulic Engineering 111(7): 1043–1056 (with discussion and closure, 113(7): 1205–1214).
Wandle SW Jr (1977) Estimating the magnitude and frequency of floods on natural-flow streams in Massachusetts, US Geological Survey Water Resources Investigations Report, pp. 77–39.
Wang QJ (1990) Estimation of the GEV distribution from censored samples by method of partial probability weighted moments. Journal of Hydrology 120(1–4): 103–114.
Wang QJ (1997) LH moments for statistical analysis of extreme events. Water Resources Research 33(12): 2841–2848.
Wang QJ (1998) Approximate goodness-of-fit tests of fitted generalized extreme value distributions using LH moments. Water Resources Research 34(12): 3497–3502.
Ward JH Jr. (1963) Hierarchical grouping to optimize an objective function. Journal of the American Statistical Association 58: 236–244.
Waylen PR, Caviedes CN (1986) El Nino and annual floods on the north Peruvian littoral. Journal of Hydrology 89(1–2): 141–156.
Webster R, Burrough PA (1972) Computer-based soil mapping of small areas from sample data, II: Classification smoothing. Journal of Soil Science 23(2): 222–234.
Whalen TM, Savage GT, Jeong GD (2002). The method of self-determined probability weighted moments revisited. Journal of Hydrology 268: 177–191.
Willmott CJ, Vernon MT (1980) Solar climates of the conterminous United States: A preliminary investigation. Solar Energy 24: 295–303.
Willshaw DJ, Malsburg CV (1976) How patterned neural connections can be set up by self-organization. Proceedings of Royal Statistical Society London B 194: 431–445.
Wiltshire SE (1986) Regional flood frequency analysis II. Multivariate classification of drainage basins in Britain. Hydrological Sciences Journal 31(3): 335–346.
Windham MP (1981) Cluster validity for fuzzy clustering algorithms. Fuzzy Sets and Systems 5(2): 177–185.
Windham MP (1982) Cluster validity for the fuzzy c-means clustering algorithm. IEEE Transactions on Pattern Analysis and Machine Intelligence 4(4): 357–363.

Winkler JA (1985) Regionalization of the diurnal distribution of summertime heavy precipitation. Preprints, Sixth Conference of Hydrometeorology, Indianapolis, IN, American Meteorological Society, pp. 9–16.
Wu S, Chow TWS (2004) Clustering of the self-organizing map using a clustering validity index based on inter-cluster and intra-cluster density. Pattern Recognition 37: 175–188.
Xie XL, Beni G (1991) A validity measure for fuzzy clustering. IEEE Transactions on Pattern Analysis and Machine Intelligence 13(8): 841–847.
Yang MS (1993) On a class of fuzzy classification maximum likelihood procedures. Fuzzy Sets and Systems 57(3): 365–375.
Zadeh LA (1965) Fuzzy sets. Information and Control 8(3): 338–353.
Zrinji Z, Burn DH (1994) Flood frequency analysis for ungauged sites using a region of influence approach. Journal of Hydrology 153: 1–21.
Zrinji Z, Burn DH (1996) Regional flood frequency with hierarchical region of influence. Journal of Water Resources Planning and Management 122(4): 245–252.

Index of Notation

$a(i)$	Average distance from the i–th feature vector to all other feature vectors in the cluster k
$b(i)$	Minimum average distance from the i-th feature vector to all the feature vectors in another cluster j
c	Number of fuzzy clusters
C_{Exp}	Expected number of clusters
$\mathrm{C}_{\max}$	Maximum number of clusters
C_k	Cluster k
c_{ij}	(i, j)th elements of cophenetic matrix
d_{ij}^{p}	(i, j)th elements of proximity matrix
$DAR(\mathrm{i})$	Drainage area ratio defined as ratio of the drainage area at site i divided by the sum of drainage areas at all the sites in a region
DB	Davies-Bouldin index
d_{ij}	Geographic distance between catchments i and j
$d_{\max}$	Maximum geographic distance between catchment pairs
$\mathbf{D}_i$	Discordancy statistic for site i in a region
D	Dunn's cluster validity Index
$d_{\omega,j}$	Topological distance between the winning node ω and its neighboring node j in the output layer of SOFM
$e_j^{D,T,M}$	Average error (in percentage) for region j with method M, probability distribution D and recurrence interval T
F	Non-exceedence probability
G	Partition comprising of clusters
$f(\cdot)$	Transformation function
HM	Heterogeneity measure
h	Order of statistical moment
H_1	Heterogeneity measure based on L-CV
H_2	Heterogeneity measure based on L-CV and L-skewness
H_3	Heterogeneity measure based on L-skewness and L-kurtosis
$h_{j,\omega}(t)$	Neighborhood function in SOFM
$\mathbf{I}$	Unit matrix
K	Number of hard clusters
l_1	First at-site L-moment or sample mean

l_2	Second at-site L-moment or L-CV
l_3	Third at-site L-moment or L-skew
l_4	Fourth at-site L-moment or L-Kurtosis
l_1^R	Regional average L-location or mean of the distribution
m	Number of output nodes in Kohonen layer of SOFM
m'	Number of output nodes in Kohonen layer of SOFM that are winners for at least one input vector
n	Number of attributes
n_i	Record length of peak flows at site i
N	Number of feature vectors or watersheds considered for cluster analysis
N_R	Number of feature vectors (sites) in cluster (or region) R
N_{sim}	Number of realizations of region obtained by Monte-Carlo simulations
N_i^f	Cardinality of a fuzzy cluster
Q_{kj}	Peak flow value at site k for year j
$\hat{\mu}_Q^j$	Mean annual flood (or mean annual peak flow) at site j
Q_{log}	logarithm of mean annual peak flow
$\hat{Q}_T^k$	Quantile estimate at site k for T year recurrence interval
$Q_M^{D,T}(i)$	T-year flood quantile estimated at site i by method M and distribution D
$\hat{q}_T^R$	Quantile of normalized regional distribution (or Growth curve ordinates estimated using index flood method) for T year recurrence interval
$\boldsymbol{r}_j$	Discrete vector denoting the position of node j in Kohonen lattice
$\mathbf{S}$	Covariance matrix in expression of discordancy measure
$s(i)$	silhouette width
$S_{k,q}$	Scatter within the k-th cluster estimated by using q-th root of the q-th moment of Euclidean distance of points in the cluster about its centroid
$t^{(i)}$	L-CV of peak flows at site i
$t_3^{(i)}$	L-skewness of peak flows at site i
$t_4^{(i)}$	L-kurtosis of peak flows at site i
t^R	Regional average L-CV
t_3^R	Regional average L-skewness
t_4^R	Regional average L-kurtosis
$\mathbf{U}$	Fuzzy partition matrix containing memberships of feature vectors in c fuzzy clusters
$\boldsymbol{u}_{\mathrm{i}}$	Vector containing L-CV, L-skew and L-kurtosis values of site i
$\bar{\boldsymbol{u}}$	Vector containing unweighted regional average L-moment ratios
u_{ik}	Fuzzy membership of feature vector k in cluster i
$\boldsymbol{V}$	Matrix containing centroids of c fuzzy clusters

$\boldsymbol{v}_i$	Centroid of fuzzy cluster i
$\boldsymbol{v}_i^{init}$	Initialized centroid of fuzzy cluster i
V	Weighted standard deviation of the at-site sample L-CVs
V_2	Weighted average distance from the site to the group weighted mean in the two dimensional space of L-CV and L-skewness
V_3	Weighted average distance from the site to the group weighted mean in the two dimensional space of L-skewness and L-kurtosis
V_{PC}	Partition coefficient
V_{PE}	Partition entropy (or classification entropy)
V_{FPI}	Fuzziness performance index
V_{NCE}	Normalized classification entropy
V_{CH}	Calinski Harabasz cluster validity Index
V_{FS}	Fukuyama and Sugeno cluster validity index
V_{XB}	Xie-Beni cluster validity index
$V_{XB,m}$	Extended FCM Xie-Beni index
V_K	Kwon's cluster validity Index
w_{ij}	Weight of connection from the input node i to the output node j in SOFM
$\boldsymbol{w}_j$	Weight vector between the output node j and the nodes in the input layer of SOFM
$\boldsymbol{x}_i$	n-dimensional rescaled feature vector used for clustering
$\bar{\boldsymbol{x}}$	Centroid of the entire set of rescaled feature vectors
$\mathbf{X}$	$n \times N$ data matrix containing set of N rescaled feature vectors
$\boldsymbol{y}_i$	n-dimensional feature vector i
$\mathbf{z}_k$	Centroid of cluster k
$\eta(t)$	Learning rate parameter
λ	Scale parameter
$\hat{\lambda}_1(k)$	First L-moment
$\bar{\lambda}$	Average of $\hat{\lambda}_1^k$ values (same as l_1^R when estimated for a region)
$\hat{\lambda}_2(k)$	Second L-moment for data at site k
$\hat{\lambda}_3(k)$	Third L-moment for data at site k
$\hat{\lambda}_1^k$	Index flood value for site k (taken as mean annual flood)
μ	Fuzzifier value in fuzzy c-means algorithm
$\hat{\mu}_Q^j$	Mean annual peak flow at site j
μ_c	Means of elements in cophenetic matrix
μ_p	Means of elements in proximity matrix
μ_V	Mean of the N_{sim} values of V
μ_{V_2}	Mean of the N_{sim} values of V_2
μ_{V_3}	Mean of the N_{sim} values of V_3
σ_V	Standard deviation of the N_{sim} values of V
π_i	Compactness of fuzzy cluster i
σ_{V_2}	Standard deviation of the N_{sim} values of V_2
σ_{V_3}	Standard deviation of the N_{sim} values of V_3

σ_i^f	Variation of fuzzy cluster i
$\hat{\sigma}_Q^j$	Standard deviation of the annual peak flows at site j
σ_j	Standard deviation of attribute j
σ	Total variation of a data set
θ	Scaling exponent
ν^2	Variance used to evaluate the precision in estimation of regional flood quantiles

Abbreviations

A	Drainage area
ASW	Average silhouette width
BMDP2M	BioMeDical computer Program
CCA	Canonical correlation analysis
CE	Classification entropy validity measure
CL	Complete linkage
CLARA	Clustering Large Application
CLARANS	Clustering Large Applications based on Randomized Search
CN	Cluster number
CPCC	Cophenetic correlation coefficient
CV	Coefficient of variation
FCM	Fuzzy c-means
FFA	Flood frequency analysis
GM2	Two parameter gamma distribution
GLO	Generalized logistic distribution.
GLS	Generalized least square
GEV	Generalized extreme value distribution
GREHYS	Groupe de recherche en hydrologie statistique
IRAS PSC	Infrared Astronomical Satellite Point Source Catalog
KL	Kohonen layer
KMA	K-means algorithm
LAT	Latitude in decimal degrees
L-CV	Coefficient of L-variation
LMRs	L-moment ratios (L-CV, L-skewness and L-kurtosis)
LN3	Three parameter log normal distribution
LONG	Longitude in decimal degrees
LP3	Log-Pearson type III distribution
MAF	Mean annual flood
MDL	Minimum Description Length cluster validity Index
MEF	Median annual flood
MOR	Method of residuals
MSE	Mean square error
OLS	Ordinary least square

P	Precipitation
PAM	Partitioning around medoids
PC	Partition coefficient validity index
PT3	Pearson type 3 distribution
RC	Runoff coefficient
RFFA	Regional flood frequency analysis
ROI	Region of Influence
S	Slope
SI	Scenario
SOFM	Self-Organizing Feature Map
STOR	Drainage area covered by lakes in percentage (same as W)
T	Return period or recurrence interval
UPGA	Unweighted pair-group average
UPGC	Unweighted pair-group centroid
USA	United States of America
USGS	United States Geological Survey
W	Wet area percentage (same as STOR)
WPGA	Weighted pair-group average
WPGC	Weighted pair-group centroid

Index

Water Science and Technology Library

1. A.S. Eikum and R.W. Seabloom (eds.): *Alternative Wastewater Treatment.* Low-Cost Small Systems, Research and Development. Proceedings of the Conference held in Oslo, Norway (7–10 September 1981). 1982 ISBN 90-277-1430-4
2. W. Brutsaert and G.H. Jirka (eds.): *Gas Transfer at Water Surfaces.* 1984 ISBN 90-277-1697-8
3. D.A. Kraijenhoff and J.R. Moll (eds.): *River Flow Modelling and Forecasting.* 1986 ISBN 90-277-2082-7
4. World Meteorological Organization (ed.): *Microprocessors in Operational Hydrology.* Proceedings of a Conference held in Geneva (4–5 September 1984). 1986 ISBN 90-277-2156-4
5. J. N.ěmec: *Hydrological Forecasting.* Design and Operation of Hydrological Forecasting Systems. 1986 ISBN 90-277-2259-5
6. V.K. Gupta, I. Rodríguez-Iturbe and E.F. Wood (eds.): *Scale Problems in Hydrology.* Runoff Generation and Basin Response. 1986 ISBN 90-277-2258-7
7. D.C. Major and H.E. Schwarz: *Large-Scale Regional Water Resources Planning.* The North Atlantic Regional Study. 1990 ISBN 0-7923-0711-9
8. W.H. Hager: *Energy Dissipators and Hydraulic Jump.* 1992 ISBN 0-7923-1508-1
9. V.P. Singh and M. Fiorentino (eds.): *Entropy and Energy Dissipation in Water Resources.* 1992 ISBN 0-7923-1696-7
10. K.W. Hipel (ed.): *Stochastic and Statistical Methods in Hydrology and Environmental Engineering.* A Four Volume Work Resulting from the International Conference in Honour of Professor T.E. Unny (21–23 June 1993). 1994
    10/1: Extreme values: floods and droughts ISBN 0-7923-2756-X
    10/2: Stochastic and statistical modelling with groundwater and surface water applications ISBN 0-7923-2757-8
    10/3: Time series analysis in hydrology and environmental engineering ISBN 0-7923-2758-6
    10/4: Effective environmental management for sustainable development ISBN 0-7923-2759-4
    Set 10/1–10/4: ISBN 0-7923-2760-8
11. S.N. Rodionov: *Global and Regional Climate Interaction: The Caspian Sea Experience.* 1994 ISBN 0-7923-2784-5
12. A. Peters, G. Wittum, B. Herrling, U. Meissner, C.A. Brebbia, W.G. Gray and G.F. Pinder (eds.): *Computational Methods in Water Resources X.* 1994 Set 12/1–12/2: ISBN 0-7923-2937-6
13. C.B. Vreugdenhil: *Numerical Methods for Shallow-Water Flow.* 1994 ISBN 0-7923-3164-8
14. E. Cabrera and A.F. Vela (eds.): *Improving Efficiency and Reliability in Water Distribution Systems.* 1995 ISBN 0-7923-3536-8
15. V.P. Singh (ed.): *Environmental Hydrology.* 1995 ISBN 0-7923-3549-X
16. V.P. Singh and B. Kumar (eds.): *Proceedings of the International Conference on Hydrology and Water Resources* (New Delhi, 1993). 1996
    16/1: Surface-water hydrology ISBN 0-7923-3650-X
    16/2: Subsurface-water hydrology ISBN 0-7923-3651-8

16/3: Water-quality hydrology ISBN 0-7923-3652-6
16/4: Water resources planning and management ISBN 0-7923-3653-4
Set 16/1–16/4 ISBN 0-7923-3654-2

17. V.P. Singh: *Dam Breach Modeling Technology*. 1996 ISBN 0-7923-3925-8
18. Z. Kaczmarek, K.M. Strzepek, L. Somlyódy and V. Priazhinskaya (eds.): *Water Resources Management in the Face of Climatic/Hydrologic Uncertainties*. 1996 ISBN 0-7923-3927-4
19. V.P. Singh and W.H. Hager (eds.): *Environmental Hydraulics*. 1996 ISBN 0-7923-3983-5
20. G.B. Engelen and F.H. Kloosterman: *Hydrological Systems Analysis*. Methods and Applications. 1996 ISBN 0-7923-3986-X
21. A.S. Issar and S.D. Resnick (eds.): *Runoff, Infiltration and Subsurface Flow of Water in Arid and Semi-Arid Regions*. 1996 ISBN 0-7923-4034-5
22. M.B. Abbott and J.C. Refsgaard (eds.): *Distributed Hydrological Modelling*. 1996 ISBN 0-7923-4042-6
23. J. Gottlieb and P. DuChateau (eds.): *Parameter Identification and Inverse Problems in Hydrology, Geology and Ecology*. 1996 ISBN 0-7923-4089-2
24. V.P. Singh (ed.): *Hydrology of Disasters*. 1996 ISBN 0-7923-4092-2
25. A. Gianguzza, E. Pelizzetti and S. Sammartano (eds.): *Marine Chemistry*. An Environmental Analytical Chemistry Approach. 1997 ISBN 0-7923-4622-X
26. V.P. Singh and M. Fiorentino (eds.): *Geographical Information Systems in Hydrology*. 1996 ISBN 0-7923-4226-7
27. N.B. Harmancioglu, V.P. Singh and M.N. Alpaslan (eds.): *Environmental Data Management*. 1998 ISBN 0-7923-4857-5
28. G. Gambolati (ed.): *CENAS. Coastline Evolution of the Upper Adriatic Sea Due to Sea Level Rise and Natural and Anthropogenic Land Subsidence*. 1998 ISBN 0-7923-5119-3
29. D. Stephenson: *Water Supply Management*. 1998 ISBN 0-7923-5136-3
30. V.P. Singh: *Entropy-Based Parameter Estimation in Hydrology*. 1998 ISBN 0-7923-5224-6
31. A.S. Issar and N. Brown (eds.): *Water, Environment and Society in Times of Climatic Change*. 1998 ISBN 0-7923-5282-3
32. E. Cabrera and J. García-Serra (eds.): *Drought Management Planning in Water Supply Systems*. 1999 ISBN 0-7923-5294-7
33. N.B. Harmancioglu, O. Fistikoglu, S.D. Ozkul, V.P. Singh and M.N. Alpaslan: *Water Quality Monitoring Network Design*. 1999 ISBN 0-7923-5506-7
34. I. Stober and K. Bucher (eds): *Hydrogeology of Crystalline Rocks*. 2000 ISBN 0-7923-6082-6
35. J.S. Whitmore: *Drought Management on Farmland*. 2000 ISBN 0-7923-5998-4
36. R.S. Govindaraju and A. Ramachandra Rao (eds.): *Artificial Neural Networks in Hydrology*. 2000 ISBN 0-7923-6226-8
37. P. Singh and V.P. Singh: *Snow and Glacier Hydrology*. 2001 ISBN 0-7923-6767-7
38. B.E. Vieux: *Distributed Hydrologic Modeling Using GIS*. 2001 ISBN 0-7923-7002-3
39. I.V. Nagy, K. Asante-Duah and I. Zsuffa: *Hydrological Dimensioning and Operation of Reservoirs*. Practical Design Concepts and Principles. 2002 ISBN 1-4020-0438-9

Water Science and Technology Library

40. I. Stober and K. Bucher (eds.): *Water-Rock Interaction.* 2002 ISBN 1-4020-0497-4
41. M. Shahin: *Hydrology and Water Resources of Africa.* 2002 ISBN 1-4020-0866-X
42. S.K. Mishra and V.P. Singh: *Soil Conservation Service Curve Number (SCS-CN) Methodology.* 2003 ISBN 1 4020 1132 6
43. C. Ray, G. Melin and R.B. Linsky (eds.): *Riverbank Filtration.* Improving Source-Water Quality. 2003 ISBN 1-4020-1133-4
44. G. Rossi, A. Cancelliere, L.S. Pereira, T. Oweis, M. Shatanawi and A. Zairi (eds.): *Tools for Drought Mitigation in Mediterranean Regions.* 2003 ISBN 1-4020-1140-7
45. A. Ramachandra Rao, K.H. Hamed and H.-L. Chen: *Nonstationarities in Hydrologic and Environmental Time Series.* 2003 ISBN 1-4020-1297-7
46. D.E. Agthe, R.B. Billings and N. Buras (eds.): *Managing Urban Water Supply.* 2003 ISBN 1-4020-1720-0
47. V.P. Singh, N. Sharma and C.S.P. Ojha (eds.): *The Brahmaputra Basin Water Resources.* 2004 ISBN 1-4020-1737-5
48. B.E. Vieux: *Distributed Hydrologic Modeling Using GIS. Second Edition.* 2004 ISBN 1-4020-2459-2
49. M. Monirul Qader Mirza (ed.): *The Ganges Water Diversion: Environmental Effects and Implications.* 2004 ISBN 1-4020-2479-7
50. Y. Rubin and S.S. Hubbard (eds.): *Hydrogeophysics.* 2005 ISBN 1-4020-3101-7
51. K.H. Johannesson (ed.): *Rare Earth Elements in Groundwater Flow Systems.* 2005 ISBN 1-4020-3233-1
52. R.S. Harmon (ed.): *The Río Chagres, Panama.* A Multidisciplinary Profile of a Tropical Watershed. 2005 ISBN 1-4020-3298-6
53. To be published.
54. V. Badescu, R.S. Cathcart and R.D. Schuiling (eds): Macro-Engineering: A Challenge for the Future. 2006 ISBN 1-4020-3739-2
55. K.-P. Seiler and J.R. Gat: *Groundwater Recharge from Run-off, Infiltration and Percolation.* 2008 ISBN 978-1-4020-5305-4
56. G. Salvadori, C. De Michele, N.T. Kottegoda and R. Renzo: *Extremes in Nature.* An Approach Using Copulas. 2007 ISBN 1-4020-4414-3
57. S.K. Jain, R.K. Agarwal and V.P. Singh: *Hydrology and Water Resources of India.* 2007 ISBN 1-4020-4414-4
58. A.R. Rao and V.V. Srinivas: *Regionalization of Watersheds.* An Approach Based on Cluster Analysis. 2008 ISBN 978-1-4020-6851-5
59. M. Shahin: *Water Resources and Hydrometeorology of the Arab Region.* 2007 ISBN 1-4020-4577-8
60. A.R. Rao and E-C. Hsu: *Hilbert-Huang Transform Analysis of Hydrological and Environmental Time Series.* 2008 ISBN 978-1-4020-6453-1
61. R.S. Govindaraju and B.S. Das: *Moment Analysis for Subsurface Hydrologic Applications.* 2007 ISBN 978-1-4020-5751-9
62. Giuseppe Rossi, Teodoro Vega and Brunella Bonaccorso (eds.): *Methods and Tools for Drought Analysis and Management.* 2007 ISBN 978-1-4020-5923-0

MIX
Papier aus verantwortungsvollen Quellen
Paper from responsible sources
FSC® C105338

If you have any concerns about our products,
you can contact us on
ProductSafety@springernature.com

In case Publisher is established outside the EU,
the EU authorized representative is:
Springer Nature Customer Service Center GmbH
Europaplatz 3, 69115 Heidelberg, Germany

Printed by Libri Plureos GmbH
in Hamburg, Germany